Green Energy and Technology

Lingfeng Wang (Ed.)

Modeling and Control of Sustainable Power Systems

Towards Smarter and Greener Electric Grids

Springer

Editor

Prof. Lingfeng Wang
University of Toledo
Electrical Engineering and Computer Science Dept.
MS 308
2801 W Bancroft St.
Toledo, OH 43606
USA
E-mail: lingfeng.wang@utoledo.edu

ISBN 978-3-642-22903-9 e-ISBN 978-3-642-22904-6

Springer Series in Green Energy and Technology ISSN 1865-3529

Library of Congress Control Number: 2011934147

Cover design: Scientific Publishing Services Pvt. Ltd., Chennai, India

Printed on acid-free paper

9 8 7 6 5 4 3 2 1

springer.com

Preface

The concept of the smart grid promises the world an efficient and intelligent approach of managing energy production, transportation, and consumption by incorporating intelligence, efficiency, and optimality into the power grid. Both energy providers and consumers can take advantage of the convenience, reliability, and energy savings achieved by real-time and intelligent energy management. To this end, the current power grid is experiencing drastic changes and upgrades. For instance, more significant green energy resources such as wind power and solar power are being integrated into the power grid, and higher energy storage capacity is being installed in order to mitigate the intermittency issues brought about by the variable energy resources. At the same time, novel power electronics technologies and operating strategies are being invented and adopted. For instance, Flexible AC Transmission Systems (FACTS) and Phasor Measurement Units (PMU) are two promising technologies for improving the power system reliability and power quality. Demand side management (DSM) will enable the customers to manage the power loads in an active fashion. As a result, modeling and control of modern power grids pose great challenges due to the adoption of new smart grid technologies. In this book, chapters regarding representative applications of smart grid technologies written by world-renowned experts are included, which explain in detail various innovative modeling and control methods.

In Chapter 1, A. K Srivastava, R. Zamora, N. N Schulz, K. G. Ravikumar, and V. M. Mohan address two important aspects of smart grid modeling and simulation including the development of a wide area monitoring and control (WAMC) testbed.

In Chapter 2, C. Cecati, C. Citro, A. Piccolo, and P. Siano propose an energy management system for smart grids using optimal power flow and integrating the demand side management mechanism and the active management schemes for the optimization of a smart grid in a competitive power market.

In Chapter 3, A. Egea-Alvarez, A. Junyent-Ferre, and O. Gomis-Bellmunt describe active and reactive power control for distributed generation and storage systems connected to the grid by means of voltage source converters.

In Chapter 4, S. T. Cha, Q. Wu, A. Saleem, J. Østergaard, and Y. Ding propose a multi-agent based controller by utilizing different resources in the distribution systems to stabilize the frequency. A test platform is established to test the proposed multi-agent based frequency controller.

In Chapter 5, A. V. Stankovic, D. Schreiber, and S. Wu present two control methods for complete harmonic elimination of a grid side inverter under unbalanced operating conditions. Simulation results show the excellent behavior of the wind power system under several fault conditions.

In Chapter 6, F. Yang, V. Donde, Z. Li, and Z. Wang present their recent research efforts by examining distribution automation and online system analysis. State-of-the-art high performance computation architectures that efficiently manage the complexity of the distribution system analysis problem on a large scale are explored.

In Chapter 7, H. Falaghi and M. K. Mahmooee model and analyze the power system operation with the Battery Energy Storage Systems (BESS) in order to manage and control emission of power plants. Also, the life-time parameters of the BESS are considered in the problem modeling.

In Chapter 8, S. Kamalasadan, J. T. Haney, and C. M. Tanton discuss the modeling and control of a PV array and PEM fuel cell based hybrid smart microgrid system and the interconnection to power grid as a smart grid option.

In Chapter 9, P. Siano, P. Chen, Z. Chen, and A. Piccolo propose a hybrid optimization method that aims at maximizing the net present value related to the investment made by wind turbine developers in an active distribution network. The proposed method combines a genetic algorithm with a multi-period optimal power flow.

In Chapter 10, Y. Cheng and M. Sahni discuss relevant issues on wind generation resources model equivalencing, requirement and assessment of voltage ride-though and reactive power, as well as inertial and frequency control of the wind generation, all from the power system planning perspective.

In Chapter 11, A. K. Sadigh and S. M. Barakati discuss multiple topologies of multilevel converters as well as their several control methods. Multilevel converters have been continuously developed in recent years due to the necessity of increase in power level of industrial applications.

In Chapter 12, M. A. Sofla, L. Wang, and R. King discuss the modeling of DC-AC converters based on the characteristics of the microgrids. A performance test of microgrids is analyzed based on power quality and stability of microgrids.

The contributors of this edited book are from both academia and industry. The book is primarily intended for professionals in areas including sustainable power systems, integration of renewable energy sources, power system monitoring and control, wide area measurement and control, power system planning and operations, power system reliability and security, microgrid modeling and control, energy storage, energy management systems, hybrid power systems, smart grid testbed, power system optimization, applications of high-performance computing in power system automation, applications of computational intelligence and intelligent control in power grids, and so forth. The book contains materials reflecting the state-of-the-art developments in modeling and control methods of modern electric grids. This book is expected to be useful not only to researchers and industry practitioners in this field, but also to the general readers who are interested in the promising applications of various smart grid technologies.

The editor would like to thank all the authors who have contributed their valuable work to this book. Thanks are also due to the reviewers who have devoted their time to reviewing the chapters. In addition, the editor is grateful for the great efforts from the staffs of the publisher in making the entire publication process a pleasant and rewarding experience.

June, 2011

L.Wang
University of Toledo, Ohio

Contents

Real Time Modeling and Control of Smart Grid Systems

Anurag K. Srivastava, Ramon Zamora, Noel N. Schulz, Krishnanjan G. Ravikumar, and Vinoth M. Mohan

Abstract. The real time modeling and simulation of smart grid behavior under the disturbance helps in the analysis and planning of system in order to minimize the possible damage. Real time modeling and simulations with hardware in the loop capabilities allows testing the smart grid control algorithms and new equipments.

This chapter addresses two aspects of smart grid modeling and simulation. First one relates to the development of wide area monitoring and control (WAMC) test bed including phasor measurements units (PMUs), synchophasor vector processor (SVP), phasor data concentrators (PDC) and real time digital simulator (RTDS). A WAMC system provides improved reliability, security and coordinated control actions to mitigate or prevent large area disturbances. PMU technology is being deployed all over the world by different utilities for wide area measurement and control actions. These units are mainly based on time synchronized measurements of voltages and currents that lead to accurate decisions and faster control actions. RTDS was used to model and simulate power system and PMUs was used to generate measurement data for power system monitoring. Measured data was sent to PDC and SVP for further analysis and control. Developed WAMC test bed can also be used for testing control algorithm and device performance. Second aspect of this chapter relates to modeling and simulation of Microgrid. Microgrid with controllers have been modeled and simulated in MATLAB/ Simulink.

Anurag K. Srivastava · Ramon Zamora
Department of Electrical Engineering and Computer Science, Washington State University, Pullman, WA 99164, 509-335-2348
e-mail: asrivast@eecs.wsu.edu

Noel N. Schulz
Department of Electrical & Computer Engineering, Kansas State University, Manhattan, KS

Krishnanjan G. Ravikumar
Schweitzer Engineering Lab, Pullman, WA

Vinoth M. Mohan
AREVA T&D, Redmond, WA

L. Wang (Ed.): *Modeling and Control*, Green Energy and Technology, pp. 1–26.
springerlink.com

1 Introduction

Smart grid activities include phasor based application for transmission system and integration of distributed generation/ demand response for distribution system or microgrid. Even though EMS/SCADA systems can monitor a wide-area system, they do so with a time skew. The introduction of phasor measurement units (PMU) have enabled wide-area monitoring with time synchronization thereby facilitating real-time monitoring of the power system. Technological development in satellite based synchronization, power communication network and digital signal processing laid the necessary foundation for the development of PMU technology. Application of PMU within wide area monitoring and control (WAMC) facilitate real time monitoring, improved state estimation, enhanced control, model validation, post event analysis and better situational awareness [1]. With advancement in phasor based applications, several techniques have been developed by researchers for better situational awareness and decision support. These techniques need to be validated in real time before real time implementation in smart grid. Smart grid devices such as PMU, PDC has to be tested under extreme operational conditions. This chapter addresses the development of a wide area monitoring and control test bed using Real Time Digital Simulator (RTDS), Synchophasor Vector Processor (SVP), Phasor Measurement Units (PMU) and Phasor Data Concentrator (PDC). Developed test bed can be used for phasor based algorithm testing as well as PMU/ PDC testing under extreme scenarios through hardware in the loop modeling and simulation. Details for test bed developments and example of PMU performance testing as well as transient stability control have been presented here.

A microgrid is an interconnection of distributed energy sources, such as microturbines, wind turbines, fuel cells and PVs integrated with storage devices, like batteries, flywheels and power capacitors on low voltage distribution systems. Intelligent control and management of power flow is required with increased penetration of distributed generation to have operational parameters within the limit. This chapter also presents modeling and simulation of microgrid and review of controller using MATLAB/ Simulink.

2 Overview of Phasor Measurement and Devices

2.1 Phasor

Phasor measurements denote a quantity in terms of a magnitude and an angle. For example, the voltages, currents, and power in a power system can be represented in terms of phasor magnitude and phasor angle, where the phasor angles are with respect to a certain reference.

Fig. 1 shows the phasor representation of a waveform. Here the phase angle is the angular difference between the sinusoidal peak and the time reference t=0. In a real power system network, instead of using an arbitrary time reference t=0, the angle from any one bus may be used as the reference and the relative phase angles at different buses are calculated using this primary bus angle as the reference.

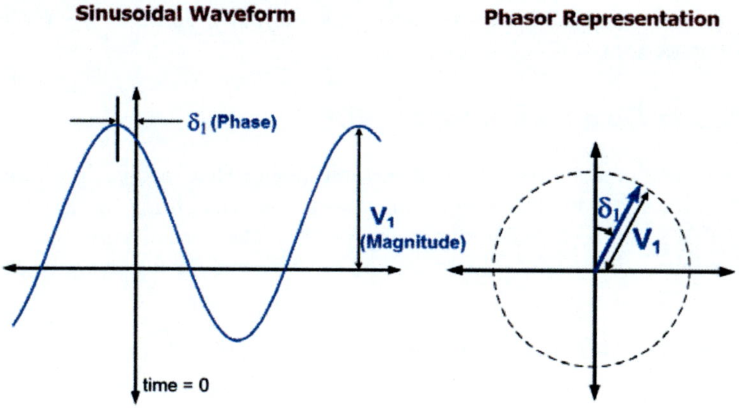

Fig. 1 Phasor representation

2.2 Phasor Measurement Unit (PMU)

The phasor measurement unit is an electronic device that receives analog current and voltage signals from CTs and PTs where these signals are further processed using special algorithms (usually Digital Fourier Transform) to compute the phasor angles and line frequencies.

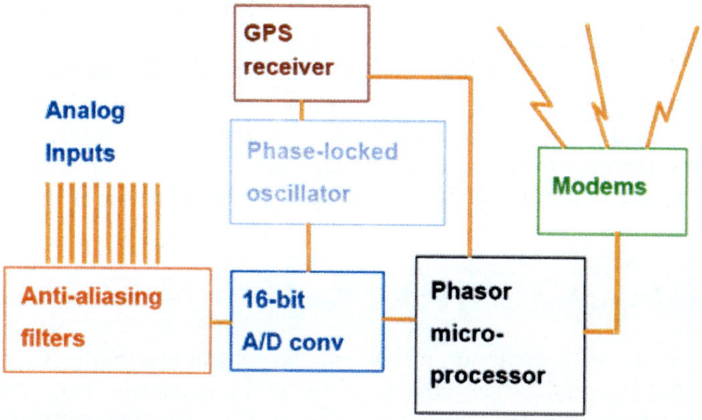

Fig. 2 PMU Block Diagram

Fig. 2 is reproduced from [2] and shows the block diagram of the PMU. Analog signals from CT and PT are fed into an anti-aliasing filter that filters out frequencies above the Nyquist rate. The GPS clock sends a pulse every second and the phase locked oscillator splits this pulse into a sequence of timing pulses. The A/D converts the analog signal into digital form and feeds it into the microprocessor which processes the input waveforms using DFT algorithms and computes the phasor measurements. These measurements normally are then sent out using TCP/IP connections.

2.3 *Phasor Data Concentrator (PDC)*

The phasor data concentrator is an electronic device that receives its inputs from the PMUs, time-tags the input signals and sorts them according to their time. Two or more PDCs can be connected to a super-PDC. The outputs from the PDCs are used to monitor the system in real time and are archived for future analyses.

3 Wide Area Monitoring and Control Testbed

3.1 *Wide Area Monitoring and Control*

The core idea of the wide area monitoring and control (WAMC) systems is the centralized/ distributed processing of the data collected from various locations of a power system, aiming at the evaluation of the actual power system operating conditions and subsequent control [3]. PMUs are being increasingly deployed by power utilities throughout the world. When optimally placed in the electricity network, these can provide more comprehensive information about the system. The PMUs turn the state estimation measurements into linear measurements and can enhance the state estimation efficiency. Wide area control can addresses automatic healing capabilities to some extent by proposing decisive smart topology changes and control actions with the goal of maintaining the integrity of the grid under adverse conditions [4].

At any time in a conventional SCADA system, only a steady-state picture of the power system at an instant is available with time skew. It is highly difficult to construct the current scenario of the power system using SCADA information as data from different points reach the control centre at different times. Time synchronization plays a big role in portraying the real situation of a power system network.

3.2 *Developed WAMC Testbed Setup*

Fig. 3 shows the wide area monitoring and control test bed that was set up at Mississippi State University. The power system was modeled in RSCAD (Real time Simulator Computer Aided Design) software that runs on RTDS (Real Time Digital Simulator). The RTDS [5] in our case is designed to emulate power system components such as CTs and PTs. The low level values from RTDS are directly fed into the SEL-421 Phasor Measurement Control Unit (PMCU) but in the case of GE relays D60 and N60, the values are amplified using the Omicron amplifier. PMCU have ability to provide control signals in addition to phasor measurements. All the PMUs are time synchronized using a time signal from the GPS clock SEL-2407. The outputs from the three PMUs are sent to the SEL-3306 Phasor Data

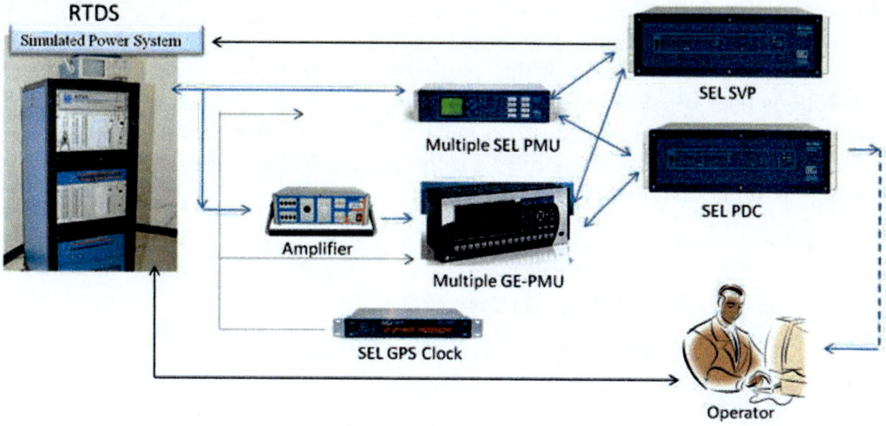

Fig. 3 Developed wide area monitoring and control testbed

Concentrator (PDC). Outputs from two PMCU were also sent to SVP to take control action. SEL-5078 synchrowave console is used to monitor the phasor values of the original power system modeled and simulated in RSCAD.

Interface between RTDS, PMCU, PMU, PDC, and SVP were established through several communication links. Table 1 shows the different cables that were used to connect the individual devices [6].

Table 1 Connection cables used in the test bed

To	From	Cable
SEL 421 (PMCU)	Computer (commands)	C234A
SEL 421 (PMCU)	PDC	C276
GE D70 (PMCU)	PDC (data)	Ethernet
GE N60 (PMCU)	PDC (data)	Ethernet
SEL-3306 (PDC)	Computer (commands)	C235
SEL-3306 (PDC)	Computer (data)	Ethernet
SEL 2407 (Clock)	SEL 421	C953
SEL 2407 (Clock)	GE relays	C953

The SEL-421 is a distance protection relay that has PMU capabilities. The PMU capability is enabled by setting the command EPMU under global enables to 'Y'. When high quality time signal is fed into the PMU, the relay bits TSOK and TIRIG would be set indicating that the relay is in high accuracy time keeping mode. The SEL-2407 clock was set to Greenwich Mean Time. To check if the relay is working in PMU mode, the command MET PM can be used. It displays the synchrophasor measurements at that particular instance. To connect to the PDC over Ethernet, an IP address first has to be assigned. The command STA is used to find the device status. The SEL-3306 has two Ethernet interfaces and the first interface was enabled. Further settings for the PDC are changed on the browser by logging on to the assigned IP address.

SEL-3306 Configuration

PDC-General-Data	
PDCSTN	MSU-1
PDCID	1
HID	Mississippi State Power Lab SEL-3306
NFREQ	60
MRATE	60
MWAITP	1000
T_OUT	5
REQPASS	N
IRIGTYPE	DEMOD
DATE_F	MDY
Network-Configuration	
Network01	
ETHEN	Y
ETDHCP	Y
ETIPAD	130.18.65.130
ETSNMK	255.255.254.0
ETGTWY	130.18.64.1
ETDNS	130.18.65.11
ETDOMA	ece.msstate.edu
ETHOST	substation

Fig. 4 SEL 3306 settings

Fig. 4 shows the settings for SEL 3306 [6]. The IP address, subnet mask, gateway address and the IP address of the host computer are to be fed into the PDC settings. Also the message rate, frequency rate, waiting period, time out, IRIG-B type, and date format are set using the browser.

The C276 serial cable from the SEL-421 is connected to the 10th serial port on the PDC. The setting for that particular port is given in Fig. 5. The serial port is enabled, the message format is set to C37.118 and the speed of the message is set at 57600. The two GE PMUs are connected to the PDC using Ethernet cables. Fig. 6 shows the Ethernet settings for the two GE PMUs.

Serial10	
SPEN	Y
SMFMT	C37.118
SNAME	10SP
SPMUID	0x0001
SPHFMT	P
SVCOMP	0.00
SICOMP	0.00
SSPEED	57600

Fig. 5 Settings for serial port connection for SEL 421 and SEL 3306

PMU-Ethernet-Configuration			
Ethernet01		Ethernet02	
EPEN	Y	EPEN	Y
EMFMT	C37.118	EMFMT	C37.118
ENAME	GE-D60 PMU	ENAME	GE-N60
EPMUID	0x0004	EPMUID	0x0005
EPHFMT	P	EPHFMT	P
EVCOMP	0.00	EVCOMP	0.00
EICOMP	0.00	EICOMP	0.00
EAD_IP	130.18.65.97	EAD_IP	130.18.65.87
ETXDP	TCP	ETXDP	TCP
ETEL	N	ETEL	N
ENDP	4713	ENDP	4713
ENCP	4712	ENCP	4712

Fig. 6 Ethernet port settings for GE PMU

The IP addresses for GE PMU were entered using the GE Enervista software. In addition to the communication settings, the PMUs have to be tuned to receive the inputs from the simulated CTs and PTs. The CT and PT ratings are entered into the Enervista software as shown in the Fig. 7 [6]. The CT and PT ratios are not practical values as several scaling factors are to be taken into account in modeling process.

The SEL-5078 synchrowave console runs on the computer and shows the phasor measurements. The computer needs to be configured to get data from the PDC.

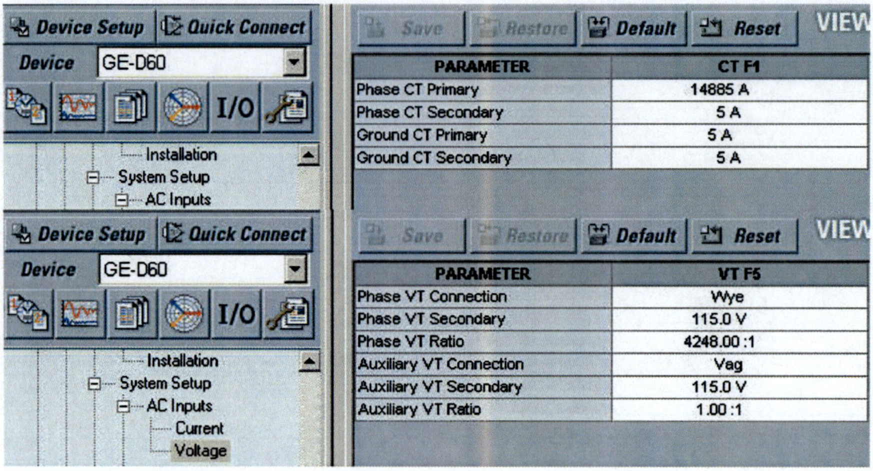

Fig. 7 CT and PT settings for GE PMU

4 Performance Testing of Phasor Devices

4.1 Power System in RSCAD

Fig. 8 shows the 8-bus system [6] that is simulated using the RTDS. The outputs from the two CTs and PTs are fetched from the RTDS device and fed into the SEL and GE relays. The 8 bus system has a three-phase source, transformer, two parallel transmission lines and a three-phase load connected through a network of 8 buses.

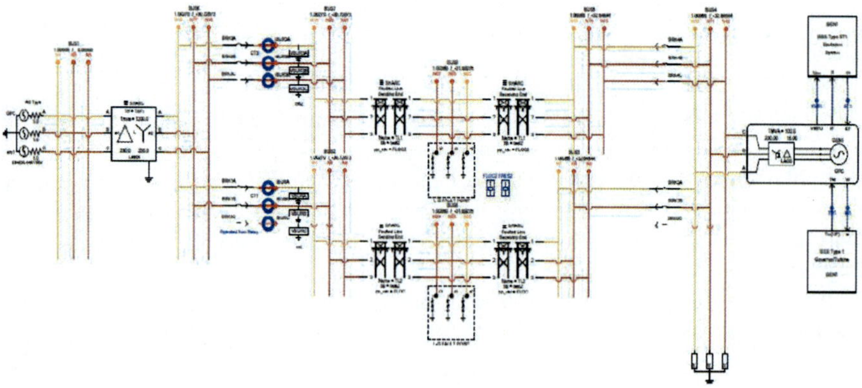

Fig. 8 Eight bus power system

When the modeled power system is simulated, the signals from the CT and PT are sent to the respective PMUs. These PMUs compute the phasor measurements, time stamp them using the GPS signal and send them to the PDC which displays the data on the screen in addition to archiving.

4.2 *Phasor Measurements Display*

The synchrowave phasor measurements can either be observed on the browser or in the synchrowave console. Fig. 9 shows the phasor measurements IA, IB, IC from the PMCUs SEL-421 and GE-D60 displayed on the browser [6].

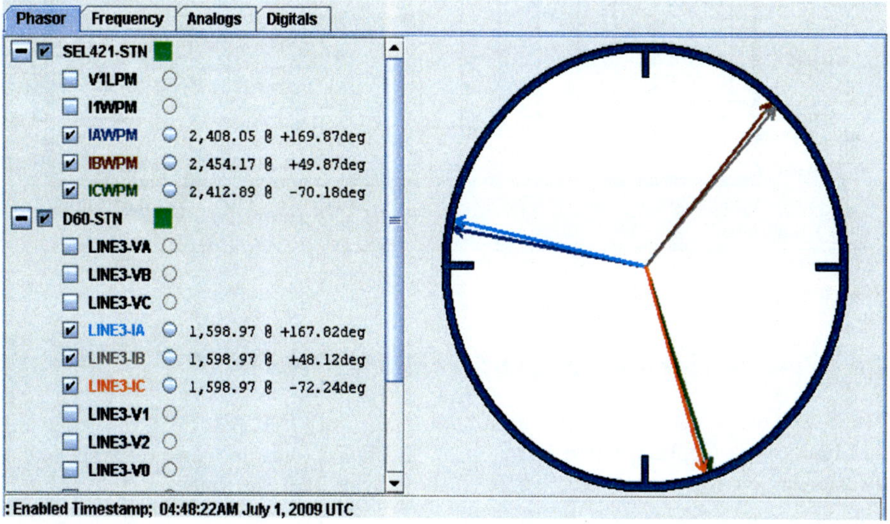

Fig. 9 Phasor measurement display from PDC

4.3 *Frequency Monitoring*

Both the PMUs come with the ability to monitor frequency and rate of change of frequency. Fig. 10 shows the frequency as observed by both PMUs. During the first half of the graph, the SEL-2407 was in holdover mode with a time quality between ±1 μs and ±100 μs. As shown in figure, during this time, the frequency measured by PMU-1 has spikes with very small magnitude while PMU-2 was producing steady frequency measurement output.

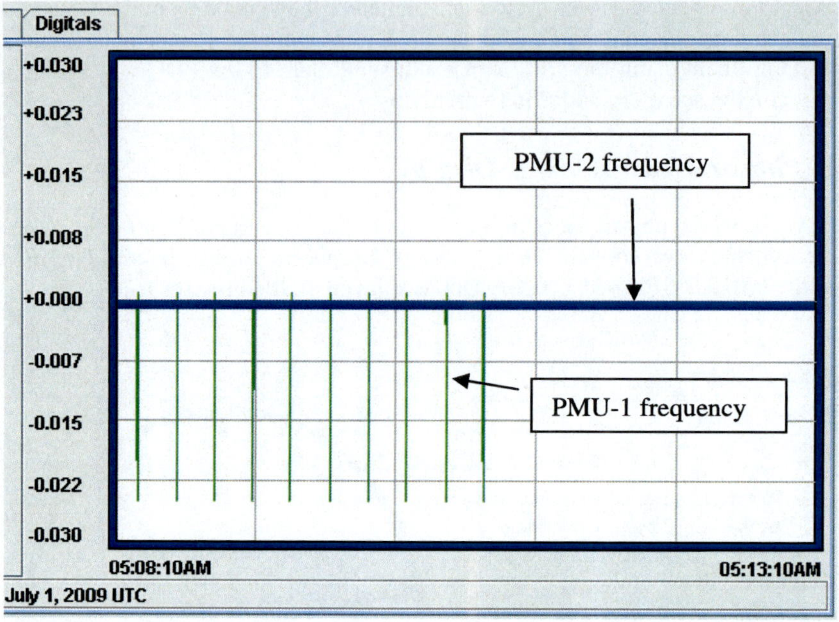

Fig. 10 Frequency monitoring using PMU

4.4 Voltage Monitoring at Off-Nominal Frequency

The 8-bus test case system was simulated using only the SEL PMCU connected and faults were applied between the bus 2 and bus 3 and the measurements made by the device were observed. Table 2 shows the average value of the voltage

Table 2 Voltage magnitude at different off-nominal frequencies

Frequency	Before fault	During fault	After fault
59.30	193064.50	30688.74	193080.20
59.60	193017.90	30670.30	193097.40
59.60	193015.00	30705.06	193032.40
59.80	192963.80	30717.07	192978.50
60.00	192940.90	30661.15	192966.64
60.00	192955.00	30655.55	192975.00
60.02	192901.70	30663.51	192918.10
60.20	192909.40	30659.04	192923.90
60.40	192878.50	30661.42	192901.60
60.70	192828.70	30674.88	192924.20

magnitudes at different frequencies measured by SEL PMCU that was outputting data at the rate of 60 messages per second. For all the different frequencies, the voltage magnitude was around 193 kV before the fault was applied. During fault, for all cases, the voltage fell to around 30 kV. After the fault was cleared, the voltage again regained its initial value of around 193 kV.

The voltage changes with frequency and exhibits transient behavior during the fault. From Table 2, it can be seen that the voltage is dropping with increase in frequency, which can be explained using the equation: $V(t) = V_{max} \cos(wt+\delta)$. As the value of frequency increases, the overall cosine value (throughout the entire time period) decreases thereby reducing the average voltage.

5 Phasor Measurement Based Control Algorithm

5.1 Test Case for PMU Based Control

The test case used in this work is shown in the Fig. 11 [7]. The power system has been modeled in RTDS with the machine and system control dynamics. Power System Stabilizer (PSS) and type ST1 excitation system available in RSCAD software are considered along with the generators. The typical exciter gain value for all the generators is 200. Generators G1, G2 and G3 are rated for 500MW, 700MW and 600MW respectively. L1 and L2 are RL type dynamic loads rated 800MW and 1000MW respectively. The RL load type is referred as dynamic load because the P and Q values can be dynamically adjusted during a simulation.

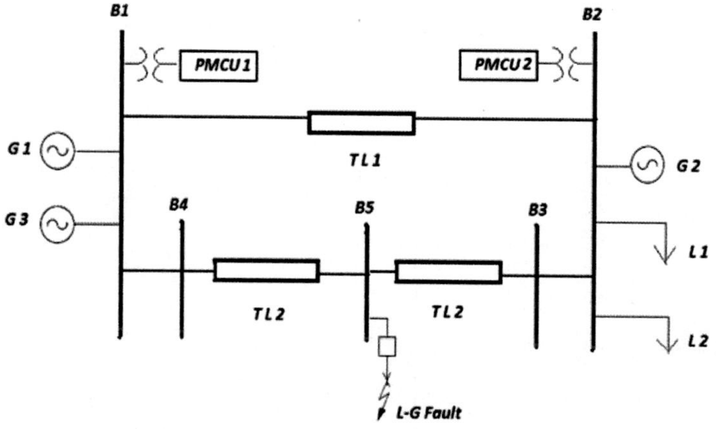

Fig. 11 Test case used for wide area control

In this test case, three generators and two dynamic loads have been used. Two Phasor Measurement and Control Units (PMCUs) have been placed at bus 1 and bus 2. The angle difference between these buses is measured for steady state condition and also for fault condition.

Basically, the PMCUs and RTDS act in close loop hardware in the loop. The PMCU measurements from RTDS are sent into the synchrophasor vector processor as input data into the wide area control algorithms. An angle difference scheme has been built in the synchrophasor vector processor, as shown in Fig. 12. This scheme has been programmed to receive bus voltage angles and calculate the angle difference.

5.2 Implementation of Control Algorithm

The wide area control algorithm has been designed and implemented in the synchrophasor vector processor. The SVP receives the phasor inputs from the PMUs and computes the difference between positive sequence voltage angles at Bus 1 and Bus 2. SVP asserts the corresponding alarm based on thresholds set at 150 degrees (0ms pickup timer) for Level 1 and 300 degrees (5ms pickup timer) for Level 2. If the angle difference exceeds the threshold, the alarm is asserted and it activates the Fast Operate (FO) function block. When the FO function block is activated, it sends the corresponding fast operate command (FOS command used for sending control signals) to the external control device (in this case, it is the PMCU 1) with the assigned IDCODE, when the Enable (EN) input is true. The angle difference schematic is shown in Fig. 12 [7].

In Fig. 12, the PADM is the Phase Angle Difference Monitoring Block which has inputs such as angle 1 and angle 2 in radians, two levels of thresholds and pickup timers along with Second of Century (SOC) and Fraction of Second (FOS) of the incoming data. The outputs of these PADM blocks are angle difference, two alarms, SOC and FOC of the output along with an indicator (OK) for valid PADM output. The Rising edge Trigger (R_TRIG) is used to enable the Fast Operate (FO) block corresponding to the alarm. The Falling Edge Trigger (F_TRIG) is used to clear the control command after the FO block sends it to the PMCU.

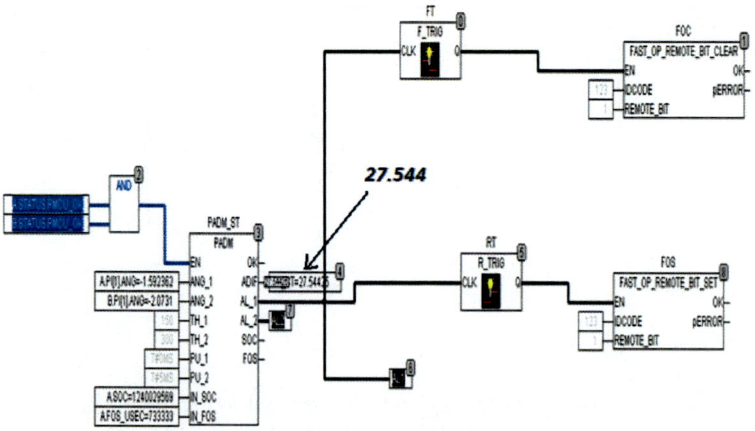

Fig. 12 Function Block Diagram Programming in SVP

Initially, the phasor voltages, currents and angles were sent to the SVP through PMCUs during the stable condition of the power system simulating in RTDS. The angle difference between Bus 1 and Bus 2 was calculated in SVP using the PMCU measurements and it was validated with the angle difference calculation in the RTDS. Fig. 12 shows the angle difference as 27.54425 degrees in the SVP which is also the same in RTDS as shown in Fig. 13.

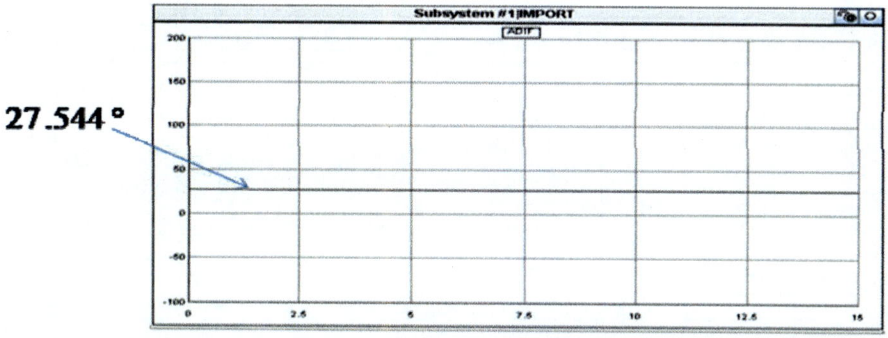

Fig. 13a Angle Difference between Bus 1 and Bus 2 terminal node voltages in RTDS during steady state

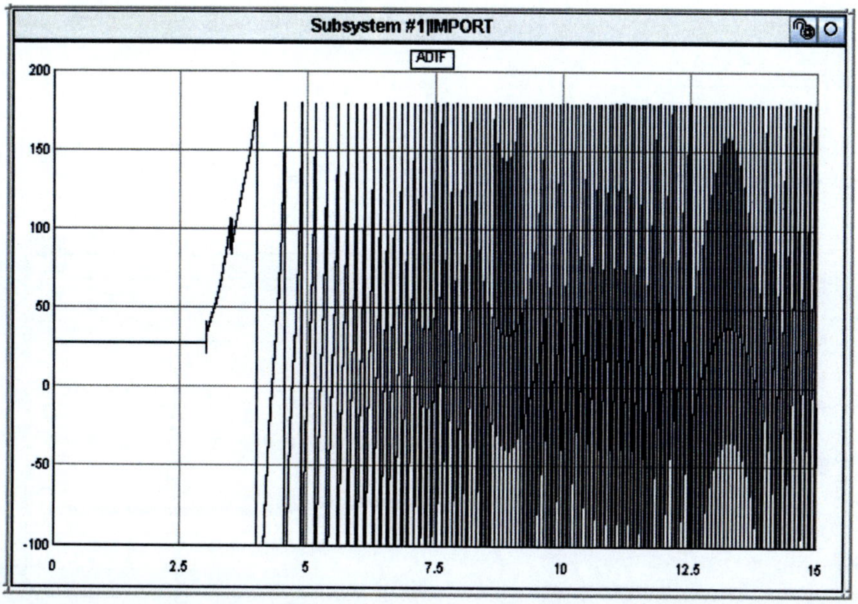

Fig. 13b Angle Difference between Bus 1 and Bus 2 during Fault condition

The system in RTDS has been simulated for 15 seconds with a time step of 100 microseconds. A fault of 28 cycles (467 ms) was applied on Bus 5 and undamped growing oscillations were observed in the system. The system's critical clearing time was observed as 450 ms cycles. In this case, the generators become unstable. The rotor angle velocity increases beyond limits and generators lose synchronism. The angle difference between Bus 1 and Bus 2 as well as transmission line real power flow are shown in Figs. 13 and 14 respectively.

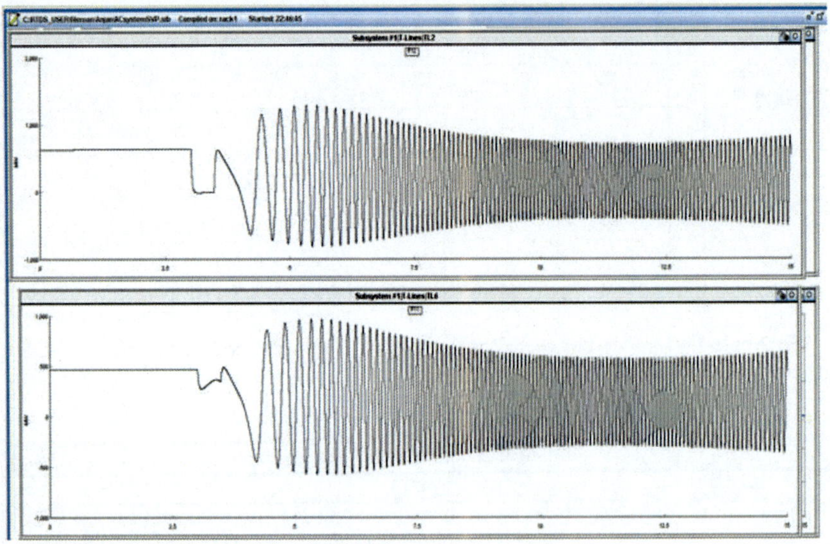

Fig. 14 Transmission line 1 and transmission line 2 real power between bus 1 and bus 2 during fault condition

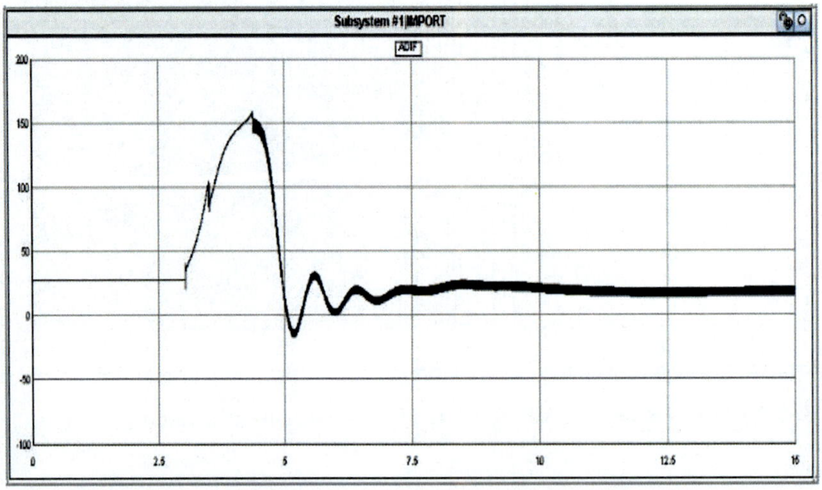

Fig. 15 Angle Difference between Bus 1 and Bus 2 after SVP action

The synchrophasor vector processor action is shown in Figs. 15-17, where the SVP calculates the angle difference and initiates the alarm, which corresponds to the activation of a control signal as shown in Fig. 16 for the system protection by shedding the load 2. The control signal is transferred from the SVP to the PMCU and the PMCU opens up a breaker, which allows load shedding.

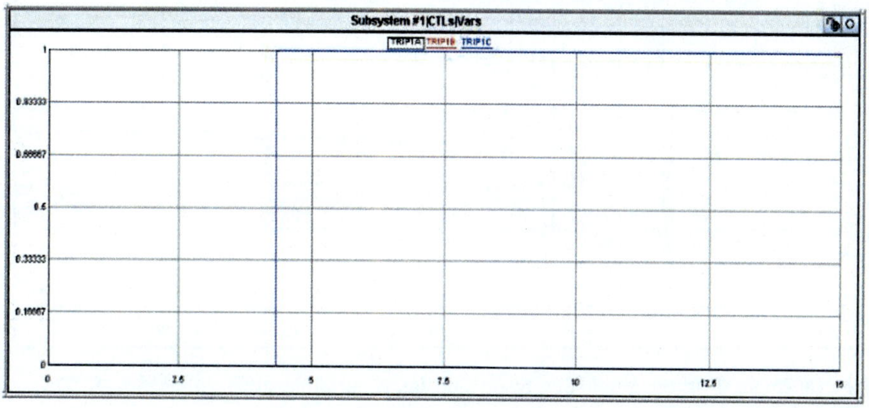

Fig. 16 Control (trip) signal for load shedding after SVP action

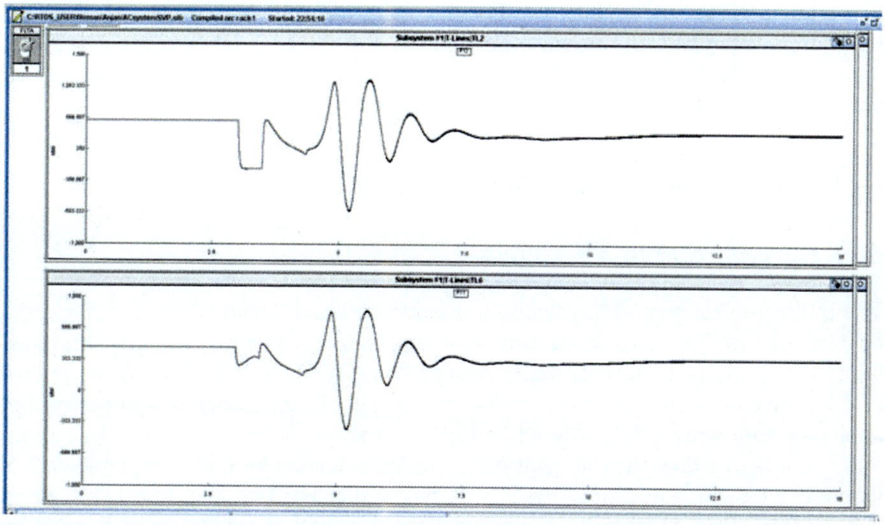

Fig. 17 Transmission line 1 and transmission tine 2 real power between bus 1 and bus 2 after SVP action

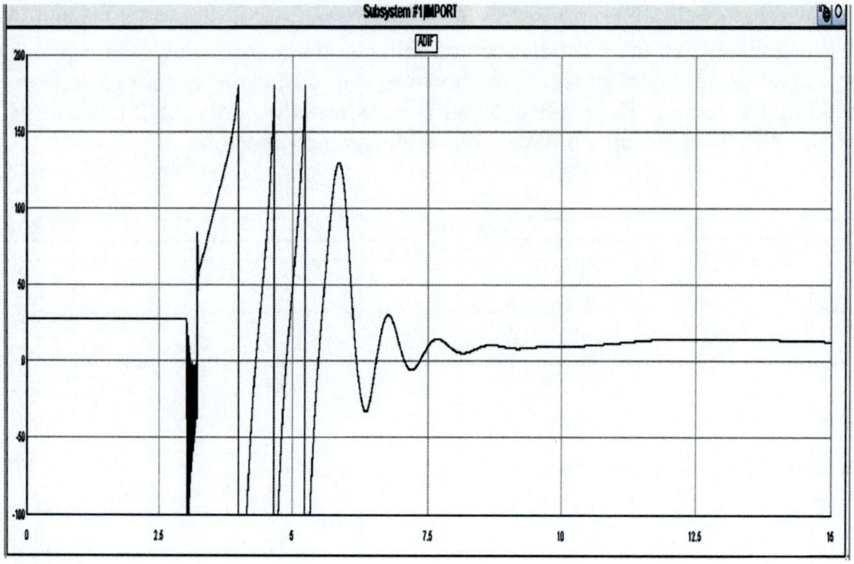

Fig. 18 Angle difference between bus 1 and bus 2 terminal node voltages in RTDS after SVP action for fault on bus 4

Similar type of fault was applied at Bus 4 for validating the operation of the SVP. The SVP action is shown in Fig. 18. Here the angle difference between the two buses does not settle down immediately, and rather it takes some time, after the action of the SVP. This test bed smart grid monitoring and control was further developed to include cyber security with enhanced communication [8].

6 Modeling and Simulation of Microgrid

6.1 Introduction to Microgrid

The interest in microgrid has been increased since last decade due to its potential benefits to provide reliable, secure, efficient, environmentally friendly, and sustainable electricity from renewable energy sources (RES) [9]. Moreover, microgrid has also been implemented in disaster management to improve power system resilience following a catastrophe or big disturbance. Many aspects of microgrid ranging from architecture to controls have been researched and implemented in laboratory test-beds and field models. A basic microgrid architecture is shown in Fig. 19 [10]. This microgrid consists of a group of radial feeders, which could be part of a distribution system or a building's electrical system. There are three sensitive-load feeders (Feeder A-C) and one non-sensitive-load feeder (Feeder D). The sensitive load feeders contain sensitive loads that must be always supplied, thus each feeder must have at least a microsource rated to satisfy the load at that feeder. On the contrary, the non-sensitive load feeder is the feeder that can be shut

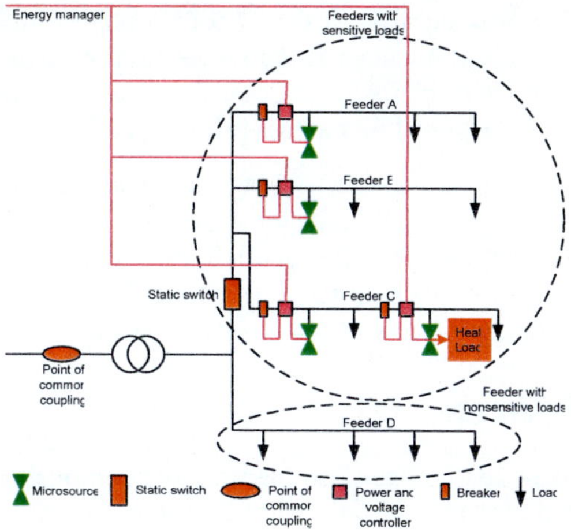

Fig. 19 Microgrid architecture

down if there is a disturbance or power quality problems on the utility side; the non-sensitive load feeder will be left to ride through the disturbance or power quality problems [11].

When there is a problem with the utility supply, Feeders A-C can island from the grid using the static switch that can separate in less than a cycle to isolate the sensitive loads from the power grid to minimize disturbance to the sensitive loads. In an islanded operation, a microgrid will work autonomously, therefore must have enough local generation to meet the demands of the sensitive loads [10,11]. Furthermore, a disturbance causing an individual feeder operation may also occur. Considering this particular scenario in the microgrid design, each sensitive-load feeder must have enough local generation to supply its own loads while the non-sensitive-load feeder will rely on the utility supply.

After any disturbance, the microgrid will reconnect to the utility and work normally as a grid-connected system. In this grid-connected, excess local power generation, if any, will supply the non-sensitive loads or charge the energy storage devices for later uses. The excess power generated by the microgrid may also be sold to the utility. In this case, the microgrid will participate in the market operation or provide ancillary services. The disconnection or reconnection processes must be specified by the point of common coupling (PCC), a single point of connection to the utility located on the primary side of the transformer. At this point, the microgrid must meet the established interface requirements, as defined in IEEE standard 1547 series [12-18]. Furthermore, the successful disconnection or reconnection processes depend upon microgrid controls. The controllers must insure that the processes occur seamlessly and the operating points after the processes are satisfied [19].

Another important characteristic of the CERTS architecture is the energy manager which is responsible to manage system operation through power dispatching and voltage setting to each microsource controller. Some possible criteria for the microgrid to fulfill this responsibility are as follows [10]:

- insure that the necessary electrical loads and heat are fulfilled by the microsources;
- insure that the microgrid satisfies operational contracts with the utility;
- minimize emissions and/or system losses; and
- maximize the operational efficiency of the microsources.

6.2 Microgrid Control

Microgrids controllers can be based on hierarchical controls as shown in Fig. 20 including centralized or decentralized controllers. The control level of hierarchical systems can be classified as follows [19-21]:

- Local controllers consisting of Microsource Controllers (MCs) and Load Controllers (LCs);
- Microgrid Cental Controllers (MGCCs); and
- Distribution Management System (DMS).

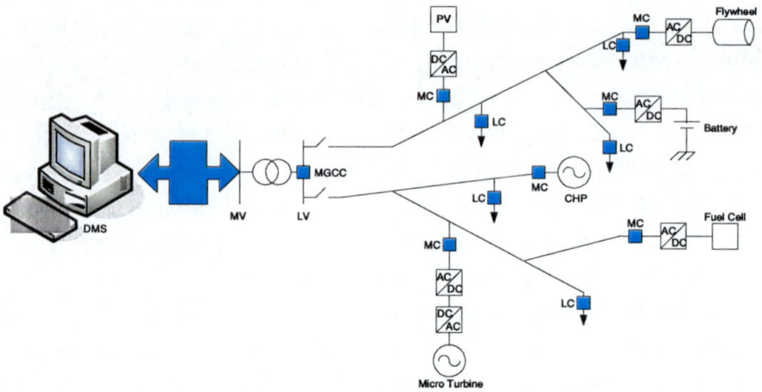

Fig. 20 Hierarchical control of microgrid

Local controls are the basic category of microgrid controls. The main usage of local controllers is to control microsources. This type of controllers is aimed to control operating points of the microsources and their power electronic interfaces without communication systems. No communication systems result in simple circuitry and low cost. The measured data for local controllers are local voltages and currents [10-11].

In most microgrid applications, local controllers will coexist with other type of controllers, while in fully islanded microgrids, as described in [22-23], the local controllers are the only required controllers. The local controllers must also ensure the "plug-and-play" function of microsources; one or several microsources must be able to seamlessly connect or disconnect to the distribution network when and where they are needed. The MCs may also be enhanced with various degrees of intelligence. In addition, LCs are installed at the controllable loads to provide load control capabilities. LCs are commonly used for demand side management. For each microgrid, there is an MGCC that interfaces between the DMS and the microgrid. The MGCC may have different roles ranging from simple coordination of the local controllers to the main responsibility of optimizing the microgrid operation. The difference between centralized and decentralized controls is defined by the centralization roles assumed by the MGCC; the level of decentralization can vary depending on the share of responsibilities assumed by the MGCC and the MCs and LCs. In a centralized control, MCs and LCs follow the orders of MGCC during grid-connected mode and have autonomy to perform their own controls during islanded mode.

In decentralized controls, the main responsibility is given to MCs to maximize their production in order to satisfy the demand and probably provide the maximum possible power export to the grid taking into account current market prices. The decentralized control is aimed to maximize autonomy of the microsources and loads. Several intelligent methods based on peer-to-peer algorithm, such as multi-agent-based [26] and gossip-based algorithms [27], may be used for decentralized controls.

DMS or Distribution Network Operator (DNO), to which several MGCCs are interfaced, has responsibility to manage the operation of medium and low voltage areas in which more than one microgrid may exist. In addition, one or more Market Operators (MO) will exist in the system if the microgrids participate in market operation. DNO and MO are not parts of microgrids but representatives of the utility.

In order to utilize renewable energy optimally without having problems related to variability and intermittency of energy and also instability of electricity, a properly designed storage system must be implemented in a local power system containing large number of small-scale RES. This optimal utilization can be fully competitive either technically or economically to the utilizations of energy from the best fossil fuels or nuclear technologies [23]. Microsources have small generating capacities and mostly require inverters to convert their output to suit power system specifications. Thus, the connections of small-size sources which are dominated by power-electronic-interfaced sources can be considered as inertia-less systems [28]. This inertia-less system cannot response to the initial or surge power or energy mismatch by using their machines' inertia as commonly found in bulk systems. Therefore, a microgrid requires energy storage systems to solve the mismatch problems.

Owing to the facts that different RES have different characteristics and the likeliness of hybrid energy sources in a microgrid, the design of versatile energy storage systems having capability to operate in wide ranges of power density and energy density is required. Since no single energy-storage technology has this capability, system will incorporate combinations of technologies such as supercapacitors, batteries, superconducting magnetic energy storage (SMES), and kinetic energy storage in flywheels [29]. The capacity of the energy storage system depends upon the characteristics of compensation being provided.

Besides optimizing system operation electrically, microgrid controls also aim to optimize production and consumption of heat, gas, and electricity in order to improve overall efficiency. Moreover, controlling a large number of microsources and storage having different characteristics will be very challenging due to the possibility of conflicting requirement and limited communication. Transitions from grid-connected to islanded modes of operation are likely to cause large mismatches between generation and loads, causing a severe frequency and voltage control problem. The "plug-and-play" capability may also create serious problem if the connection and disconnection processes involve big number of microsources at the same time.

6.3 Simulation Results

The microgrid model was developed as given in Fig. 21. The microgrid consists of several microsources: a PV Array of 10 kWp, and a wind turbine of 100 kVA, and sensitive loads of 109 kW and 1.1 kVAR, two batteries as energy storage elements, and a three phase voltage source representing the grid including its loads and the non-sensitive load of the microgrid. The non sensitive load is placed on the grid side of the circuit breaker. In the islanded operation, the microgrid only supplies the sensitive loads while the non sensitive loads will be supplied by the grid.

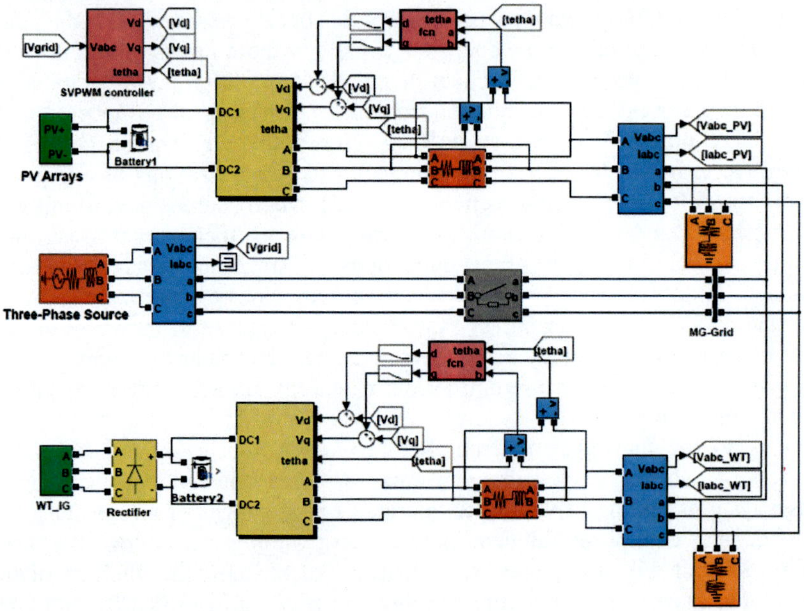

Fig. 21 Microgrid test case system in MATLAB

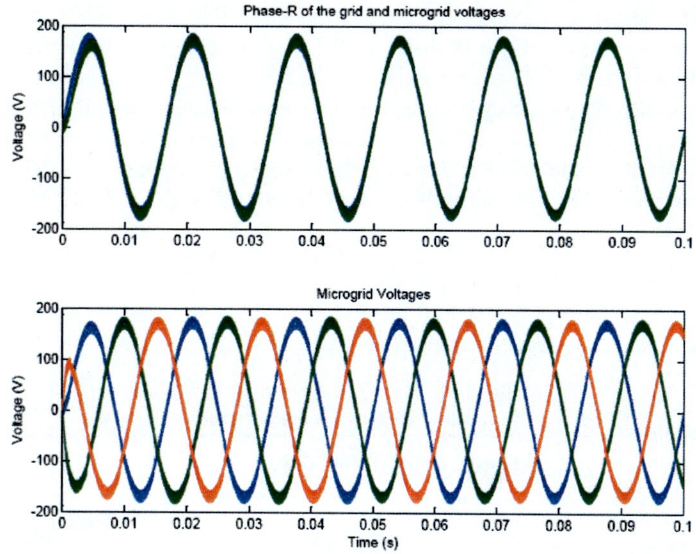

Fig. 22 Phase-R voltages of grid and microgrid and microgrid voltages at the point of connection to the grid

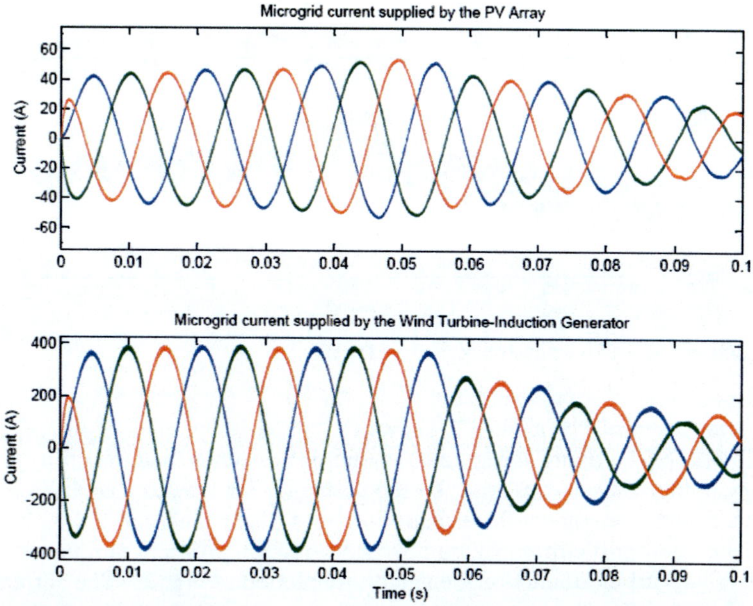

Fig. 23 Microgrid currents

The simulation was conducted in Matlab/Simulink. The wind turbine model used in the simulation is the built-in model in SimPowerSystem toolbox of Matlab/Simulink and the PV array model is based on [30]. The simulation was conducted to assess the basic operation of the system in islanded and grid-connected. First, the microgrid operates in islanded mode. After 3 cycles (50 ms) of the islanded operation, the breaker closes and the microgrid connects to the utility. For the grid connected operation, both microgrid and utility must fulfill the following requirements:

- The voltage magnitude is equal,
- The frequency is equal, and
- The phase sequence is the same.

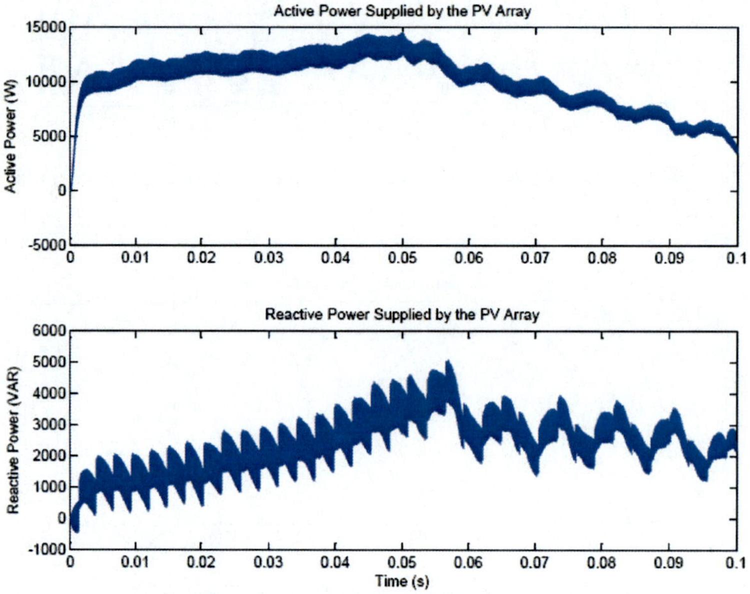

Fig. 24 Active and reactive power supplied by the PV Array

The simulation result in Fig. 22 shows the phase-R voltages of the microgrid and utility [23]. The figure shows that the microgrid voltage and the utility voltage are overlapping. This means that the requirements for interconnection are satisfied. The figure also shows three-phase voltages of the microgrid, which is balanced. The microgrid current by pointing contribution current from the PV array and the Wind Turbine-Induction Generator are plotted in Fig. 23. The currents decrease when the microgrid is operated in grid-connected mode due to the contribution of the main grid to supply the local loads. The connection process from stand-alone to grid-connected operation is seamless and has no overshoot current. Furthermore, Figs. 24 and 25 show active and reactive power supplied by

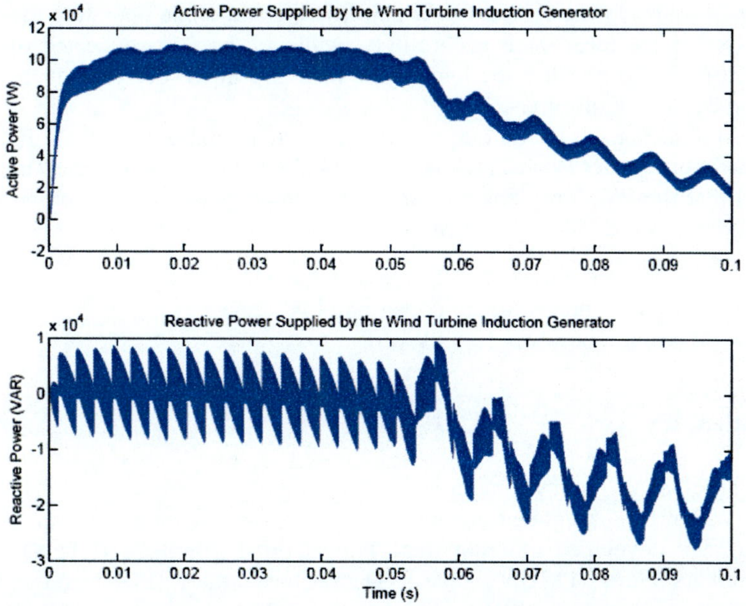

Fig. 25 Active and reactive power supplied by the Wind Turbine-Induction Generator

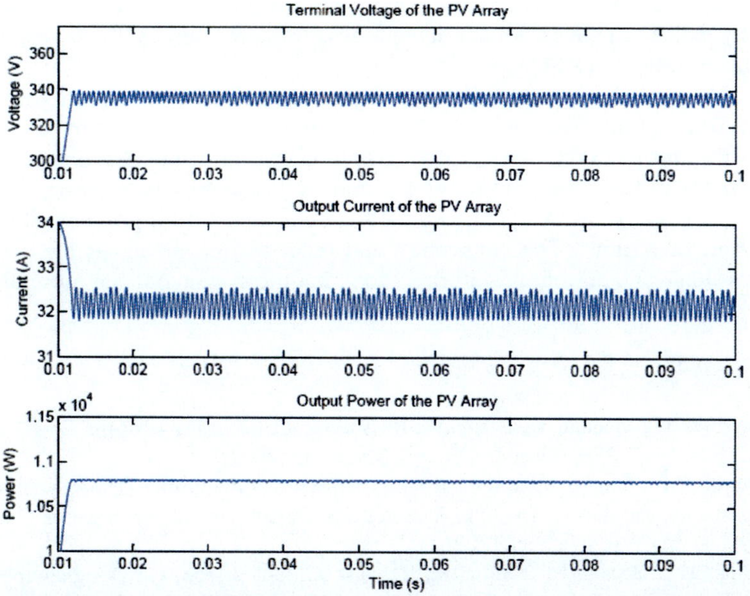

Fig. 26 Output voltage, current, and power of the PV Array

both microsources. The figures show that in the stand-alone operation the microsources share the local loads accordingly, while in the grid-connected operation the main grid also supplies the local loads resulting in the decrease of the power supplied by the microsources.

In addition, Fig. 26 shows output voltage, current, and power of the PV array. The Maximum Power Point Tracking (MPPT) algorithm applied in this simulation ensures that the PV Array operates at its maximum power point. The maximum power point has the following quantities:

- Voltage at the maximum power point is 335 V,
- Current at the maximum power point is 32 A,
- Maximum power is 10.8 kWp.

7 Summary

Modeling of smart grid system requires modeling of transmission and distribution system with newly integrated components. The wide-area monitoring and control test bed was developed utilizing Real Time Digital Simulator (RTDS), Phasor Measurement Units (PMU's), Phasor Data Concentrator (PDC) and Synchrophasor Data Concentrator (SVP). The frequency and phasor measurements were observed for different time accuracies and during off-nominal frequencies. The simulation results of wide area control using SVP were presented. The modeling of the test system in RTDS and the control algorithm in the SVP were discussed and analyzed. An angle difference scheme was implemented for the power system test case developed in RSCAD.

This chapter also presented modeling and simulation of microgrid using MATLAB/Simulink. Microgrid can provide reliable power delivery due to its ability to island from the main grid and to connect and disconnect one or several microsources in the "plug-and-play" function. The model has two microsources, two batteries, and local loads. The model has been verified for grid connected and stand-alone operations. The connection and reconnection processes are seamless and the voltage and current waveforms show that the system operate normally.

References

[1] Novosel, D., Madani, V., Bhargava, B., Vu, K., Cole, J.: Dawn of the Grid Synchronization. IEEE PES Power and Energy Magazine (2008)

[2] Nuqui, R.F.: State Estimation and Voltage Security Monitoring Using Synchronized Phasor Measurements. Doctoral Dissertation, Department of Electrical & Computer Engineering, Virginia Tech (2001)

[3] Zima, M., Larsson, M., Korba, P., Rehtanz, C., Andersson, G.: Design Aspects for Wide-Area Monitoring and Control Systems. Proc. of IEEE 93, 980–996 (2005)

[4] Tholomier, D., Schmitt, L., Giri, J.: Impact of Recent Blackout: Return of Experience of Experience on Utility Operational IT Infrastructures. In: iREP Symposium- Bulk Power System Dynamics and Control - VII, Revitalizing Operational Reliability, Charleston, SC, USA (2007)

[5] Srivastava, A.K., Schulz, N.N.: Applications of Real Time Digital Simulator in Power System Education and Research. In: ASEE Annual Conference and Exposition, Austin, TX (2009)
[6] Mohan, V.: Advancements in Power System Monitoring and Inter-Operability. M.S. Thesis, Mississippi State University (2009)
[7] Ravikumar, K.G.: Distributed simulation of power systems using Real Time Digital Simulator. M.S. Thesis, Mississippi State University (2009)
[8] Reddi, R.M., Srivastava, A.K.: Real Time Test Bed Development for Power System Operation, Control and Cyber Security. In: North American Power Symposium, Arlington, TX (2010)
[9] Lasseter, R.H., Piagi, P.: Microgrid: A conceptual solution. In: Proc. of the 35th IEEE Annual PESC, vol. 6, pp. 4285–4290 (2004)
[10] Lasseter, R., Akhil, A., Marnay, C., Stephens, J., Dagle, J., Guttromson, R., et al.: The CERTS microgrid concept. White Paper on Integration of distributed energy resources. Report prepared for Transmission Reliability Program, Office of Power Technologies, U.S. Department of Energy (2002)
[11] Lasseter, R.H.: MicroGrids. In: Proc. of IEEE PES Winter Meeting, vol. 1, pp. 305–308 (2002)
[12] IEEE Standard 1547^{TM}, IEEE Standard for Interconnecting Distributed Resources with Electric Power Systems (2003)
[13] IEEE Standard P1547.1^{TM}, IEEE Standard Conformance Test Procedures for Equipment Interconnecting Distributed Resources with Electric Power Systems (2005)
[14] IEEE Standard 1547.2^{TM}, Application Guide for IEEE Std 1547, Interconnecting Distributed Resources with Electric Power Systems (2008)
[15] IEEE Standard 1547.3^{TM}, Guide for Monitoring, Information Exchange, and Control of Distributed Resources Interconnected with Electric Power Systems (2007)
[16] IEEE Standard P1547.4^{TM}, Guide for Design, Operation, and Integration of Distributed Resource Island Systems with Electric Power Systems (2010)
[17] P1547.6^{TM}, Recommended Practice for Interconnecting Distributed Resources with Electric Power Systems Distribution Secondary Networks (2010)
[18] P1547.7^{TM}, Guide to Conducting Distribution Impact Studies for Distributed Resources Interconnection (2010)
[19] Hatziargyriou, N., Dimeas, A., Tsikalakis, A.: Centralized and decentralized control of microgrids. Int. J. Dist. Energy Resources 1, 197–212 (2005)
[20] Peças, L.J.A., Moreira, C.L.F., Resende, O.: Control strategies for microgrids black start and islanded operation. Int. J. Distr. Energy Resources 2, 211–231 (2006)
[21] Peças, L.J.A., Moreira, C.L., Madureira, A.G., Resende, F.O., Wu, X., Jayawarna, N., Zhang, Y., Jenkins, N., Kanellos, F., Hatziargyriou, N.: Control strategies for MicroGrids emergency operation. In: Int. Conf. Future Power Systems (2005)
[22] Piagi, P., Lasseter, R.H.: Autonomous control of microgrids. In: IEEE PES General Meeting, Montreal (2006)
[23] Zamora, R., Srivastava, A.K.: Controls for Microgrids with Storage: Review, Challenges, and Research Needs. Journal of Renewable and Sustainable Energy Reviews 14, 2009–2018 (2010)
[24] Hiskens, I.A., Fleming, E.M.: Control of inverter-connected sources in autonomous Microgrids. In: Proc. of American Control Conference, Seattle, vol. 1, pp. 586–590 (2008)

[25] Pedrasa, M.A., Spooner, T.: A survey of techniques used to control microgrid generation and storage during island operation. In: Australian Universities Power Engineering Conference, Melbourne, Australia (2006)
[26] Logenthiran, T., Srinivasan, D., Wong, D.: Multi-agent coordination for DER in MicroGrid. In: Proc. of IEEE Inter. Conf. on Sustainable Energy Technology, vol. 1, pp. 77–82 (2008)
[27] De Brabandere, K., Vanthournout, K., Driesen, J., Deconinck, G., Belmans, R.: Control of microgrids. In: Proc. of IEEE PES General Meeting (2007)
[28] Chakraborty, S., Weiss, M.D., Simoes, M.G.: Distributed intelligent energy management for a single phase high-frequency AC microgrid. IEEE Trans. Ind. Electron 54 (2007)
[29] Hall, P.J., Bain, E.J.: Energy-storage technologies and electricity generation. Energy Policy 36, 4352–4355 (2008)
[30] Zamora, R.: Power Electronic Interfaces for Stand-Alone Photovoltaic Systems. M.S. thesis, Univ. of Arkansas, Fayetteville, AR (2008)

Smart Grids Operation with Distributed Generation and Demand Side Management

C. Cecati, C. Citro, A. Piccolo, and P. Siano

Abstract. The integration of Distributed Generation (DG) based on renewable sources in the Smart Grids (SGs) is considered a challenging task because of the problems arising for the intermittent nature of the sources (e.g. wind or sun), only partly predictable. Another important issue concerning the design of SGs is how to support the consumers' participation in the electricity market minimizing the costs of the global energy consumption. This chapter proposes an Energy Management System for SGs using Optimal Power Flow and integrating the demand side management mechanism and the active management schemes for the optimization of a SG in a competitive power market. The idea is that they can achieve a better integration of different types of DG improving flexibility while reducing the costs of energy for customers. The efficiency of the proposed technique is verified on a 19-bus 11 kV distribution network.

1 Introduction

The increasing diffusion of Distributed Generation (DG) based on renewable sources is significantly affected by the discontinuous and partially predictable behaviour of the renewable energy resources like the wind or the sun.

Network stresses due to recurrent over voltage/current as well as power system instability conditions are limiting factors to a large penetration of renewable DG because of the increasing costs of power system upgrades needed for improving network capacity of integration and flexibility [1].

Many papers dealing with the problem of the grid integration of large amounts of renewable DG [2]-[4] have been proposed as well as many solutions for the electrical energy efficiency raise [5]-[7]. Some papers deal also with the participation of end consumers in the electricity market in order to minimize the costs of energy consumption [8]-[10].

C. Citro · A. Piccolo · P. Siano
Department of Industrial Engineering, University of Salerno, Fisciano (SA), Italy
e-mail: costantinocitro@gmail.com, {psiano,apiccolo}@unisa.it

C. Cecati
Department of Electrical and Information Engineering, University of L'Aquila, Italy
e-mail: carlo.cecati@univaq.it

L. Wang (Ed.): *Modeling and Control*, Green Energy and Technology, pp. 27–46.
springerlink.com

Other papers foresee the birth of new investors in the energy system capable of purchasing energy when the price is low, reselling it during peak loads, thus profiting by the open energy market [11], [12]. All these new opportunities require an active network management based on a real-time pricing of the electricity, which undoubtedly will be a key feature of the Smart Grid (SG) in which energy prices will frequently change during the day according to the availability of energy produced by renewable sources [13], [14].

The term smart grid makes reference to a fully automated electric power system controlling and optimizing the operation of all its interconnected elements in order to guarantee the most efficient and safely operations during the electricity generation, transmission and distribution processes [15].

The interaction mechanisms between the grid system and the individual end users will change and demand side participation in competitive power markets will represent a strategic resource to manage. In this scenario, end users will play a strategic role by modifying their energy consumption, thus giving an important contribution to the balance between demand and supply [16].

These features attracted significant research attention during the last decade and several studies demonstrated that the growth of the SG will greatly depend on the availability of sophisticated the Demand Side Management (DSM) systems [17]. DSM concept includes a set of strategies adopted by grid operators in order to influence the users' electricity demand , favoring the higher electrical energy demand when it is mostly available with the aim of achieving energy cost reduction for both consumers and grid operators and allowing the best use of renewable energy sources.

DSM includes mechanisms of Price Responsive Demand and Demand Response Programs, the latter defined as the customer's ability to modify his own electricity demand in response to signals forecasted by the system when reliability is not guaranteed. Essentially, it refers to curtailment service programs actualized by paying the end-users to take their electrical load off when the system is deficient in capacity or operating reserves. Price-responsive demand refers to the changes applied by consumers to their electric load profile in response to energy market price signals to improve the economic efficiency of the energy consumption. It is a mechanism increasing the economic efficiency of an electricity markets by encouraging the energy load demand when the real-time price is low and discouraging it when it is high. In this way the peak demand as well as the need of additional generation and transmission infrastructures can be significantly reduced [18]-[20].

It's worth noting that customer's response depends on the load type that can be fixed, curtailable or shiftable. While fixed loads are inelastic to price, curtailable loads are inessential loads that can be shed (but not shifted) due to high prices or incentives. On the contrary, shiftable loads can be moved to other periods during the day.

The actualization of these new mechanisms is possible thanks to the new technologies. In fact, a SG includes complex communication systems allowing real time information exchanges about electrical energy price and availability.

This chapter is based on the idea that a combination Active Management Schemes (AMS) and DSM can improve variable renewable energy exploitation by allowing a better integration of different types of DG, such as diesel generators (DGens), wind turbines (WTs), photovoltaic generators (PV) and other sources. By supporting the electricity market in a SG, this combination can also increase flexibility in electricity consumption while reducing the costs of energy for customers.

On these basis, an Energy Management System (EMS) making use of an Optimal Power Flow (OPF) that integrates AMS and DSM and aims at maximizing the social welfare in a competitive power market is proposed.

The efficiency of the proposed method, is successful demonstrated on a 19-bus 11 kV radial distribution network.

In the following, Sections 2 presents a scheme to actualize the active control of a SG, Section 3 describes the mathematical problem formulation, Section 4 presents different case studies. A discussion on the presented results is given in Section 5 while conclusions are drawn in Section 6.

2 The Smart Grid Infrastructure

Active power networks require new control methodologies and techniques needed for achieving a dynamical control of all the interconnected elements, thus optimizing network operations and ensuring regulatory status of the whole electrical system. A possible scenery for the actualization of this infrastructure is described in Figure 1.

The main elements of the system are:

- Remote Terminal Units (RTUs);
- Supervisory Control And Data Acquisition (SCADA);
- Energy Management System (EMS);
- Advanced Metering Infrastructure (AMI);
- State Estimation Algorithms (SEA);
- Generation and Load Forecast System (GLFS).

The SCADA transmits the measurement data provided by remote collecting data devices (RTUs) placed in strategic positions along the grid to the EMS.

In order to obtain a full control of the grid, real time information of the state of the network are needed to identify the current and voltage profiles at all its nodes. To achieve this goal the use of SEA, i.e. a set of methods and mathematical procedures for evaluating the state of a network by making use of remote measurements of apposite parameters is required. Moreover, as load flow and voltage profiles estimation suffers from a substantial inaccuracy and in order to obtain an accurate state estimation, adequate monitoring systems [21] as well as an AMI, based on wireless and/or wired automatic metering systems (smart meters) with bidirectional capabilities, are needed [22]-[25].

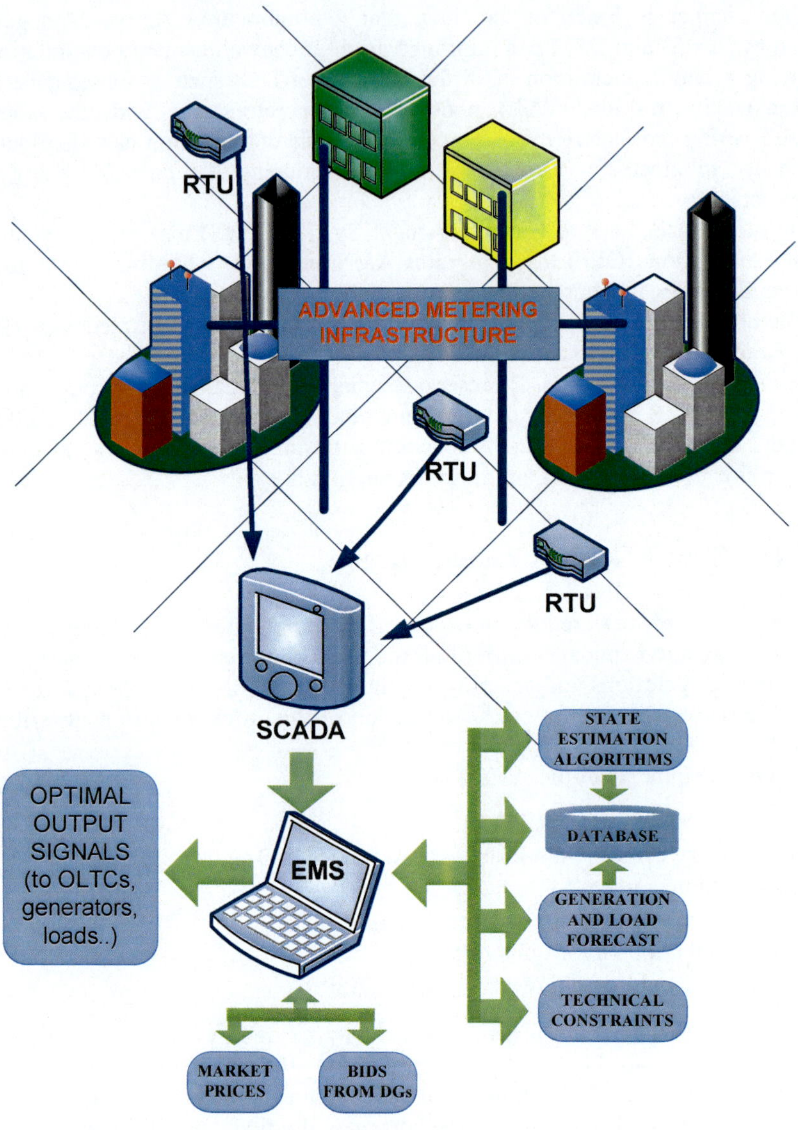

Fig. 1 Energy Management System in the Smart Grid infrastructure.

Optimization, monitoring and control of the SG performances are entrusted to a suite of hardware/software applications that constitutes the so-called EMS [26]. After evaluating the data coming from SCADA using SEA [27] it determines those actions required for achieving the optimum state of the grid, setting the optimal combination of control variables including the reference values for the active and reactive power dispatching and the position of the OLTC transformer in the substation.

The proposed EMS, making use of SEA and GLFS [28], evaluates current and voltage profiles at all nodes and determines those actions needed for maintaining the optimum state of the network. In order to determine these actions, the EMS execute an OPF with the aim of maximizing the social welfare in a competitive power market. In particular, the complementary operations executed by the EMS are:

- a first one day-ahead schedule of DGs according to the market prices;
- a real time intra-day optimization operation that every *m* minutes modifies the scheduling in order to consider the operation requirements.

3 Mathematical Problem Formulation

Recently, the OPF problem has received a particular attention from the perspective of global optimization methods including genetic algorithms [29]-[30], simulate annealing [31], tabu search [32], nonlinear programming solvers [33] and so on, however, these techniques require very high speed computers especially for large-scale systems. The optimization approach proposed in this chapter, i.e. Sequential Quadratic Programming (SQP), provides very good results, comparable with the performances of interior point method solution of OPF relaxation [34], without requiring huge amount of computational resources.

Nevertheless, the proposal of a new solution methodology offering relative advantages if compared with other optimization methods is outside the scope of this chapter. The major aim and novelty of this study is, instead, to evaluate the benefits of DSM and AMS, such as on-load tap-changers (OLTCs) regulation and the WTs power factor control [35], integrated in the OPF for the optimization of a SG in a competitive power market.

In the assumed open market environment structured as a pool, the energy trading takes place over regular intervals of time. Each source or load, every *m* minutes, e.g. 15 minutes, bids for the production or the consumption for the next hour in *m*-minute intervals. The Distribution System Operator (DSO) optimizes the market operation sending dispatch signals to the local DG controllers and load controllers, as well as to the OLTC. The objective consists in supplying at lower cost without any security violations all pool demand [36], maximizing during each time interval the social welfare:

$$\begin{aligned}&\text{maximize}\left[\mathbf{f}(\mathbf{x},\mathbf{u})=\sum_{c=1}^{N_c}B_c(\mathbf{d})-\sum_{g=1}^{N_g}C_g(\mathbf{s})\right]\\&\text{subject to}\\&\mathbf{h}(\mathbf{x},\mathbf{u},\mathbf{s},\mathbf{d})=0\\&\mathbf{g}(\mathbf{x},\mathbf{u},\mathbf{s},\mathbf{d})\le 0\end{aligned} \tag{1}$$

Where $\mathbf{x}$ is the vector of dependent variables, containing the amplitudes and angles of the buses voltages, $\mathbf{u}$ is the vector of control variables, including the secondary voltage of the OLTC and the active and reactive power injected or absorbed by generators and loads.

N_c is the set of pool load buses;

N_g is the set of pool generator buses;

$\mathbf{d} = [\mathbf{d}_p, \mathbf{d}_q]$ is the demand vector;

$B_c(\mathbf{d}) = B_c(\mathbf{d}_p, \mathbf{d}_q) = B_{pc}(\mathbf{d}_p) + B_{qc}(\mathbf{d}_q)$ is the benefit of consumer c.

$\mathbf{s} = [\mathbf{s}_p, \mathbf{s}_q]$ is the supply vector;

$C_g(\mathbf{s}) = C_g(\mathbf{s}_p, \mathbf{s}_q) = C_{pg}(\mathbf{s}_p) + C_{qg}(\mathbf{s}_q)$ is the cost of supplier g. Subscript p and subscript q signify a relation to active and reactive power, respectively. In the pool model, production costs and benefit functions are assumed as quadratic functions of active and reactive power of pool loads and generators, as follows:

$$C_{pg}(\mathbf{s}_p) = a_{0g} + a_{1g} s_{pg} + a_{2g} s_{pg}^2 \tag{2}$$

$$C_{qg}(\mathbf{s}_q) = \alpha_{0g} + \alpha_{1g} \left| s_{qg} \right| + \alpha_{2g} s_{qg}^2 \tag{3}$$

$$B_{pc}(\mathbf{d}_p) = b_{0c} + b_{1c} d_{pc} + b_{2c} d_{pc}^2 \quad \phi_g^- < \phi_g < \phi_g^+ \tag{4}$$

$$B_{qc}(\mathbf{d}_q) = -B_{q0c} \left[d_{qc} - \gamma d_{pc} \right]^2 \tag{5}$$

The price-dependent load is modeled with a consumer benefit function, $B_c(\mathbf{d})$, concave and increasing, with $\mathbf{d}$ including both the real and reactive power demand [36].

In order to integrate the simulation of reactive power exchange, market price-dependent reactive loads should be considered. As reactive power acts more as a service which enables the consumption of real power, a benefit function different from the real power benefit equation should be determined.

Accordingly, the benefit of the reactive power can be considered as the avoidance of its shifting from a given desired level for a specified active power consumption. Desired reactive power demand is that required by the load at the given load level and can be defined as a function of the real power demand $d_{qdesired} = f(d_p)$. Assuming that the magnitude of the function increases with d_p as $d_{qdesired} = \gamma d_p$ and considering a concave function for B_q as $k(x) = k(d_q - d_{qdesired})^2$ follows Eq. (5).

In the objective function, the active and reactive power production costs of the generators, as well as the costs of the power imported/exported from/to the HV/MV transformer referred to as the slack bus, i.e. the boundary between the MV distribution network and the HV transmission network, are considered.

In order to maximize the social welfare the nonlinear programming formulation of the OPF described in [37]-[40] has been modified in order to include the AMS. The OPF has been solved by SQP and a nonlinear programming solver was used. SQP applies the equivalent of a Newton step to the KKT conditions of the nonlinear programming problem and this leads to a fast rate of convergence [41], [42].

3.1 Discrete Variables Handling

Using both discrete and continuous controls converts the OPF into a mixed discrete-continuous optimization problem. Since in the considered mixed-integer nonlinear problem the sole integer variable is the OLTC transformer tap, the solution of the mixed-integer nonlinear programming problem was implemented in a two-stage approach [43]. First, a solution over the full range of variables is generated while assuming that all variables are continuous. Then, the discrete variable was moved to the nearest discrete setting, and treated as fixed in a second-stage solution. The second solution was compared to the first finding a negligible difference between these two solutions: this would imply that the effects of the OLTC discretization is negligible [43]. In fact, it is largely agreed that the round-off technique is generally suitable for discrete variables with small steps (e.g., OLTC transformer ratio and phase shifter angle) but requires some caution for discrete variables with larger steps (e.g. shunt compensation banks, network switching) [44]-[46].

3.2 Constraints

The equality constraints $\mathbf{h}(\mathbf{x},\mathbf{u},\mathbf{s},\mathbf{d})$ represent the static load flow equations such as Kirchhoff current law $\forall b \in B$, where B is the set of busses (indexed by b) and Kirchhoff voltage law $\forall l \in L$, where L is the set of lines (indexed by l) [33].

The inequality constraints $\mathbf{g}(\mathbf{x},\mathbf{u},\mathbf{s},\mathbf{d})$ are listed in the following:

- *Active and reactive power constraints for the interconnection to external network (slack bus)* $\forall n \in N$:

$$
\begin{aligned}
s_{pn}^{-} \le s_{pn} \le s_{pn}^{+} \\
s_{qn}^{-} \le s_{qn} \le s_{qn}^{+}
\end{aligned}
\tag{6}
$$

where N is the set of external sources (indexed by n), s_{pn} and s_{qn} are the active and reactive power outputs of n, respectively and s_{pn}^{-} / s_{pn}^{+} and s_{qn}^{-} / s_{qn}^{+} are the min/max values they can assume.

- *Active and reactive power constraints for generators:* $\forall g \in N_g$:

$$\begin{aligned} s_{pg}^{-} \leq s_{pg} \leq s_{pg}^{+} \\ s_{qg}^{-} \leq s_{qg} \leq s_{qg}^{+} \end{aligned} \tag{7}$$

where s_{pg} and s_{qg} are the active and reactive power outputs of g , respectively and s_{pg}^{-} / s_{pg}^{+} and s_{qg}^{-} / s_{qg}^{+} are the min/max values they can assume.

- *Active and reactive power constraints for consumers load,* $\forall c \in N_c$:

$$\begin{aligned} d_{pc}^{-} \leq d_{pc} \leq d_{pc}^{+} \\ d_{qc}^{-} \leq d_{qc} \leq d_{qc}^{+} \end{aligned} \tag{8}$$

where d_{pc} and d_{qc} are the active and reactive consumed by consumer c , respectively and d_{pc}^{-} / d_{pc}^{+} and d_{qc}^{-} / d_{qc}^{+} are the min/max values they can assume.

- *Voltage level constraints* $\forall b \in B$ *:*

$$V_b^{-} \leq V_b \leq V_b^{+} \tag{9}$$

where V_b is the voltage at *b*, V_b^{+} and V_b^{-} are the max/min values it can assume.

- *Flow constraints for lines and transformers* $\forall l \in L$ *:*

$$\sqrt{\left(f_l^P\right)^2 + \left(f_l^Q\right)^2} \leq f_l^{+} \tag{10}$$

where f_l^P and f_l^Q represent the active and reactive power injection onto *l* respectively and f_l^{+} the maximum power flow on *l*.

The additional constraints derived from the AMS are related to the coordinated on-load tap-changer voltage, the WTs and DGens power factor angles.

- *Coordinated on-load tap-changer voltage constraint:*

$$V_{OLTC}^{-} < V_{OLTC} < V_{OLTC}^{+} \tag{11}$$

where V_{OLTC} is the secondary voltage of the OLTC, $V_{OLTC}^{-} / V_{OLTC}^{+}$ are the max/min values it can assume.

- *Coordinated generator reactive power constraints,* $\forall g \in N_g$ *:*

$$\phi_g^{-} < \phi_g < \phi_g^{+} \tag{12}$$

where ϕ_g is the power factor angles of *g*, ϕ_g^{-} / ϕ_g^{+} are the max/min values it can assume.

4 Case Study

Study System

The proposed technique was applied to the 19-bus 11 kV radial distribution system, shown in Fig. 2.

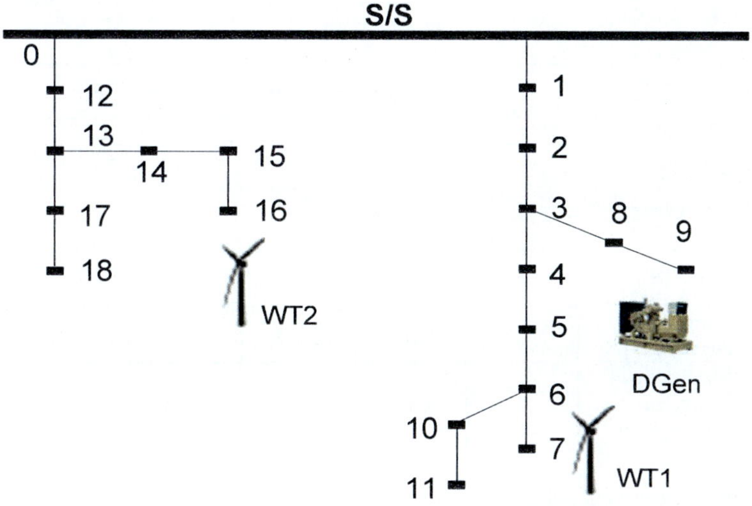

Fig. 2 Test network

The phasor dynamic models for the WTs, the DGens, the OLTC and the other distribution system elements have been implemented using Matlab® SimPowerSystems®.

The two feeders are supplied by a 6 MVA 33/11 kV transformer, the tap position allows nine different voltages with a step $\Delta U = 0.0235$ p.u. Voltage limits are taken to be ±10% of nominal and feeder thermal limits are 1.5 MVA (81 A/phase). The loading at each bus is assumed to track a load curve [27]: discrete load bands across a year have been considered: maximum, normal, medium and minimum load. The loading levels for each band are given in Table 1.

In the test network two wind turbines (WT1 and WT2) are connected at node 7 and 16 respectively, while a high cost DGen generating a maximum active power of about 600 kW is connected at bus 9. Each WT can generate about 1.05 MW of active power at wind speed of 12 m/s, operating within a power factor varying between 0.85 leading and lagging. The WTs generated active and reactive power dependence on the wind speed is given in Table 2. It has been assumed that each consumer has both fixed and curtailable price responsive loads and that the SG, by means of the slack bus connecting it to the transmission system, participates in the open market as a prosumer, buying or selling active and reactive power by/to the grid.

As regards with the loads, it has been assumed that each consumer, excepting the slack bus, has a load equally divided among fixed load and the price responsive load. The suppliers and customers cost coefficients are given in tables 3 and 4, respectively.

Table 1 Maximum Network Loading

Load Band	Active Power [MW]	Reactive Power [MVAR]
Minimum	1.32	0.71
Medium	1.88	1.02
Normal	2.29	1.23
Maximum	2.68	1.45

Table 2 WTs generated active and reactive power

Wind velocity [m/s]	0	6	8	10	12
Active Power [MW]	0	0.11	0.30	0.60	1.05
Maximum Reactive Power [MVar]	0	0.07	0.16	0.32	0.47

Table 3 Suppliers characteristics

Supplier	a_{2g} [€/MWh^2]	a_{1g} [€/MWh]	a_{0g} [€]	α_{2g} [€/MVarh^2]	α_{1g} [€/MVarh]	α_{0g} [€]
Slack bus (Nor/Max load)	60	220	0	0	20	0
Slack bus (Min/Med load)	52	180	0	0	20	0
Wind turbine	0	100	0	0	10	0
Diesel Generator	80	240	0	-	-	-

Table 4 Customer characteristics

Customer bus	b_{2g} [€/MWh2]	b_{1g} [€/MWh]	b_{0g} [€]
Slack bus	40	160	0
3	35	400	0
9-12	40	380	0
17	45	360	0
2-4-5-6-7-8-10-11-13-14-15-16-18-19	50	450	0

Simulation Results

The operation of the considered EMS endowed with AMS and DSM has been evaluated considering discrete load and wind speed states, respectively varying from maximum to minimum load and from 0 m/s to 12 m/s. The corresponding objective function and control variables are evaluated and discussed in the following.

Table 5 Objective Function [Euro]

Wind Speed [m/s]	Load			
	Minimum	Medium	Normal	Maximum
0	192	236	161	131
6	236	285	249	240
8	295	374	356	380
10	330	424	417	461
12	372	469	505	531

As expected, the objective function, shown in Table 5, increases proportionally to the wind speed and tends to increase proportionally to the load value too, except from minimum to normal load, mainly due to the different cost of the active power imported from the grid as defined in Table 3.

All variable loads are supplied at their maximum values when the wind speed equals 12 m/s or in correspondence of the minimum load condition, as in these cases the network constraints are not limiting the supplied active and reactive power. This is mainly achieved thanks to both the DGen operation and the AMS, such as the coordinated voltage regulation of OLTC and the power factor control of WTs. In fact, as evidenced by simulation results, the AMS are able to increase the total absorbed power by variable loads. For instance, in case of maximum load, the power factor control of WTs can increase up to 10% the total absorbed power by loads at buses 3, 12 and 17, particularly if compared to the case with only the coordinated voltage regulation of OLTC.

On the contrary, in correspondence of a wind speed of 0 m/s and 6 m/s and of maximum load, while all the other variable loads are supplied at the desired active and reactive power levels, loads at buses 3, 12 and 17 can absorb reduced active and reactive power as shown in Tables 6 and 7.

While the active power absorbed by buses 3 and 12 is reduced only during normal and maximum load conditions, that absorbed by bus 17 is reduced also in the case of medium load.

Table 6 Active power absorbed at bus 3 and at bus 12 [kW]

Wind Speed [m/s]	Load			
	Bus 3		Bus 12	
	Normal	Maximum	Normal	Maximum
0	329.75	445.38	476.28	355.09
6	487.5	426.99	487.5	512.58
8	483.42	568.75	487.5	568.75
10	487.5	568.75	487.5	568.75
12	487.5	568.75	487.5	568.75

Table 7 Active power absorbed at bus 17 [kW]

Wind Speed [m/s]	Load			
	Minimum	Medium	Normal	Maximum
0	284.37	375.91	243.75	284.37
6	284.37	406.25	371.04	284.37
8	284.37	371.73	243.75	451.9
10	284.37	406.25	487.5	368.87
12	284.37	406.25	487.5	568.75

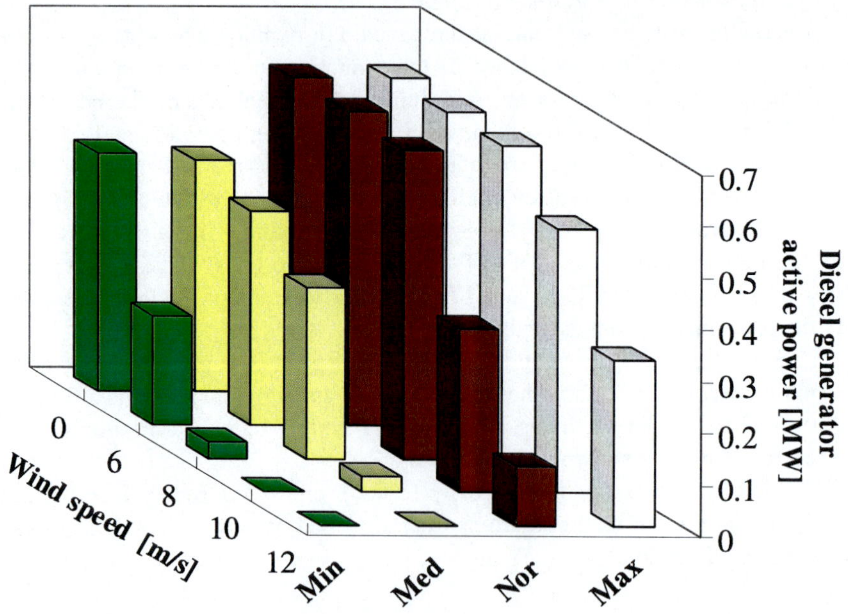

Fig. 3 Diesel generator generated active power

In correspondence of a wind speed of 12 m/s the objective function reaches its maximum values, varying from 372 € (minimum load) to 531 € (maximum load) with an increase of 159 € or of 43%.

As shown in Fig. 3, in this case, the DGen supplies active power only during maximum and normal loads, with about 116 kW and 321 kW, respectively.

The active power generated by the DGen increases with the load value and tends to supply the loads to their maximum desired value in order to maximize the social welfare. Tendentiously, the lower the wind speed, the higher the active power generated by the DGen, while it is reduced to zero in correspondence of minimum/medium load and maximum wind power production. A maximum generated power of about 606 kW is generated mainly for wind speeds below 10 m/s and maximum and normal load conditions.

It's worth noting that when the wind speed is below the cut in value, the load at bus 17 is the most limited one absorbing about 284 kW, with a reduction of about 50% if compared to the desired power level. As the variable loads operate at fixed power factor, the absorbed reactive power exhibits the same trend. Both thermal and voltage constraints limit, in fact, the power absorbed by loads at buses 3, 12 and 17 for wind speeds varying from 0 m/s to 10 m/s. Fundamentally, the thermal constraint on the wire 0-12 limits the power absorbed by loads at buses 12 and 17, while the power absorbed by the load at bus 3 is restricted by the thermal constraint on the wire 0-1 when the power generated by the WTs is zero. In contrast, in all other cases, the voltage constraint at bus 3 limits the power absorbed by the load at bus 3, while the voltage constraint at bus 17 limits the power absorbed by the loads at buses 12 and 17.

It's worth noting that, due to the implemented AMS and, in order to reduce the effects of voltage drops on the wires and relieve the voltage constraints, the WTs supply always positive (capacitive) reactive power for positive wind speeds. The reactive power output from the WTs decreases with the active power output because of the fixed relation existing between the generated active power and the maximum reactive power that can be generated, with a maximum value of about 0.48 MVAR, in correspondence of a wind speed of 12 m/s.

The active and reactive power imported from the grid tend to increase for decreasing wind speed as shown in Fig. 4 and Fig. 5.

This is always true for the active power except when the wind speed varies between 0 m/s and 6 m/s in correspondence of both normal and minimum load. In the case of normal load, the imported active power increases of about 30 kW even if the active power generated by the WTs increases of about 220 kW.

This is mainly due to the corresponding increase of the power absorbed by the variable loads as the active power supplied by the DGen remains unchanged. In the case of minimum load, the increase of the active power produced by the WTs is, instead, counterbalanced by a corresponding decrease of the power generated by the DGen as in this case all the variable load is supplied at its maximum value.

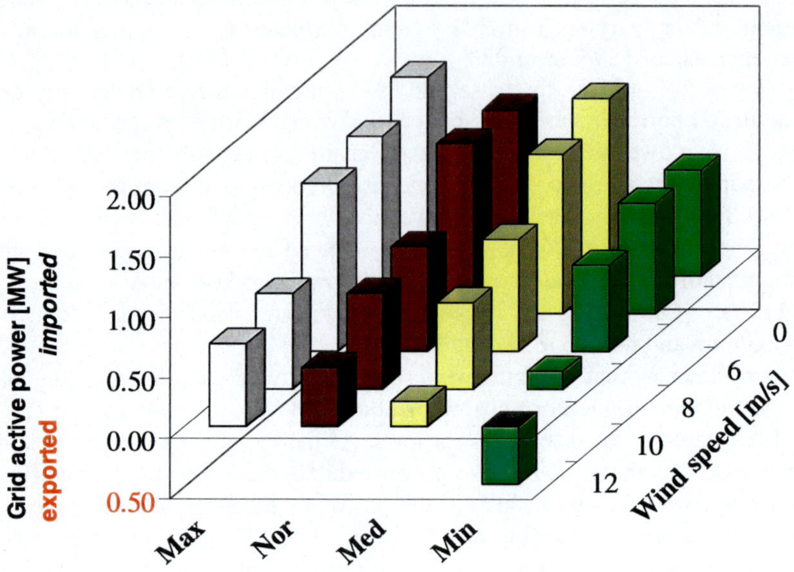

Fig. 4 Active power imported/exported from the grid

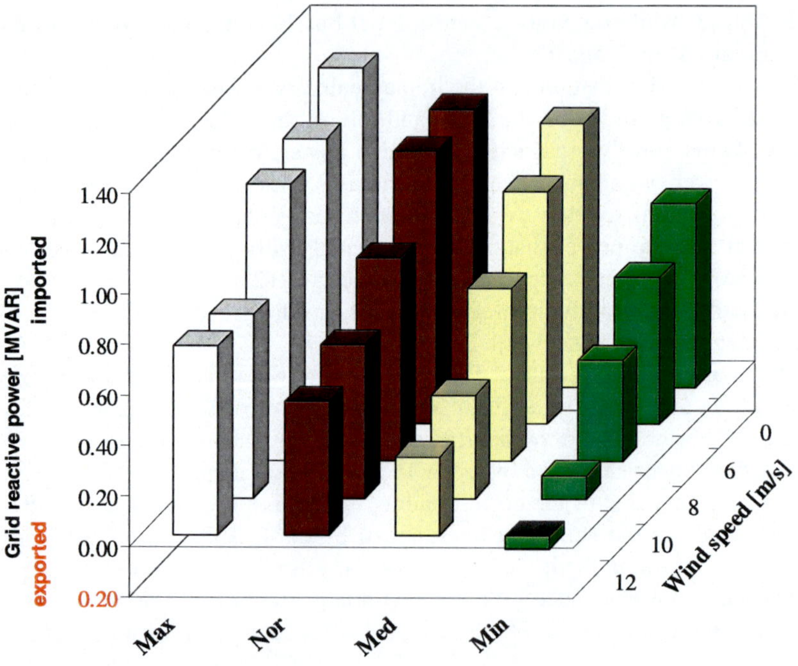

Fig. 5 Reactive power imported/exported from the grid

The relation between the tendencies of the imported/exported active and reactive power in relation with the load value is more complex as it depends much more on the active/reactive power generated by the WTs and the DGen and on the active/reactive power absorbed by the variable loads. The total costs related to the active and reactive power also vary according to the active power/reactive supplied by the WTs and the DGen and on the active/reactive power absorbed by the variable loads.

It's worth noting that, in correspondence of a wind speed of 12 m/s and minimum load, both active and reactive power are exported to the transmission grid, causing an increase in both active and reactive power costs and consumers surplus, if compared to the case of a wind speed of 10 m/s. The increased cost is due to the increased active and reactive power provided by the WTs, while the increased consumer surplus is due to the active power absorbed by the slack bus participating in the open market as a prosumer, buying and selling active and reactive power to the grid. In this case, in fact, the slack bus acts as a consumer as the costs of active and reactive power generated by the WTs are lower than the bid made for buying active and reactive power from the SG.

The use of dynamic models for WTs and DGen offers the opportunity of analyzing the dynamic performances of the local control systems and the transient shape of system quantities. Considering a condition of maximum load, Fig. 6 shows the variation of the current injected by WT2 and DGen due to the wind speed variations occurring every 5 minutes. Wind speed variations start at t=100 s and simulate a constant increase of the wind speed from 0 m/s to 12 m/s going through 8 m/s, 10 m/s and 12 m/s every 5 minutes.

It has been assumed that within 5 minutes new reference values can be sent by the EMS for the active and reactive dispatching of the DGen and the WTs and for setting the OLTC regulation. Indeed, the measured timings demonstrated that the proposed method is computationally-efficient enough to reflect the real-time requirements. In order to solve a single OPF with the considered network, it takes less than 3 minutes measuring the CPU time consumption with reference to a personal computer with a Intel® Core™ i7 processor and 8 Gb RAM. As this time closely depends on the performance of the computation machine that executes the algorithm, it can be easily reduced with more powerful computing machines.

The compliance with the statutory voltage and current limits are fully satisfied, also during transients. In particular, both the currents injected by WT2 and the current flowing on the wire 0-12 never exceeds the feeder thermal limit (81 A/phase) that limits the power transfer for wind speeds below 8 m/s.

As shown in Fig. 6, the proposed control method is able to deal with a strong correlation between the current injected by the WT, the DGen and the line constraint that, when active, limits the power absorbed by the variable loads.

Results evidenced that the proposed optimization approach provides very good results, comparable with those of interior point method solution of OPF relaxation [26]. Moreover, to validate the results the method was also coded in the AIMMS optimization modeling environment [40, 47]. The non-linear programming solver CONOPT was employed.

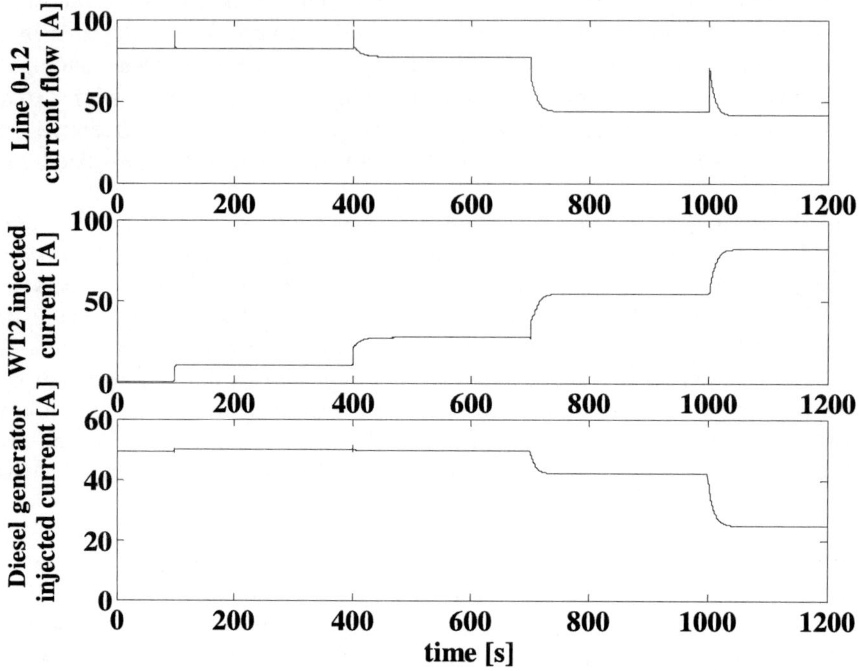

Fig. 6 Cooperation between WT2 and diesel generator while adhering to the capacity limitation of the line 0-12

The method is scalable and can be used in EMS in order to manage large networks since it can cope with a larger number of control variables, as evidenced by other simulations with larger networks not discussed in this chapter.

5 Discussion

The implementation of the proposed method based on active distribution management and the functionality of the electricity market with an active customer connection point can improve use of energy and a renewable energy exploitation. In fact, the obtained results show that introducing a combination of AMS and DSM results in an increase of the surplus of both energy producers and customers, in the enhancement of the social value of wind generation to the system and in the reduction of standby generation.

Therefore, AMS and DSM increase social welfare while reducing costs and improving reliability and the interconnection of renewable DG in distribution systems regardless of whether it is in the system.

However, even though the base infrastructure already exists, the transition from automation to active control and the SG implementation is challenging. It will, in fact, require substantial investments in huge hardware and software resources [35], [48].

Moreover, large amounts of voltage and power flow monitoring equipments should be installed across distribution networks, and a reliable communication infrastructure and communication standards for SCADA systems and for the integration of renewable energies such as IEC 61400 and IEC 61850 [49] need to be provided.

In addition to conventional energy metering, different kinds of innovative sophisticated functions based on local intelligence and power electronic devices will be essential to enable the interactive, intelligent customer to be active in network management and in the electricity market.

The customer gateway connected with DSO, service providers and electricity energy market players will be based on advanced AMI. Real-time electronically controlled devices, appliances and solutions allowing customers to dynamically react to changing prices should be introduced. These devices and solutions should be based on open standards, user friendly and should achieve a compromise optimal solution between energy or cost savings, customers' standard of comfort and equipment duration. [50]-[52].

International directives or national regulatory acts are currently encouraging these kinds of investments and large-scale projects on SG, and smart metering infrastructures are initiated by utilities in the recent years [53]-[58].

6 Conclusion

This chapter presents a new active energy management system endowed with an OPF integrating demand side management and active management schemes for the optimization of a SG in a competitive power market is proposed. Simulations results evidence that it allows a better integration of different types of DG, such as DGens and WTs, and significantly increase the electricity consumption flexibility, improving the benefits for customers.

References

1. Karkkainen, S.: Integration of Demand Side Management, Distributed Generation, Renewable Energy Sources and Energy Storages. Technical Research Centre of Finland (VTT)
2. Sioshansi, R., Short, W.: Evaluating the impacts of real-time pricingon the usage of wind generation. IEEE Trans. Power Syst. 24, 516–524 (2009)
3. Sioshansi, R.: Evaluating the impacts of real-time pricing on the cost and value of wind generation. IEEE Trans. Power Syst. 25, 741–748 (2010)
4. Vittal, V.: The impact of renewable resources on the performance and reliability of the electricity grid. Bridge 40(1), 5–12 (2010)
5. Amin, S.M., Wollenberg, B.F.: Toward a smart grid: Power delivery for the 21st century. IEEE Power Energy Mag. 3(5), 34–41 (2005)
6. Vojdani, A.: Smart integration. IEEE Power Energy Mag. 6(6), 71–79 (2008)

7. Ipakchi, A., Albuyeh, F.: Grid of the future. IEEE Power Energy Mag. 7(2), 52–62 (2009)
8. Carrión, M., Philpott, A.B., Conejo, A.J., Arroyo, J.M.: A stochastic programming approach to electric energy procurement for large consumers. IEEE Trans. Power Syst. 22(2), 744–754 (2007)
9. Gómez-Villalva, E., Ramos, A.: Optimal energy management of an industrial consumer in liberalized markets. IEEE Trans. Power Syst. 18, 716–723 (2003)
10. Philpott, A.B., Pettersen, E.: Optimizing demand-side bids in day-ahead electricity markets. IEEE Trans. Power Syst. 21, 488–498 (2006)
11. Carrión, M., Conejo, A.J., Arroyo, J.M.: Forward contracting and selling price determination for a retailer. IEEE Trans. Power Syst. 22(4), 2105–2114 (2007)
12. Gabriel, S.A., Conejo, A.J., Plazas, M.A., Balakrishnan, S.: Optimal price and quantity determination for retail electric power contracts. IEEE Trans. Power Syst. 21, 180–187 (2006)
13. Borenstein, S.: The long-run efficiency of real-time electricity pricing. Energy J. 26(3), 93–116 (2005)
14. Holland, S.P., Mansur, E.T.: The short-run effects of time-varying prices in competitive electricity markets. Energy J. 27(4), 127–155 (2006)
15. Smart Grids European Technology Platform SmartGrids, `http://www.smartgrids.eu`
16. Smart Grids European Technology Platform, Vision and Strategy for Europe's. Electricity Networks of the Future, European Commission (2006)
17. Thornley, V., Kemsley, R., Barbier, C., Nicholson, G.: User perception of demand side management. In: CIRED Seminar 2008: SmartGrids for Distribution, Paper No.70, June 2008, pp. 1–4 (2008)
18. Chao, H.-p.: Price responsive demand management for a smart grid world
19. Nilsson, H.: The many faces of demand-side management. IET - Power Eng. Journ. 8(5), 207–210 (1994)
20. Bossi, C., Ciapessoni, E., Cirio, D., L'Abbate, A., Martini, L., Mauri, G., Moneta, D., Tornelli, C.: Smart Grid: tecnologie, funzionalità ed iniziative di dimostrazione in corso in Europa. In: Contributo ERSE al Piano di Azione Tecnologica del Major Economies Forum (2010)
21. Siano, P., Chen, P., Chen, Z., Piccolo, A.: Evaluating maximum wind energy exploitation in active distribution networks. IET Gener. Transm. Distrib. 4(5), 598–608 (2010)
22. Muscas, C., Pilo, F., Pisano, G., Sulis, S.: Optimal allocation of multichannel measurement devices for Distribution State Estimation. IEEE Trans. on Instrumentation and Measurement 58(6) (June 2009)
23. Fabro, M., Roxey, T., Assante, M.: No Grid Left Behind. IEEE Security & Privacy 8(1), 72–76 (2010)
24. Russell, B.D., Benner, C.L.: Intelligent Systems for Improved Reliability and Failure Diagnosis in Distribution Systems. IEEE Trans. on Smart Grids 1(1), 48–56 (2010)
25. Neenan, B., Hemphill, R.C.: Societal Benefits of Smart Metering Investments. Elsevier - The Electricity Journal 21(8), 32–45 (2008)
26. Gungor, V.C., Lu, B., Hancke, G.P.: Opportunities and Challenges of Wireless Sensor Networks in Smart Grid. IEEE Trans. on Ind. Electron, Early Access
27. Muller, H., Rudolf, A., Aumayr, G.: Studies of distributed energy supply systems using an innovative energy management system. In: Power Industry Computer Applications, PICA 2001, pp. 87–90 (2001)

28. Kamireddy, S., Schulz, N.N., Srivastava, A.K.: Comparison of state estimation algorithms for extreme contingencies. In: 40th North American Power Symposium, NAPS 2008, pp. 1–5 (2008)
29. Ackermann, T.: Wind Power in Power Systems. John Wiley & Sons, Ltd, Chichester
30. Lai, L.L., Ma, J.T., Yokoyama, R., Zhao, M.: Improved Genetic Algorithms for Optimal Power Flow under both normal and contingent operation states. Journal of Electric Power and Energy Systems 19(5), 287–292 (1997)
31. Bakirtzis, A.G., Biskas, P.N., Zoumas, C.E., Petridis, V.: Optimal Power Flow by Enhanced Genetic Algorithm. IEEE Trans. on Power Syst. 17, 229–236 (2002)
32. Chen, L., Suzuki, H., Katou, K.: Mean-field theory for optimal power flow. IEEE Trans. on Power Syst. 12, 1481–1486 (1997)
33. Kulworawanichpong, T., Sujitjorn, S.: Optimal Power Flow using tabu search. IEEE Power Engineering review, 37–40 (2002)
34. Yan, W., Liu, F., Chung, C.Y., Wong, K.P.: A Hybrid Genetic Algorithm-Interior Point Method for Optimal Reactive Power Flow. IEEE Trans. on Power Syst. 21(3), 1163–1169 (2006)
35. Capitanescu, F., Wehenkel, L.: A new heuristic approach to deal with discrete variables in optimal power flow computations. In: IEEE Bucharest Power Tech Conference, Bucharest, Romania, June 28 - July 2, pp. 1–6 (2009)
36. Webera, J.D., Overbye, T.J., DeMarco, C.L.: Modeling the consumer benefit in the optimal power flow. Decision Support Systems 24(3-4), 279–296 (1999)
37. Harrison, G.P., Piccolo, A., Siano, P., Wallace, A.R.: Exploring the Trade-offs Between Incentives for Distributed Generation Developers and DNOs. IEEE Trans. Pwr. Sys. 22(2), 821–828 (2007)
38. Harrison, G.P., Piccolo, A., Siano, P., Wallace, A.R.: Hybrid GA and OPF evaluation of network capacity for distributed generation connections. Electrical Power Systems Research 78(3), 392–398 (2008)
39. Piccolo, A., Siano, P.: Evaluating the Impact of Network Investment Deferral on Distributed Generation Expansion. IEEE Trans. Pwr. Sys. 24, 1559–1567 (2009)
40. Siano, P., Ochoa, L.F., Harrison, G.P., Piccolo, A.: Assessing the Strategic Benefits of Distributed Generation Ownership for DNOs. IET Gener. Transm. Distrib. 3, 225–236 (2009)
41. Liu, H., Shen, Y., Zabinsky, Z.B., Liu, C.-C., Courts, A., Joo, S.-K.: Social Welfare Maximization in Transmission Enhancement Considering Network Congestion. IEEE Transactions on Power Systems 23(3), 1105–1114 (2008)
42. Biegler, L.T., Grossmann, I.E.: Retrospective on optimization. Computers & Chemical Engineering 28(8), 1169–1192 (2004)
43. Capitanescu, F., Wehenkel, L.: A new heuristic approach to deal with discrete variables in optimal power flow computations. In: 2009 IEEE Bucharest Power Tech Conference, Bucharest, Romania, June 28-July 2, pp. 1–6 (2009)
44. Tinney, W.F., Bright, J.M., Demaree, K.D., Hughes, B.A.: Some deficiencies in Optimal Power Flow. IEEE Trans. on Power Syst. 3, 676–683 (1988)
45. Papalexopoulos, A.D., Imparato, C.F., Wu, F.W.: Large-Scale Optimal Power Flow: Effects of Initialization, Decoupling & Discretization. IEEE Trans. on Power Syst. 4, 748–759 (1989)
46. Liu, E., Papalexopoulos, A.D., Tinney, W.F.: Discrete Shunt controls in a Newton Optimal Power Flow. IEEE Trans. on Power Syst. 7, 1519–1528 (1992)
47. Bisschop, J., Roelofs, M.: AIMMS – The user's guide (Paragon Decision Technology) (2006)

48. Roytelman, I., Ganesan, V.: Coordinated Local and Centralized Control in Distribution Management Systems. IEEE Tran. on Pwr Del. 15(2), 718–724 (2000)
49. Timbus, A., Larsson, M., Yuen, C.: Active Management of Distributed Energy Resources Using Standardized Communications and Modern Information Technologies. IEEE Trans. on Ind. Electron 56(10), 4029–4037 (2009)
50. Liserre, M., Sauter, T., Hung, J.Y.: Energy Sources into the Smart Power Grid through Industrial Electronics. Industrial Electronics Magazine 4(1), 18–37 (2010)
51. Järventaustaa, P., Repoa, S., Rautiainena, A., Partanenb, J.: Smart grid power system control in distributed generation environment. Annual Reviews in Control (article in Press)
52. Intelligent energy systems, A White Paper with Danish perspectives, `http://www.ea-energianalyse.dk/reports/`
53. Directive 2009/72/EC of the European Parliament and of the COuncil of July 13, 2009 concerning common rules for the internal market in electricity and repealing Directive 2003/54/EC (2009)
54. Directive 2006/32/EC of the European Parliament and of the Council of April 5, 2006 on energy end-use efficiency and energy services and repealing Council Directive 93/76/EEC (2006)
55. ERCEG, Position Paper on Smart Grids, Ref: E09-EQS-30-04, December 10 (2009)
56. ERCEG, Smart metering with a focus on electricity regulation, Ref_ R07-RMF-04-03, October 31 (2007)
57. US Departement of Energy - DOE, Smart Grid System Report (July 2009)
58. Cavoukian, A., Polonetsky, J., Wolf, C.: Smart Privacy for the Smart Grid: embedding privacy into the design of electricity conservation. Identity in the Information Society, published online on April 20 (2010)

Active and Reactive Power Control of Grid Connected Distributed Generation Systems

Agustí Egea-Alvarez, Adrià Junyent-Ferré, and Oriol Gomis-Bellmunt

Abstract. The present chapter describes active and reactive power control for distributed generation and storage systems connected to the grid by means of voltage source converters. Renewable generation and storage systems connected to a three-phase three-wire grid are considered. The different system components are described and modelled. The overall active and reactive power control scheme based on the instantaneous power theory is described. The fundamental necessary control blocks are detailed including the phase locked loop, current controllers, current references calculation and DC bus voltage controllers. Simulations results are provided to exemplify the described control approach.

1 Introduction

The need for more reliable and flexible power systems along with the great potential of modern control and communication systems and power electronics, has led to development of the smart grid concept [1]. Modern grids will be required to be active and to adapt to a number of fault events ensuring the system optimum performance during and after faults occur. Furthermore, modern grids will have to integrate the increasing penetration of renewable energy of intermittent nature.

Power electronics are the enabling technology to convert classical power systems into smart grids, since they allow controlling the power flows and

Agustí Egea-Alvarez · Adrià Junyent-Ferré · Oriol Gomis-Bellmunt
Centre d'Innovació Tecnològica en Convertidors Estàtics i Accionaments (CITCEA-UPC), Universitat Politècnica de Catalunya UPC, Av. Diagonal, 647, Pl. 2. 08028 Barcelona, Spain
e-mail: {agusti.egea,adria.junyent,gomis}@citcea.upc.edu

Oriol Gomis-Bellmunt
IREC Catalonia Institute for Energy Research, Barcelona, Spain
e-mail: ogomis@irec.cat

L. Wang (Ed.): *Modeling and Control*, Green Energy and Technology, pp. 47–81.
springerlink.com

bus voltages in the milliseconds range. In particular, AC-DC converters with bidirectional power capability are the key elements in microgrids [2, 3] and distributed generation systems. The different configurations of variable speed wind turbines need power converter structures based on two AC-DC converters. Each converter needs active and reactive power control capability in order to extract the optimum power from the wind turbine while exchanging the appropriate reactive power with the power grid. Similarly, photovoltaic systems need a AC-DC inverter to inject the generated power into the grid. Storage systems based on batteries also require bidirectional AC-DC converters to charge the battery or to inject power to the grid. An example scheme of a microgrid comprising these elements is sketched in Figure 1.

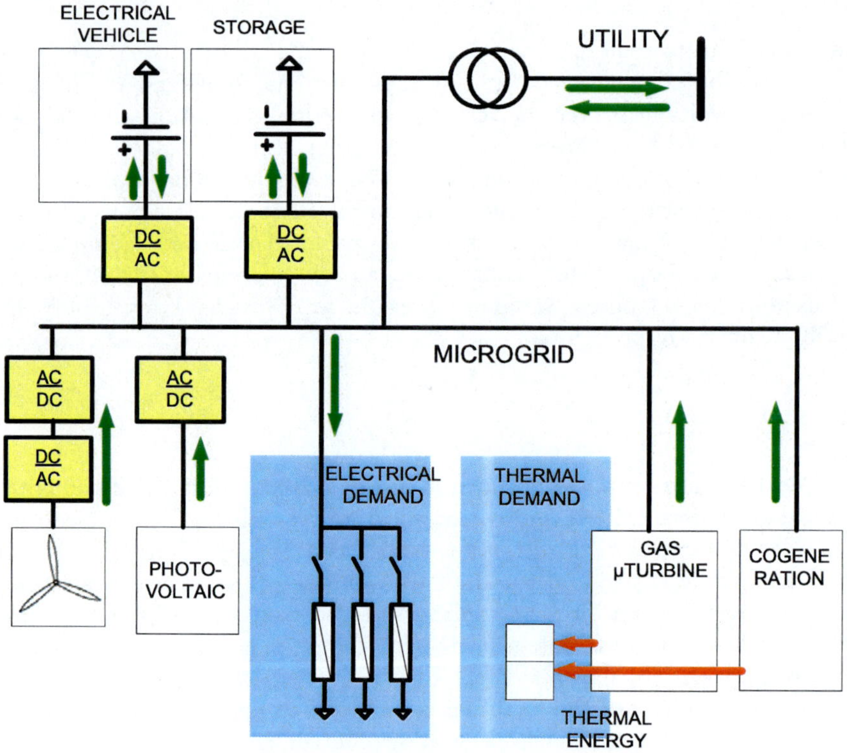

Fig. 1 Example microgrid with generation and storage systems

There exist a wide variety of converter topologies[4, 5, 6] to connect renewable generation or microgrid units to the main grid. Inverters to inject power to the grid can be classified according to different criteria. Regarding the kind of semiconductor used, inverters can be classified as:

- Inverters based on Insulated-Gate Bipolar Transistors (IGBT) or similar technologies that can provide fast switching and modulate any desired voltage. The resulting converters are the so-called Voltage Sourced Converters (VSC) which can control independently active and reactive power, can provide black start capability and inject reduced harmonic currents allowing to use lighter filters. The high switching frequency at which they are operated implies higher losses, which is the main drawback of this technology.
- Inverters based on thyristor or similar technologies that require the grid to be operated. The resulting converters are the so-called Line Commutated Converters (LCC) which can control active power while consuming non-controllable reactive power, require the grid to be operated and require large filters for the important harmonic currents they generate. The main advantage is that they are available for higher voltage and power and that they produce less losses since they commutate at low frequency (grid frequency).

Inverters can be also classified depending on their number of levels. In low voltage applications, it is common to use only two-levels, when the voltage increases, several semiconductors should be placed in series. In this case, multilevel technologies can be used, using these multiple levels to modulate voltages with fewer harmonic content. The different multilevel converter topologies can be classified[7] in multilevel configurations with diode clamps, bidirectional switch interconnection, flying capacitors or cascaded H-bridge converters.

Concerning the system requirements that have to be provided by power electronics, voltage fault ride-through capability for wind farms (and possibly also for other renewable sources or for microgrids in the near future) is one of the most relevant demands to renewable generation or microgrids. This requirement specifies that the concerned system cannot disconnect whenever a fault occurs, and it is specially important when the penetration of such nodes (wind power for example) is significant enough to produce stability problems in the main grid. Furthermore, modern microgrids and renewable energy sources should operate as conventional power plants providing ancillary services, as voltage and frequency support to the main grid.

A number of control schemes have been proposed in the literature to address the control of power converters. Feedback controlled converters present a series of advantages compared to open loop controlled converters. Namely adaptability and robustness to disturbances on the grid and to different operation points, fast response and higher stability have proven necessary in most applications and have made feedback control almost unavoidable. The feedback control techniques for power converters can be roughly classified between linear control theory based designs and nonlinear controller designs [8, 9].

The linear theory designs are based on the averaged model of the converter, which considers the control action to be able to change continuously despite the discrete number of possible switching states of the converter. This makes it possible to apply the well known linear control theory techniques to design and analyze the system, providing a lot of information on how will the system behave under different circumstances. This also makes it necessary to use a Pulse Width Modulation (PWM) technique, such as the sinusoidal PWM and the space vector PWM [10, 5], to transform the voltage output reference from the current controller into the switching signals sent to the actual converter switching devices. For the computation of the controller output, usually a variable transformation matrix is applied to the measured magnitudes. By applying the so known Park reference transformation matrix using a reference angle matching the grid angle (the so called synchronous reference frame), both voltage and current magnitudes become constant in steady state under grid balance conditions, making it possible to use classical Proportional Integrator (PI) regulators on the control loops hence simplifying a lot the design of the controllers. As this advantage is lost under unbalanced conditions, some authors have suggested either to use an enhaced double synchronous reference frame [11] or to use the Clarke transformation instead (the so called stationary reference frame) which requires Proportional Resonant (PR) regulators but enables proper operation of the system under such condition [12].

On the other hand, nonlinear based designs usually consider the discrete nature of the converter state so that the controller outputs the switching command for the converter devices instead of using PWM. This makes it possible to obtain a faster response and is said to be less dependant on the system parameters, but makes the system harder to study due to the higher complexity and the lower availability of the analysis tools for nonlinear systems. Among the nonlinear techniques the most well known are those based on the so called Direct Power Control (DPC) [13]. One important disadvantage of this methods are the non-constant switching frequency, which makes it harder to calculate to losses of the converter and to properly design the switching noise filters. To avoid these drawbacks, other techniques which require more computational power have recently gained popularity such as the current predictive control [14].

The present chapter is centered in the so-called two-level Voltage Source Converter (VSC). The system comprising a renewable generation or storage unit and its connection to the grid by means of a power converter is analyzed. A control scheme able to control independently active and reactive power is fully described and justified. Simulation results are included to illustrate the dynamic performance of the control scheme.

2 System Description

The system under analysis is sketched in Figure 2. The considered grid is a three-phase three-wire grid. The two level Voltage Source Converter (VSC)

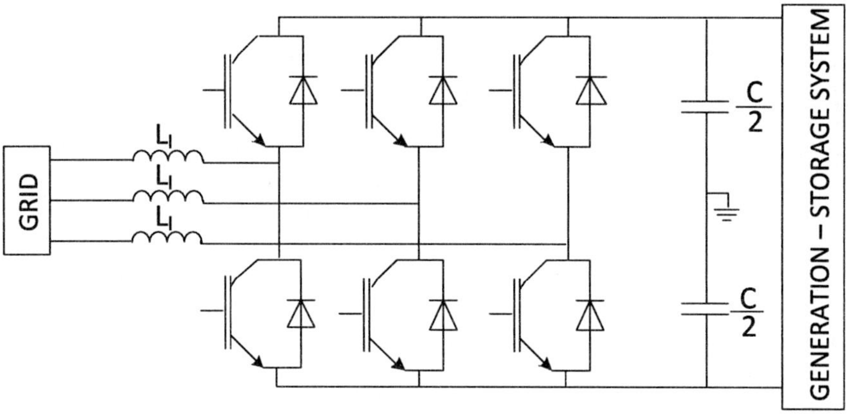

Fig. 2 System under analysis comprising the VSC converter and the three-phase utility grid

exchanges power between the AC side and the DC side. It is composed of three branches with two IGBT (Isolated Gate Bipolar Transistors), whose middle point is connected to the grid by means of inductances. Inductances allow to connect smoothly the converter to the grid. Appropriate modulation of the IGBT switching allow to generate the desired three-phase voltages on the AC side to control the active and reactive power flow.

The generation or storage source is connected to the DC side. In photovoltaic systems or batteries the source is directly of DC nature, although it has to be transformed in some applications using DC-DC converters. For wind power, the generation is of variable AC frequency, but a rectifier transforms it to DC before injecting the power to the grid. The DC side can be modelled as a DC voltage source (Figure 3) or alternatively as a current source connected to a shunt capacitor (Figure 4). The AC side can be modelled by the utility grid Thevenin equivalent or in an extremely simplified system by an AC voltage source. It is important to remark that the VSC converter alone can be considered an AC voltage source but considering also the inductances it can be considered an AC current source and therefore it can be connected to the grid (voltage source) without problem.

3 Control Scheme for P and Q Control

3.1 Clarke Transformation

The instantaneous power theory [15, 16] in the $\alpha\beta0$ frame, which will be later described, is based on the transformation of three phase instantaneous electrical quantities expressed in the *abc* reference frame to an $\alpha\beta0$ orthogonal reference frame using the Clarke transformation [17].

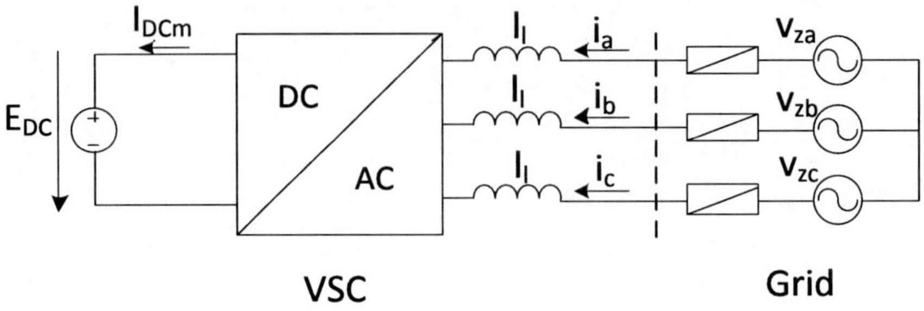

Fig. 3 VSC converter with the DC side modelled as a voltage source

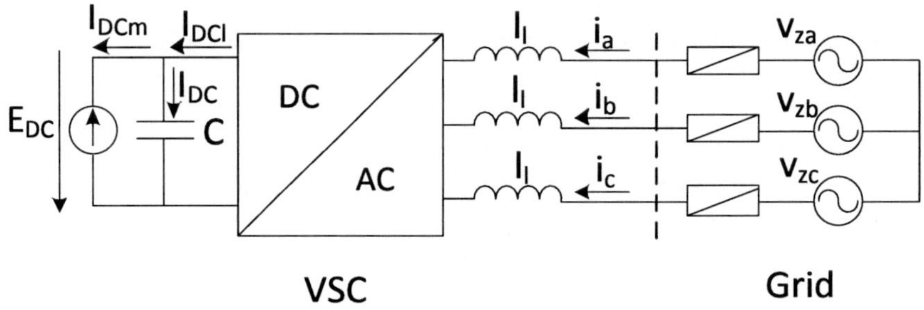

Fig. 4 VSC converter with the DC side modelled as a current source and a shunt capacitor

The so-called *Clarke transformation* is defined as

$$[x_{\alpha\beta 0}] = [T_{\alpha\beta 0}]\,[x_{abc}] \tag{1}$$

and its inverse,

$$[x_{abc}] = [T_{\alpha\beta 0}]^{-1}\,[x_{\alpha\beta 0}] \tag{2}$$

where x_{abc} is a vector with the three phase quantities in the *abc* frame and $x_{\alpha\beta 0}$ is a vector with the transformed quantities in the $\alpha\beta 0$ frame.

Expressions (1) and (2) can also be written as

$$\begin{bmatrix} x_\alpha \\ x_\beta \\ x_0 \end{bmatrix} = \frac{2}{3}\begin{bmatrix} 1 & -\frac{1}{2} & -\frac{1}{2} \\ 0 & \frac{-\sqrt{3}}{2} & \frac{\sqrt{3}}{2} \\ \frac{1}{2} & \frac{1}{2} & \frac{1}{2} \end{bmatrix}\begin{bmatrix} x_a \\ x_b \\ x_c \end{bmatrix} \tag{3}$$

$$\begin{bmatrix} x_a \\ x_b \\ x_c \end{bmatrix} = \begin{bmatrix} 1 & 0 & 1 \\ -\frac{1}{2} & -\frac{\sqrt{3}}{2} & 1 \\ -\frac{1}{2} & \frac{\sqrt{3}}{2} & 1 \end{bmatrix}\begin{bmatrix} x_\alpha \\ x_\beta \\ x_0 \end{bmatrix}$$

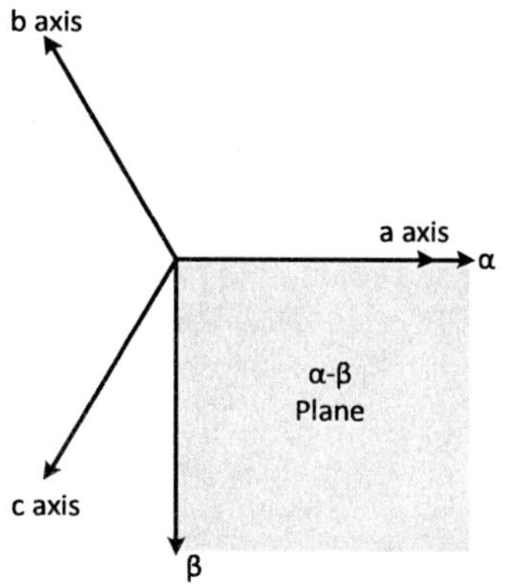

Fig. 5 $\alpha\beta$ plane representation

where x_a, x_b, x_c is the instantaneous values of voltage or current in the *abc* frame and x_α, x_β and x_0 are the transformed quantities in the $\alpha\beta0$ frame.

The Clarke transformation can be also seen as a geometric transformation as illustrated in Figure 5.

Figure 6 shows an example of three-phase voltages in the *abc* and $\alpha\beta0$ frames.

3.1.1 Instantaneous Power Theory in the $\alpha\beta0$ Frame

Instantaneous power theory [16] for balanced and unbalanced systems was presented by Akagi in 1983 [15]. Although the theory is applicable to balanced and unbalanced three-phase systems, in the present chapter balanced voltage systems are considered.

The instantaneous voltages and currents of a balanced three-phase system can be expressed as

$$\begin{aligned} x_a(t) &= \sqrt{2}X cos\left(\omega t + \phi\right) \\ x_b(t) &= \sqrt{2}X cos\left(\omega t + \phi - \frac{2\pi}{3}\right) \\ x_c(t) &= \sqrt{2}X cos\left(\omega t + \phi + \frac{2\pi}{3}\right) \end{aligned} \tag{4}$$

Transforming the *abc* (4) electrical quantities to $\alpha\beta0$ using (1)

$$\begin{aligned} x_\alpha &= \sqrt{2}X\cos(\omega t + \phi) \\ x_\beta &= -\sqrt{2}X\sin(\omega t + \phi) \\ x_0 &= 0 \end{aligned} \tag{5}$$

where it can be noted that $x_0 = 0$, since the system is balanced systems.

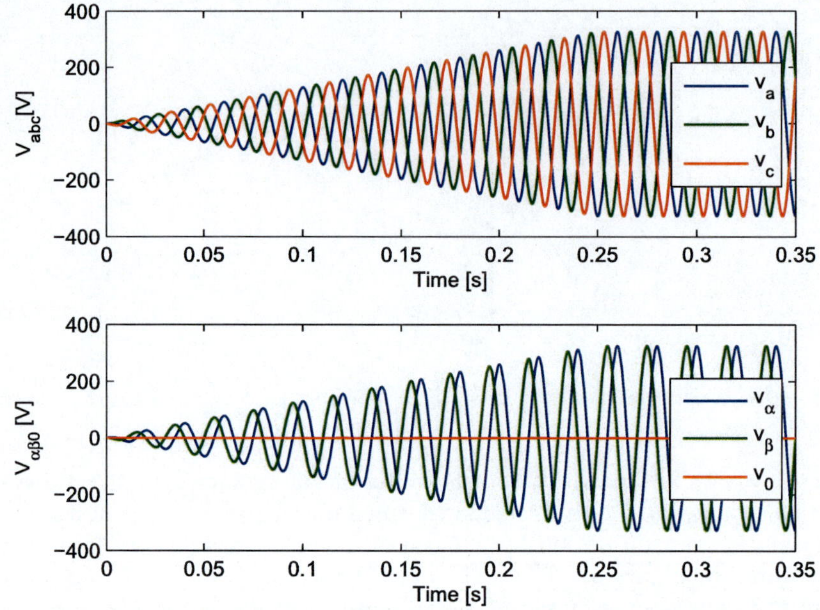

Fig. 6 Example of three-phase voltages in the *abc* and $\alpha\beta0$ frames

Defining voltage and current phasors as $\sqrt{2}\underline{V}^{\alpha\beta} = v_\alpha - jv_\beta$ and $\sqrt{2}\underline{I}^{\alpha\beta} = i_\alpha - ji_\beta$, the power expression in $\alpha\beta0$ can be deduced from the three phase power expression in *abc*

$$\underline{S} = P + jQ = 3\underline{V}^{\alpha\beta}\underline{I}^{\alpha\beta*} = 3\left(\frac{v_\alpha - jv_\beta}{\sqrt{2}}\right)\left(\frac{i_\alpha + ji_\beta}{\sqrt{2}}\right) \tag{6}$$

Rearranging expression (6), it is possible to decouple active and reactive power as

$$P = \frac{3}{2}(v_\alpha i_\alpha + v_\beta i_\beta) \tag{7}$$

$$Q = \frac{3}{2}(v_\alpha i_\beta - v_\beta i_\alpha) \tag{8}$$

obtaining the expression of active and reactive power as functions of voltages and currents in the $\alpha\beta0$ frame.

3.2 Park Transformation

The quantities in the $\alpha\beta 0$ are useful in a number of applications but have the same oscillating nature as the quantities in the *abc* frame. For the controller design it is useful to have constant quantities. This can be achieves by using the *Park transformation* [18] and the so-called synchronous reference frame.

The Park transformation is given by

$$[x_{qd0}] = [T_{qd0}]\,[x_{abc}] \tag{9}$$

and its inverse

$$[x_{abc}] = [T_{qd0}]^{-1}\,[x_{qd0}] \tag{10}$$

where x_{abc} is a vector with the three phase quantities in the *abc* frame and x_{qd0} is a vector with the transformed quantities in the *qd*0 frame.

The transformation matrix $T(\theta)$ can be written as

$$T(\theta) = \frac{2}{3}\begin{bmatrix} cos\,(\theta) & cos\left(\theta - \frac{2\pi}{3}\right) & cos\left(\theta + \frac{2\pi}{3}\right) \\ sin\,(\theta) & sin\left(\theta - \frac{2\pi}{3}\right) & sin\left(\theta + \frac{2\pi}{3}\right) \\ \frac{1}{2} & \frac{1}{2} & \frac{1}{2} \end{bmatrix} \tag{11}$$

and its inverse

$$T^{-1}(\theta) = \begin{bmatrix} cos\,(\theta) & sin\,(\theta) & 1 \\ cos\left(\theta - \frac{2\pi}{3}\right) & sin\left(\theta - \frac{2\pi}{3}\right) & 1 \\ cos\left(\theta + \frac{2\pi}{3}\right) & sin\left(\theta + \frac{2\pi}{3}\right) & 1 \end{bmatrix} \tag{12}$$

The Park transformation can be also seen as a geometric transformation which combines the Clarke transformation and a rotation as illustrated in Figure 7.

Figure 8 shows an example of three-phase voltages in the *abc* and *qd*0 frames. Note that by choosing the right θ angle, constant values are obtained.

3.2.1 Instantaneous Power Theory in the Synchronous Reference Frame

As mentioned earlier, to obtain constant steady state quantities, the angle θ employed in the Park transformation corresponds to the electrical voltage angle. Replacing θ for the electrical angle $\theta = \omega t + \phi_0$, and transforming *abc* voltages and currents to the *qd*0 frame the following voltage and current phasors can be defined, similarly to the Clarke transformation case, as

$$\underline{V}^{qd} = \frac{v_q - jv_d}{\sqrt{2}} \tag{13}$$

$$\underline{I}^{qd} = \frac{i_q - ji_d}{\sqrt{2}} \tag{14}$$

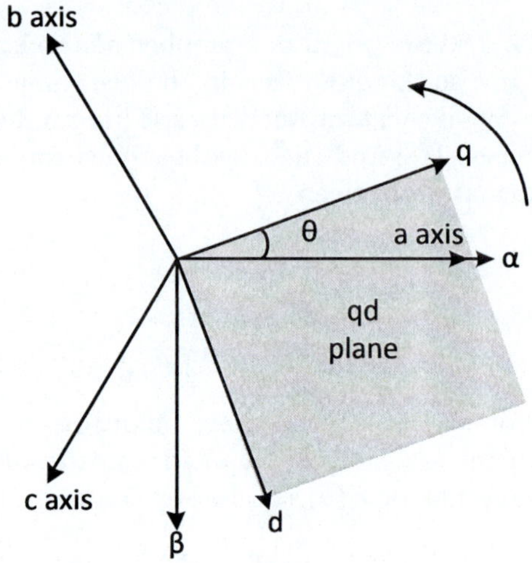

Fig. 7 *qd* plane representation

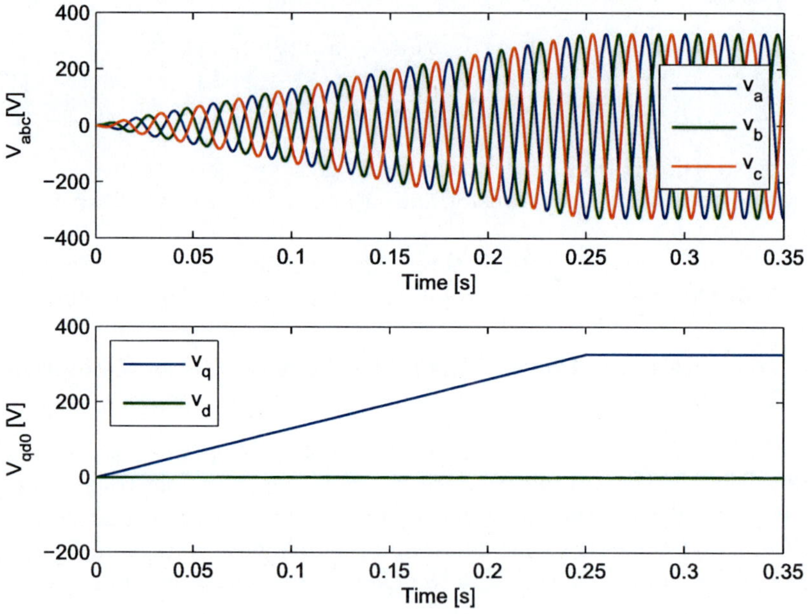

Fig. 8 Example of three-phase voltages in the *abc* and *qd*0 frames

The power of a three phase system yields

$$\underline{S} = P + jQ = 3\underline{V}^{qd}\underline{I}^{qd*} = 3\left(\frac{v_q - jv_d}{\sqrt{2}}\right)\left(\frac{i_q + ji_d}{\sqrt{2}}\right) \tag{15}$$

Reordering expression (15), active and reactive power can be expressed as [16]

$$\begin{aligned} P &= \frac{3}{2}\left(v_q i_q + v_d i_d\right) \\ Q &= \frac{3}{2}\left(v_q i_d - v_d i_q\right) \end{aligned} \tag{16}$$

obtaining the expression of active and reactive power as functions of voltages and currents in the $qd0$ frame.

3.3 VSC Converter Modelling

Although the VSC converter is based on the discrete states of the switching of the IGBTs, for control design purposes it is convenient to derive a more simplified equivalent model. A simplified model can be derived decoupling the DC and AC parts of the converter as illustrated in Figure 9. The DC side is modelled as a current source and a capacitor, while the AC side is modelled with AC voltage sources.

The current source in the DC side reflects the active power exchanged between the AC and the DC side and assures the system power balance. The DC current of the source can be computed neglecting converter losses as

$$I_{DCl} = \frac{P_{ac}}{E_{DC}} \tag{17}$$

where E_{DC} is the DC bus voltage and P_{ac} is the active power exchanged between the VSC converter and the grid in the ac side.

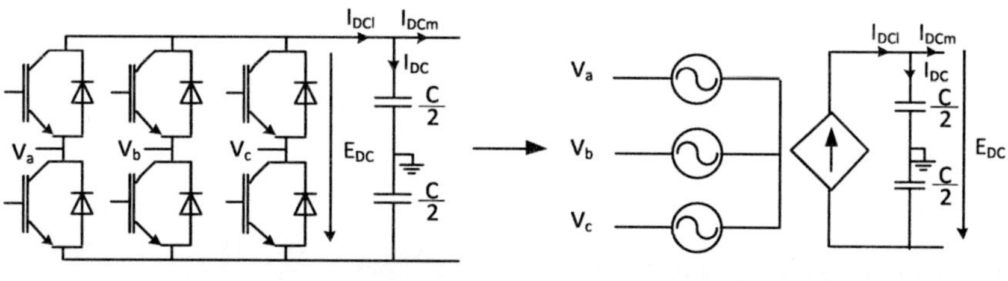

Fig. 9 Voltage Source Converter model

The DC side capacitor voltage can be obtained from the equation

$$I_{DC} = C\frac{dE_{DC}}{dt} \rightarrow I_{DCl} - I_{DCm} = C\frac{dE_{DC}}{dt} \tag{18}$$

Integrating this expression, the DC voltage can be expressed as a function of the current as

$$E_{DC} = E_{DC0} + \frac{1}{C}\int_0^t I_{DC} = E_{DC0} + \frac{1}{C}\int_0^t (I_{DCl} - I_{DCm}) \tag{19}$$

3.4 Voltage Equations in the Synchronous Reference Frame

The voltage equation can be obtained by using the equivalent scheme illustrated in Figure 10. The voltage equations yield

$$\begin{bmatrix} v_{za} \\ v_{zb} \\ v_{zc} \end{bmatrix} - \begin{bmatrix} v_{la} \\ v_{lb} \\ v_{lc} \end{bmatrix} - (v_{l0} - v_{z0}) \begin{bmatrix} 1 \\ 1 \\ 1 \end{bmatrix} = \begin{bmatrix} r_l & 0 & 0 \\ 0 & r_l & 0 \\ 0 & 0 & r_l \end{bmatrix} \begin{bmatrix} i_a \\ i_b \\ i_c \end{bmatrix} + \begin{bmatrix} l_l & 0 & 0 \\ 0 & l_l & 0 \\ 0 & 0 & l_l \end{bmatrix} \frac{d}{dt} \begin{bmatrix} i_a \\ i_b \\ i_c \end{bmatrix} \tag{20}$$

where v_{za}, v_{zb} and v_{zc} are the three-phase instantaneous grid voltages in the *abc* frame, v_{la}, v_{lb} and v_{lc} are the three-phase instantaneous converter voltages in the *abc* frame, i_a, i_b and i_c are the three-phase instantaneous currents in the *abc* frame, r_l is the inductance equivalent resistance, and l_l is the inductance value. $v_{l0} - v_{z0}$ is the voltage difference between the converter and the grid neutral. When no neutral conductor is present, it can be stated that:

$$v_{l0} - v_{z0} = \frac{1}{3}\left[1\ 1\ 1\right] \cdot \left(v_z^{abc} - v_l^{abc}\right) \tag{21}$$

where it can be noted that for three wire balanced systems, $v_{l0} - v_{z0} = 0$.

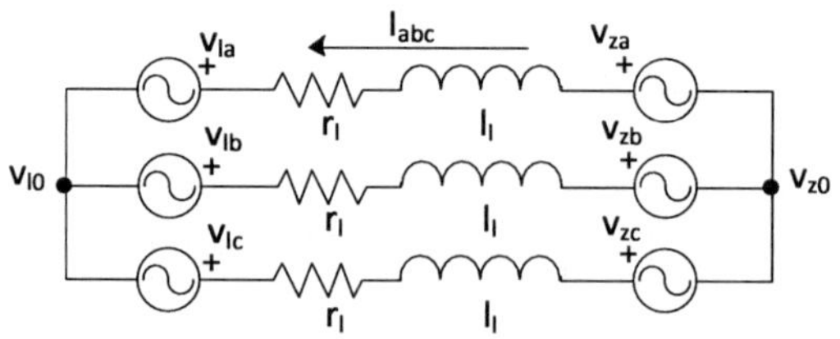

Fig. 10 Equivalent model of the AC side of a VSC converter

Applying the Park transformation (9) to (20) and taking into account that for three-wire systems $i_0 \equiv 0$, voltage equations can be expressed

$$\begin{bmatrix} v_{zq} \\ v_{zd} \end{bmatrix} - \begin{bmatrix} v_{lq} \\ v_{ld} \end{bmatrix} = \begin{bmatrix} r_l & -l_l\omega_e \\ l_l\omega_e & r_l \end{bmatrix} \begin{bmatrix} i_q \\ i_d \end{bmatrix} + \begin{bmatrix} l_l & 0 \\ 0 & l_l \end{bmatrix} \frac{d}{dt} \begin{bmatrix} i_q \\ i_d \end{bmatrix} \tag{22}$$

where v_{zq} and v_{zd} are the grid qd voltages, v_{lq} and v_{ld} are the converter qd voltages, i_q and i_d are qd currents and ω_e is the electrical angular velocity.

3.5 General Control Scheme

The VSC permits to control two electrical variables in the $qd0$ frame allowing to separately control active and reactive power. The reactive power reference can be obtained from a higher level control system (grid operator) or set to a given value. Active power reference depends on the nature of the source connected in the DC side:

- For renewable energy systems, it is adjusted to regulate the DC bus voltage and to ensure the power balance, i.e. the power injected into the grid has to be the same as the generated power. The general control scheme for this configuration is shown in Figure 11.
- For storage systems, it is adjusted to charge the battery or to inject power to the grid depending on the operation of the energy management system of the microgrid or system where the storage system is connected. The general control scheme for this configuration is shown in Figure 12.

In the present section renewable generation is considered, since it is a more complex case since a DC voltage controller is needed. For storage systems it is enough to remove the DC voltage controller and establish directly the active power reference.

The control scheme is based in a two level cascaded control system, the lower level controller allows to regulate the AC current in the q and d components, while the higher level controller deals with the regulation the DC bus voltage.

The controllers deal with currents and voltages in the $qd0$ reference frame rotating to adjust the electrical grid angle. For this reason, a Phase Locked Loop (PLL) to track the grid angle is required.

3.6 Current References Computation

The current references i_d^* and i_q^* to obtain the desired active and reactive powers P^* and Q^* can be obtained from the instantaneous power theory presented in (16),

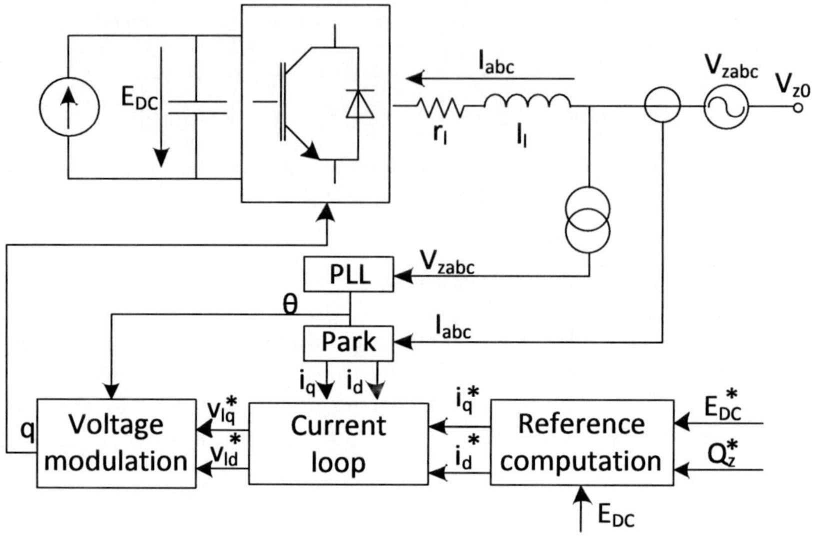

Fig. 11 Grid converter control general scheme for renewable energy generation systems

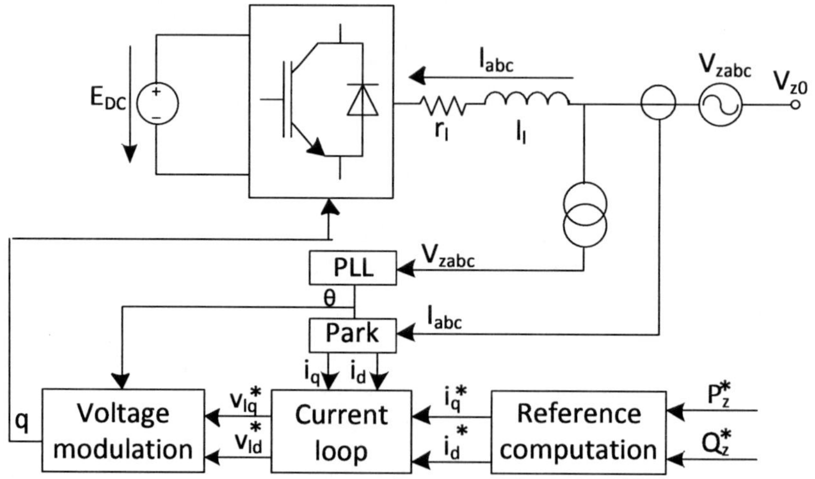

Fig. 12 Grid converter control general scheme for storage generation systems

$$P^* = \frac{3}{2}\left(v_{zq}i_q^* + v_{zd}i_d^*\right)$$
$$Q^* = \frac{3}{2}\left(v_{zq}i_d^* - v_{zd}i_q^*\right) \quad (23)$$

The Phase Locked Loop system described later, not only computes the electrical grid angle but also ensures that $v_d = 0$. Substituting it in (23), the current references i_d^* and i_q^* can be derived as

$$i_q^* = \frac{2}{3}\frac{P*}{v_{zq}} \quad (24)$$
$$i_d^* = \frac{2}{3}\frac{Q*}{v_{zq}} \quad (25)$$

The computed reference current has to be limited according to the physical limitations of the converters. The limitation can be done differently depending on the priorities between active and reactive power:

- Prioritizing i_q (active power) and providing only reactive power if there is enough available current.
- Prioritizing i_d (reactive power) and providing only active power if there is enough available current.
- Maintaining the angle between P and Q and reducing both of them equally.

3.7 DC Voltage Regulator

The DC voltage regulator is required to control the voltage of the DC bus ensuring power balance between the generation source and the power injected to the grid. The output of the DC voltage controller will provide the i_q^* reference for the current loop.

The proposed control scheme is sketched in Figure 13, where it can be seen that the controlled quantity is E^2 and a feed-forward scheme is used to improve the system response. This is a common practice, since E^2 is proportional to the energy stored in the capacitor, and the output of the controller is the active power injected to the capacitor P_C^*. Therefore, the power reference for the power converter will be $P^* = P_C^* + P_{DC}$, where P_{DC} is the measured power before the capacitor.

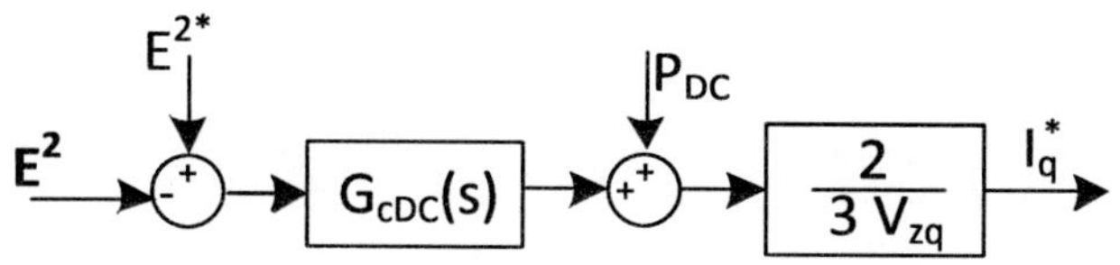

Fig. 13 Voltage controller

The control variable $W = E^2$ can be used to design the DC voltage control. The capacitor power P_C can be expressed in the Laplace domain as

$$P_C(s) = \frac{1}{2} sCW(s) \tag{26}$$

The DC voltage control can be implemented using different control approaches. Among them, P or PI controllers can be used in combination with the previously described power feed-forward scheme.

3.7.1 P Controller

A P controller can be used for the controller G_{cDC}

$$G_{c_{DC}}(s) = Kp_{DC} \tag{27}$$

The closed loop transfer function can be computed as

$$\frac{W(s)}{W^*(s)} = \frac{1}{\frac{sC}{2Kp_{DC}} + 1} \tag{28}$$

The proportional gain can be calculated as

$$Kp_{DC} = \frac{C}{2\tau_E} \tag{29}$$

where τ_E is the desired DC voltage response time constant.

Although the closed loop transfer function suggests that the system is a classical first order system with no steady state error, any perturbation provokes steady state error. This can be corrected using a PI controller thus increasing the system type.

3.7.2 PI Controller

A PI controller can be used for the controller G_{cDC}

$$G_{c_{DC}}(s) = Kp_{DC} + \frac{Ki_{DC}}{s} \tag{30}$$

The closed loop transfer function can be computed as

$$\frac{W(s)}{W^*(s)} = \frac{sKp_{DC} + Ki_{DC}}{\frac{1}{2}s^2C + sKp_{DC} + Ki_{DC}} \tag{31}$$

which is of the form

$$\frac{W(s)}{W^*(s)} = \frac{2s\xi_E\omega_E + \omega_E^2}{s^2 + 2s\xi_E\omega_E + \omega_E^2} \tag{32}$$

The PI controller gains can be calculated as

$$K_{p_{DC}} = C\xi_E \omega_E \tag{33}$$

$$K_{i_{DC}} = \frac{C\omega_E^2}{2} \tag{34}$$

where ξ_E is the desired damping ratio of DC voltage loop, and ω_E is the desired angular velocity of the voltage loop.

It is important to remark that the DC voltage loop must be much slower than the inner current controller in order to assure stable system response.

3.8 Current Loop Control

By assuming $v_{zd} = 0$ (this is done by the PLL as described in Section 3.9), the voltage equations from (37) can be written as

$$\begin{bmatrix} v_{zq} \\ 0 \end{bmatrix} - \begin{bmatrix} v_{lq} \\ v_{ld} \end{bmatrix} = \begin{bmatrix} r_l & -l_l\omega_e \\ l_l\omega_e & r_l \end{bmatrix} \begin{bmatrix} i_q \\ i_d \end{bmatrix} + \begin{bmatrix} l_l & 0 \\ 0 & l_l \end{bmatrix} \frac{d}{dt} \begin{bmatrix} i_q \\ i_d \end{bmatrix} \tag{35}$$

where it is clear that there exist a coupling between the q and d components of voltages and currents. In order to control the i_q and i_d there are mainly two different control approaches:

- Multi-variable control, controlling the q and d components with a single two dimension controller.
- Decoupling and independently controlling q and d components.

The present chapter uses the second approach of decoupling and controlling i_q and i_d separately.

The q and d components can be decoupled using

$$\begin{bmatrix} v_{lq} \\ v_{ld} \end{bmatrix} = \begin{bmatrix} -\hat{v}_{lq} + v_{zq} - l_l\omega_e i_{ld} \\ -\hat{v}_{ld} + l_l\omega_e i_{lq} \end{bmatrix} \tag{36}$$

where $\hat{v}_{lq}$ and $\hat{v}_{ld}$ are the outputs of the current controllers and v_{lq} and v_{ld} are the voltages to be applied by the converter.

Substituting in the voltage equations,

$$\begin{bmatrix} \hat{v_{lq}} \\ \hat{v_{ld}} \end{bmatrix} = \begin{bmatrix} r_l & 0 \\ 0 & r_l \end{bmatrix} \begin{bmatrix} i_q \\ i_d \end{bmatrix} + \begin{bmatrix} l_l & 0 \\ 0 & l_l \end{bmatrix} \frac{d}{dt} \begin{bmatrix} i_q \\ i_d \end{bmatrix} \tag{37}$$

Applying the Laplace transformation, the transfer function between the controller voltages and converter currents can be derived as

$$\frac{i_q(s)}{\hat{v}_{lq}(s)} = \frac{1}{l_l s + r_l} \tag{38}$$

$$\frac{i_d(s)}{\hat{v}_{ld}(s)} = \frac{1}{l_l s + r_l} \tag{39}$$

The controller can be designed using the Internal Model Control technique [19], resulting the following controller,

$$G_{ciq}(s) = G_{cid}(s) = \frac{K_p s + K_i}{s} \tag{40}$$

where the constants can be calculated as

$$K_p = \frac{l_l}{\tau} \tag{41}$$

$$K_i = \frac{r_l}{\tau} \tag{42}$$

where τ is the closed loop time constant of the electrical system. This constant must be chosen considering the converter physical restrictions. It is usual to define it a number of times (10 for example) faster than the converter switching frequency.

The implementation of the overall current controller is skected in Figure 14.

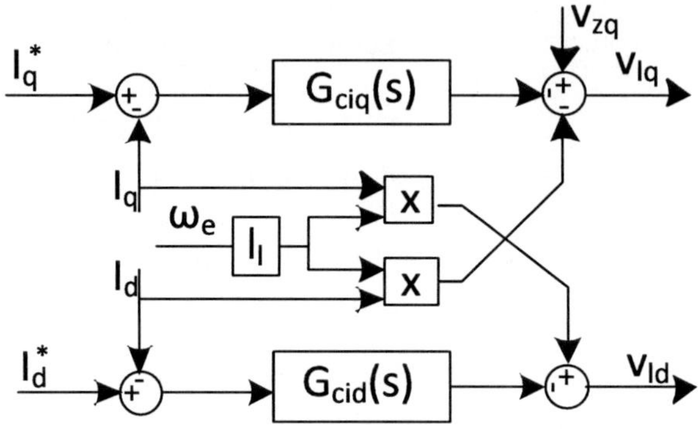

Fig. 14 Current controller

3.9 Phase Locked Loop

A phase locked loop (PLL) is used to determine the angle and the angular velocity of the electrical network. A three-phase PLL consists in a feedback of the d-axis voltage component filtered by a PI controller. The output of the controller corresponds to the angular velocity ω_e of the electrical grid and the integration of this signal corresponds the grid angle θ_e. A typical PLL scheme is illustrated in Figure 15.

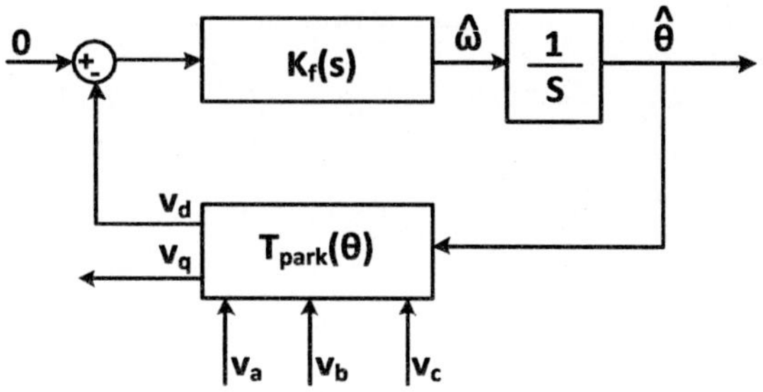

Fig. 15 Phase Locked Loop

To study the design of the PLL controller, the system can be linearized by assuming the angle error to be small. The following second order system is obtained [20]

$$\frac{\hat{\theta}(s)}{\theta(s)} = \frac{2\xi\omega_n s + \omega_n^2}{s^2 + 2\xi\omega_n + \omega_n^2} \tag{43}$$

where $\hat{\theta}(s)$ is the estimated grid angle and $\theta(s)$ is the real grid angle.

The PLL controller can be defined as

$$K_f(s) = K_p \left(\frac{\frac{1}{\tau_{PLL}} + s}{s} \right) \tag{44}$$

where τ_{PLL} is the PLL time constant.

The controller parameters K_p and τ_{PLL} can be computed using expressions [20]

$$\omega_n = \sqrt{\frac{K_p E_m}{\tau_{PLL}}} \tag{45}$$

$$\xi = \frac{\sqrt{\tau_{PLL} K_p E_m}}{2} \tag{46}$$

where E_m is the admitted peak voltage value, ξ is the damping ratio, ω_n is the electrical angular velocity.

A example of the initial transient of a PLL is illustrated in Figure 16.

3.10 Voltage Modulation

The VSC converter can apply the referenced voltages by modulating them using Pulse Width Modulation (PWM). There are different techniques to implement PWM [4, 6, 5] in power converters. The present chapter employs Space Vector PWM (SVPWM), which is the most widely used technique.

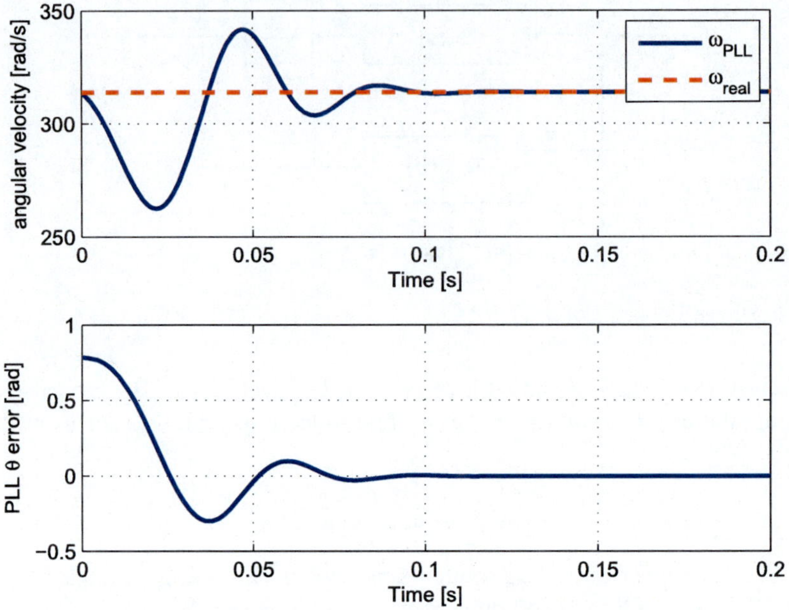

Fig. 16 PLL initial transient

Space Vector Pulse With Modulation (SVPWM) is based on space vector projection of the AC side converter voltages. Considering a two-level three-phase VSC power converter composed of three branches with two switches per branch, the obtained switching states are shown in Table 1 [6, 21].

With the proposed configuration, $2^3 = 8$ switching states are obtained. Representing the voltage vector obtained for the 8 switching states in a $\alpha\beta$ axis its vectors compose a voltage hexagon (Figure 17). The area limited between two vectors is called sector. Vectors v_0 and v_7 are denominated zero vectors, while vectors from v_1 to v_6 are denominated active vectors.

Table 1 SVPWM switching states and voltages [6, 21]

Vector	q1	q2	q3	$V_{l0}-V_{la}$	$V_{l0}-V_{lb}$	$V_{l0}-V_{lc}$	$V_{z0}-V_{la}$	$V_{z0}-V_{lb}$	$V_{z0}-V_{lc}$
0	0	0	0	$-E_{DC}/2$	$-E_{DC}/2$	$-E_{DC}/2$	0	0	0
1	1	0	0	$E_{DC}/2$	$-E_{DC}/2$	$-E_{DC}/2$	$2E_{DC}/3$	$-E_{DC}/3$	$-E_{DC}/3$
2	1	1	0	$E_{DC}/2$	$E_{DC}/2$	$-E_{DC}/2$	$E_{DC}/3$	$E_{DC}/3$	$-2E_{DC}/3$
3	0	1	0	$-E_{DC}/2$	$E_{DC}/2$	$-E_{DC}/2$	$-E_{DC}/3$	$2E_{DC}/3$	$-E_{DC}/3$
4	0	1	1	$-E_{DC}/2$	$E_{DC}/2$	$E_{DC}/2$	$-2E_{DC}/3$	$E_{DC}/3$	$E_{DC}/3$
5	0	0	1	$-E_{DC}/2$	$-E_{DC}/2$	$E_{DC}/2$	$-E_{DC}/3$	$-E_{DC}/3$	$2E_{DC}/3$
6	1	0	1	$E_{DC}/2$	$-E_{DC}/2$	$E_{DC}/2$	$E_{DC}/3$	$-2E_{DC}/3$	$E_{DC}/3$
7	1	1	1	$E_{DC}/2$	$E_{DC}/2$	$E_{DC}/2$	0	0	0

Using lineal SVPWM techniques the synthesizable voltage vector inside a sector has a maximum value limited by the DC bus voltage. For this reason it is only possible to obtain voltages inscribed in a circle of a maximum voltage of $\sqrt{3}E_{DC}/3$. Higher voltages (always inside the hexagon) can be obtained using over-modulation techniques at the cost of injecting harmonics.

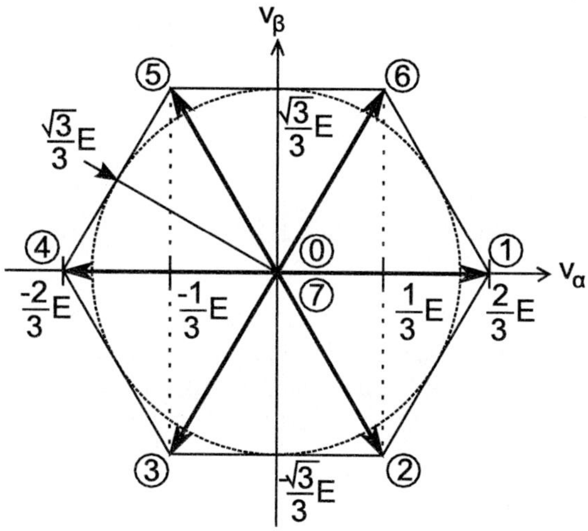

Fig. 17 Voltage vector hexagon [6]

SVPWM is based on the application of the adjacent active vectors and zero vectors during a given time. This is done inside each switching period of duration $T_s = 1/f_s$. Therefore, the higher the switching frequency f_s, the better the modulated voltage will result.

To calculate the connection time for each vector, the first step is to compute the voltage module and angle as

$$V_{svm} = \sqrt{v_\alpha^2 + v_\beta^2} \tag{47}$$

$$\theta_{svm} = \arctan \frac{v_\beta}{v_\alpha} \tag{48}$$

If the angle is not located in the first sector, it can be reduced to the first sector using

$$\theta_{sec1} = \theta_{svm} - \frac{\pi}{3}(n-1) \tag{49}$$

where n is the sector where the desired voltage is.

The modulation times t_1 and t_2 for the vectors 1 and 2 (or the corresponding vectors moved to the first sectors) can be expressed as [21]

$$t_1 = \frac{\sqrt{3}V_{svm}}{2E_{DC}} T_s \sin\left(\frac{\pi}{3} - \theta_{sec1}\right) \quad (50)$$

$$t_2 = \frac{\sqrt{3}V_{svm}}{2E_{DC}} T_s \sin(\theta_{sec1}) \quad (51)$$

The rest of the period is distributed between the zero vectors in equal part

$$t_{v_0} = t_{v_7} = \frac{T - t_1 - t_2}{2} \quad (52)$$

4 Simulation Results

The VSC converter control described in the previous section has been simulated with Matlab Simulink. Different scenarios have been considered:

1. Storage system: VSC converter connected to a DC voltage source.
2. Renewable generation system: VSC connected to a DC current source emulating a renewable source and a shunt capacitor.
3. Renewable generation system under voltage sags.

The simulations have been performed considering an AC grid of 400 V and 50 Hz, a DC bus voltage reference of 800 V and a power converter of 10 kVA of apparent power. The inductor resistance is of 0.5 Ω and the inductance 5.4 mH.

4.1 Storage System Simulation

A battery system of constant DC voltage is assumed. Alternatively a battery with a given V-I characteristic could be used. In order to test the VSC converter response, changes in active and power references are done as described in Table 2. The current controllers time constant is of $\tau = 10$ ms.

Reference and measured values of active and reactive power are illustrated in Figure 18. These power values change according to the reference values described in Table 2. Active power increases at $t = 0.3s$ and $t = 0.8s$ and decreases at $t = 0.5s$. The power converter can supply reactive power independently the active power. Between $t = 0.3s$ and $t = 0.5s$ the system

Table 2 Active and reactive power references

Time instant	Active power [kW]	Reactive power [kVAr]
$t = 0$	-3	0
$t = 0.3$	-6	-5
$t = 0.5$	-1	0
$t = 0.8$	-7	2
$t = 0.9$	-7	-7

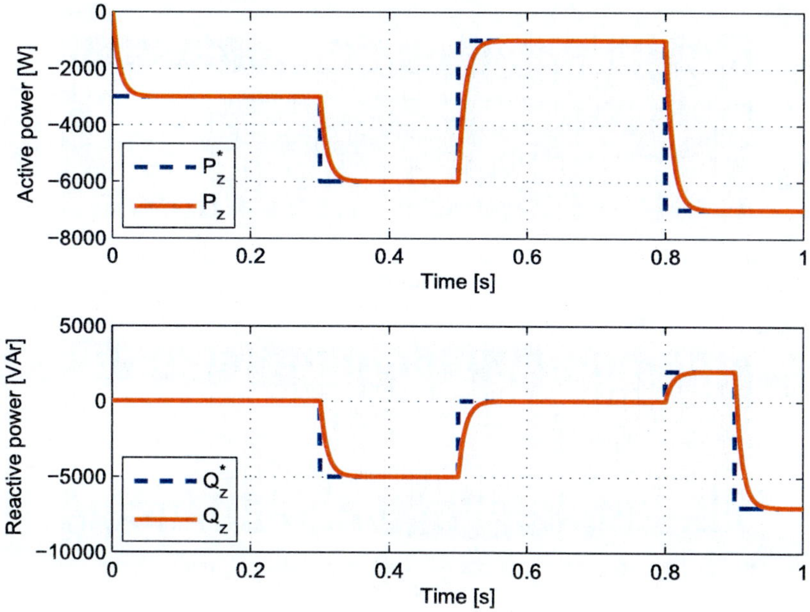

Fig. 18 Reference and measured active and reactive power P_z Q_z injected to the grid

is generating reactive power and between $t = 0.8s$ and $t = 0.9$ the power converter is consuming reactive power.

The grid and converter voltages are plotted in Figure 19.It can be observed that voltages applied for the power converter increase the magnitude between $t = 0.3-0.5s$ and $t = 0.9-1s$ due to the reactive power demand. The currents in the qd frame along with the reference currents are plotted in Figure 20. It can be observed in Figure 20 and Figure 18 the independent control of active and reactive power. A detail of the i_q current evolution is shown in Figure 21. It can be seen that the response matches with the desired time constant of $\tau = 10$ ms.

4.2 Distributed Generation System Simulation

In this simulation, the distributed generation system injects the power to the electrical grid, therefore the VSC has to control the DC voltage in order to guarantee power balance.

The presented simulation evaluates the controllers behaviour for variations of the generated power. The generation power is changed by changing the current of the DC source according to the data of Table 3. The reactive power

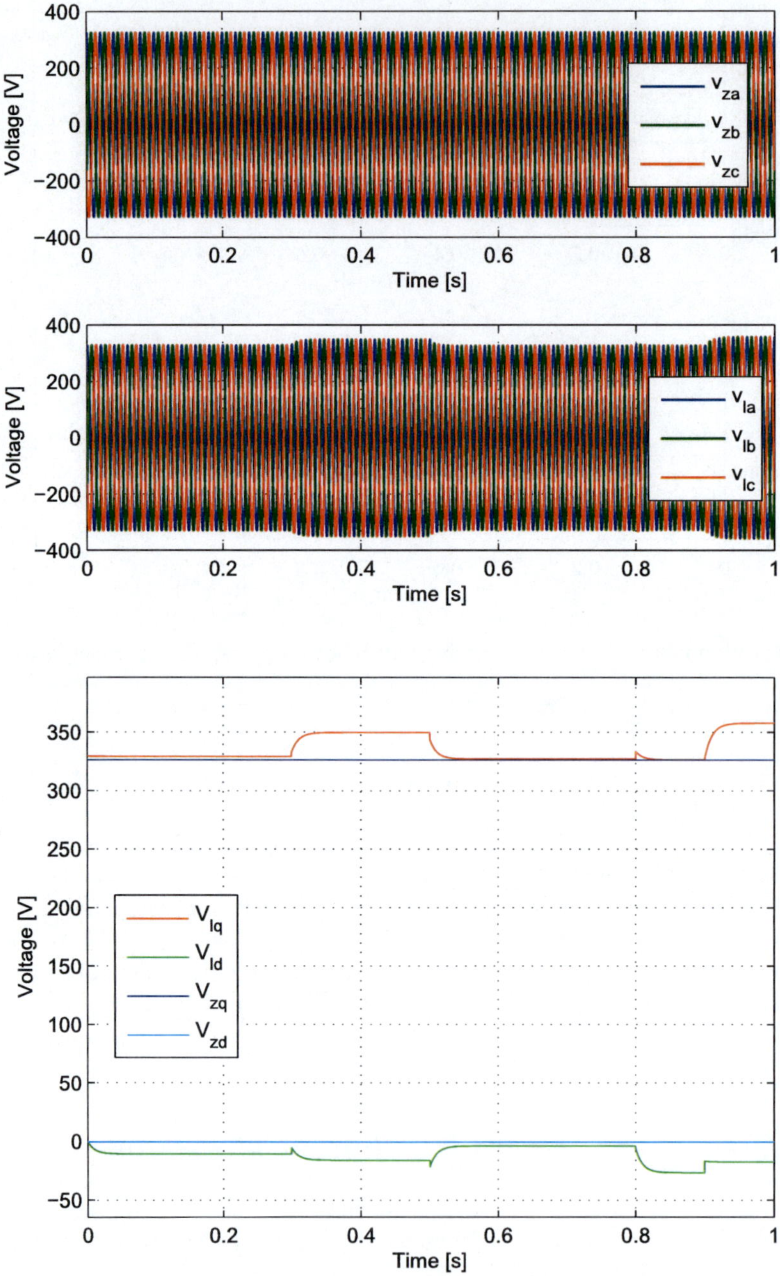

Fig. 19 Grid voltages in *abc* and *qd* frames

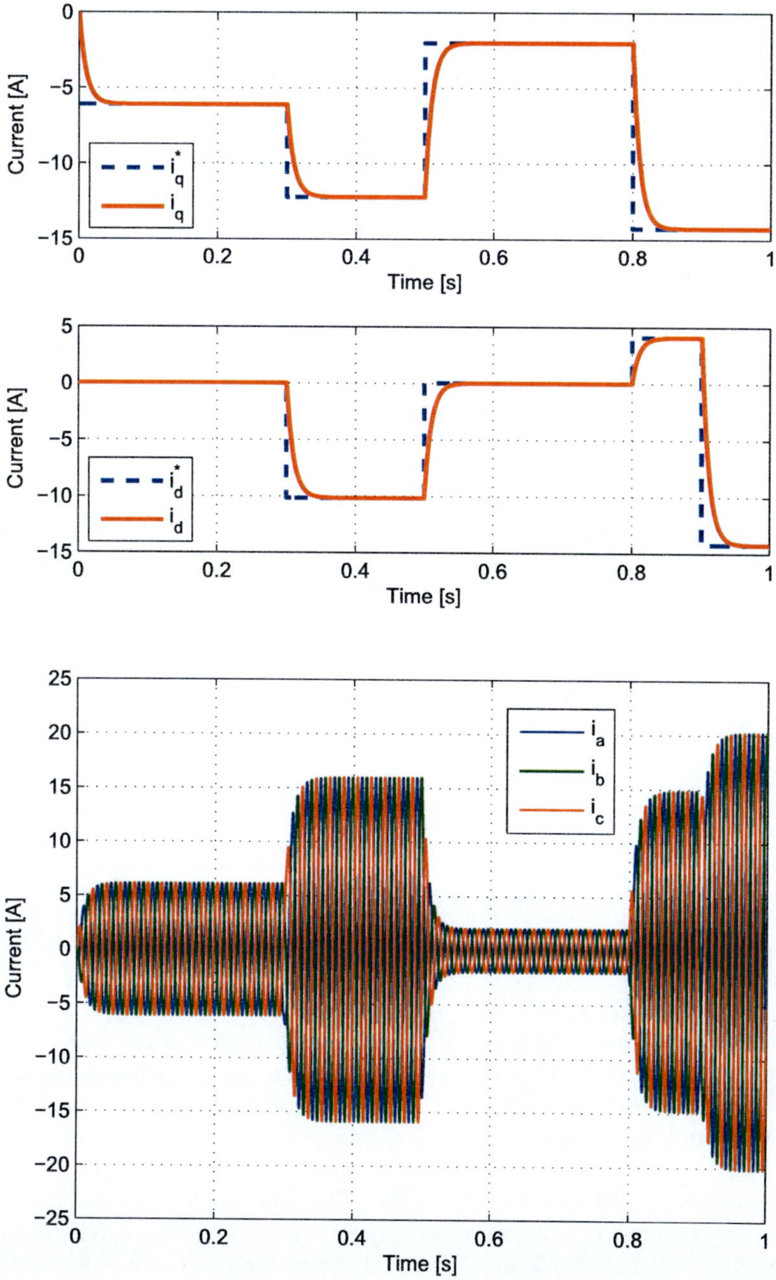

Fig. 20 Measured and reference currents in the *abc* and *qd* frames

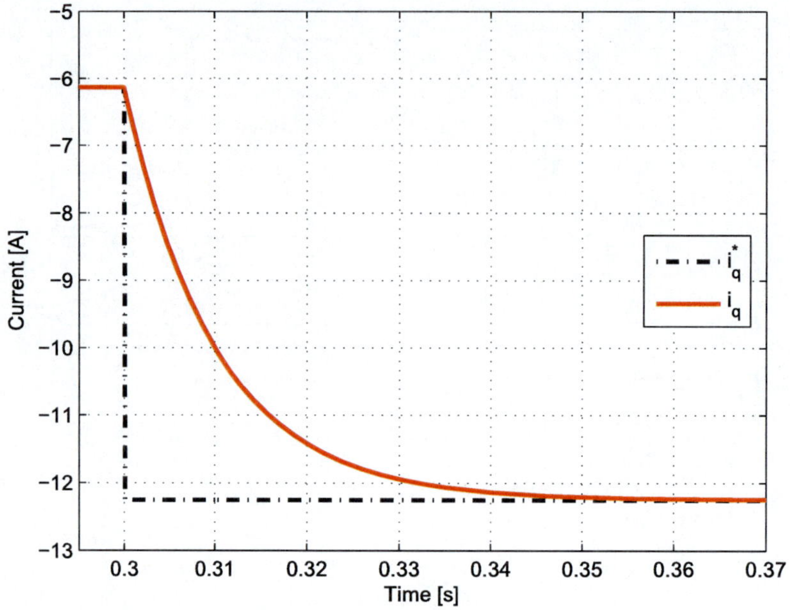

Fig. 21 Detail of the electrical i_q response

Table 3 DC currents reactive power reference for the distributed generation system simulation

Time instant	DC Current [A]	Reactive power reference [kVAr]
$t = 0$	3	0
$t = 0.3$	10	0
$t = 0.5$	5	-5
$t = 0.8$	7.5	-5
$t = 0.9$	10	0

set points are also described in Table 3. The time constant of the current loop is set to $\tau = 1$ ms, while the natural frequency of the DC voltage regulator is $\omega_n = 418.88\frac{rad}{s}$ (15 times slower than the current loops, as recommended in cascaded controllers) and $\xi = 0.707$.

Figure 22 shows the evolution of the current in the *qd* and *abc* reference frames. Current i_q changes each time that DC current changes according to Table 3. The current i_d changes according to reactive reference, as shown in Table 3. At $t = 0.3$, $t = 0.8$ and $t = 0.9$ the transferred active power suffers an increase of power due to the increase of the DC current.

The voltages in the *qd* and *abc* reference frames are plotted in Figure 23. The power converter voltage increases when the system injects reactive power between $t = 0.5$ and $t = 0.8$.

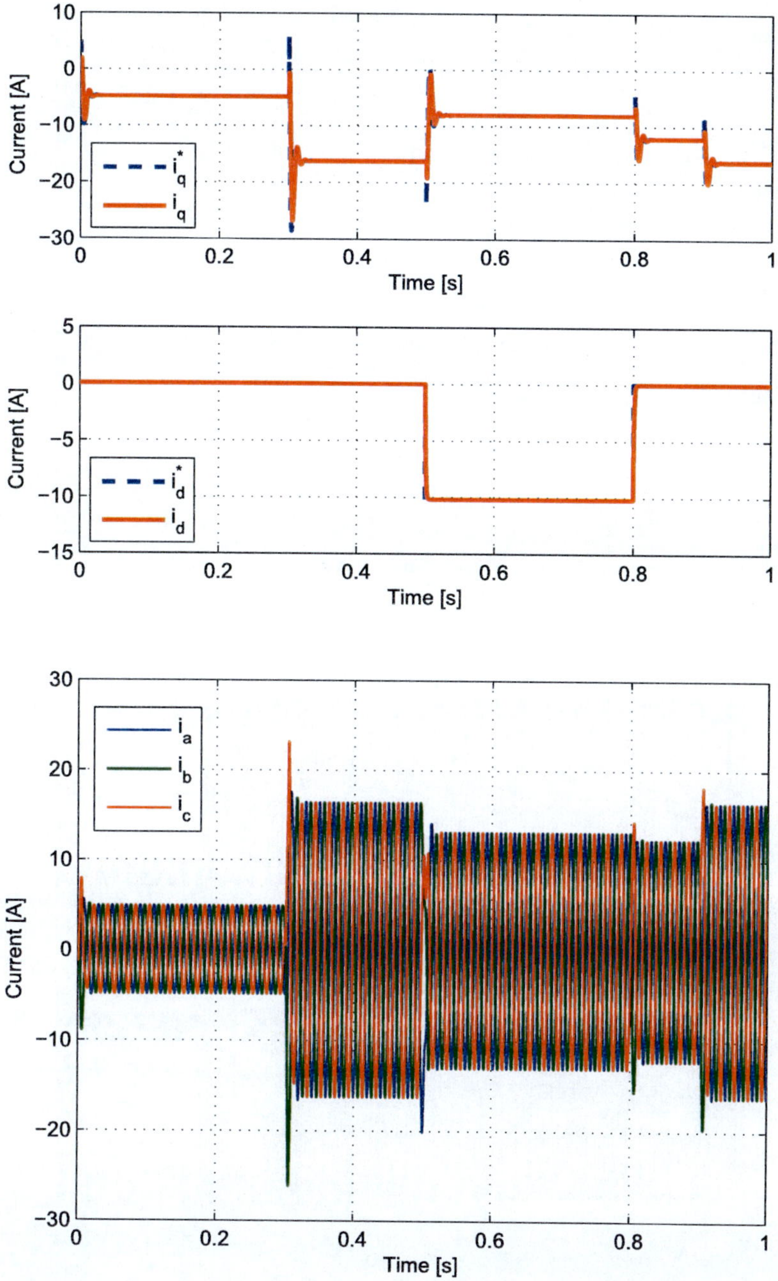

Fig. 22 Current and reference currents in the *abc* and *qd* frames

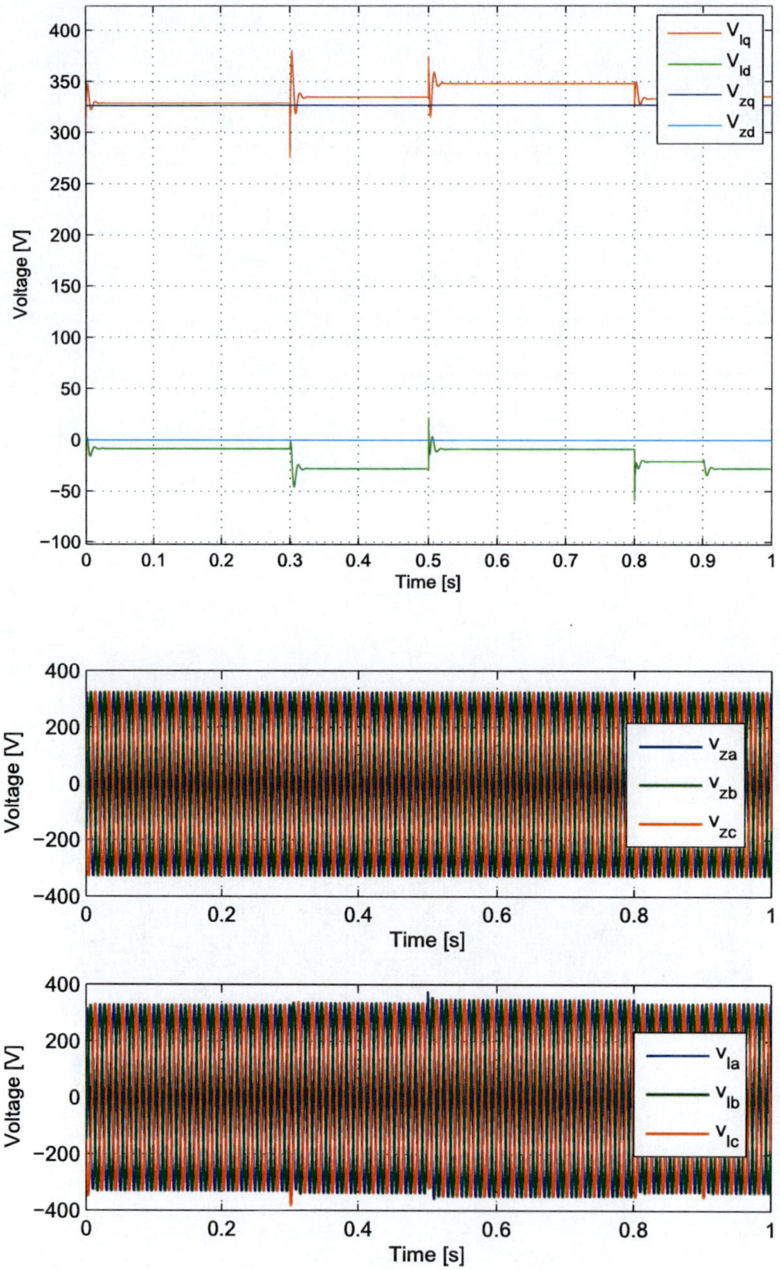

Fig. 23 Grid and converter voltages in the *abc* and *qd* frames

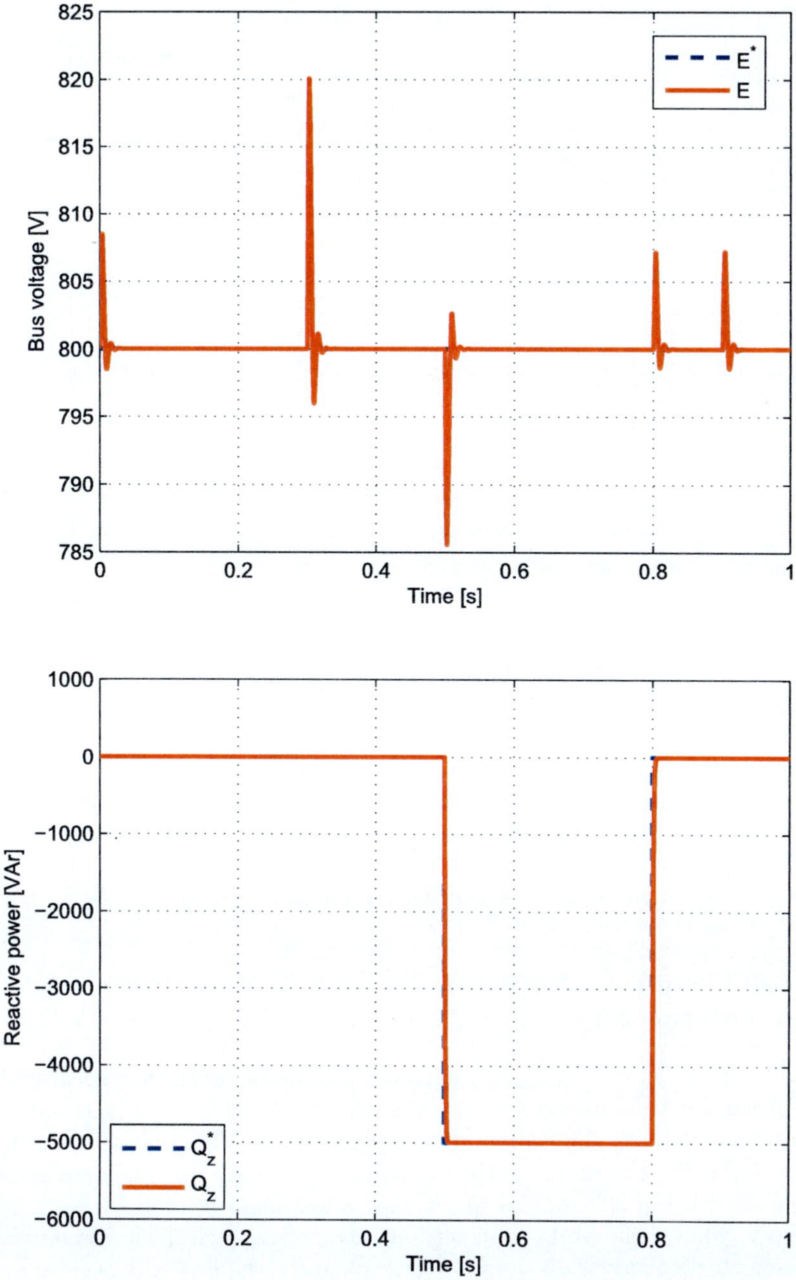

Fig. 24 DC bus voltage E_{DC} and reactive power response

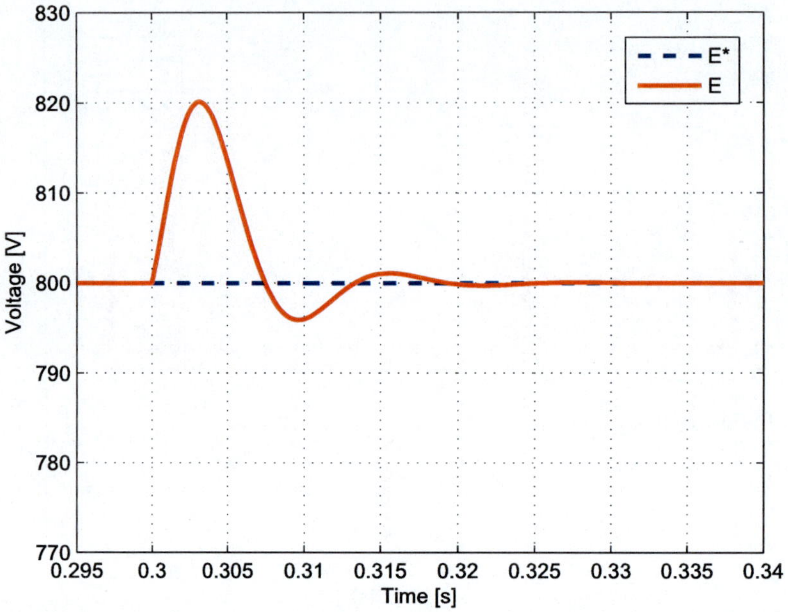

Fig. 25 Detail of the E_{DC} transient transpose

The DC voltage evolution is shown in Figure 24. Each time that DC current changes the voltage controller takes a few milliseconds to regulate the bus voltage to the reference value. The peak E_{DC} voltage is higher for larger injected DC currents. Figure 25 shows a detail of the E_{DC} bus voltage transient. Analysing the plotted voltage response, it can be observed that the angular velocity of the system is approximately $\omega_n = 418\frac{rad}{s}$.

4.3 Renewable Generation System Simulation during a Voltage Sag

Most of electrical faults in the grid produce a voltage sag in the electrical grid. The VSC can remain connected during these faults, which is usually known as ride-through capability. This simulation illustrates the VSC behavior during a voltage sag using the same controllers as in Section 4.2. In this simulation a voltage sag with of 30% occurs at $t = 0.02$ s and lasts 100 ms.

Figure 26 shows the voltage in *qd* and *abc* reference frames. When the voltage sag starts at $t = 0.02$ the voltage applied for the power converter decreases to maintain the injected power previous to the voltage sag. Figure 27 show the currents in *qd* and *abc* reference frames. The current increases its magnitude to maintain the power injected to the AC grid constant and keep the bus voltage at the reference value. Figure 28 shows the DC bus voltage

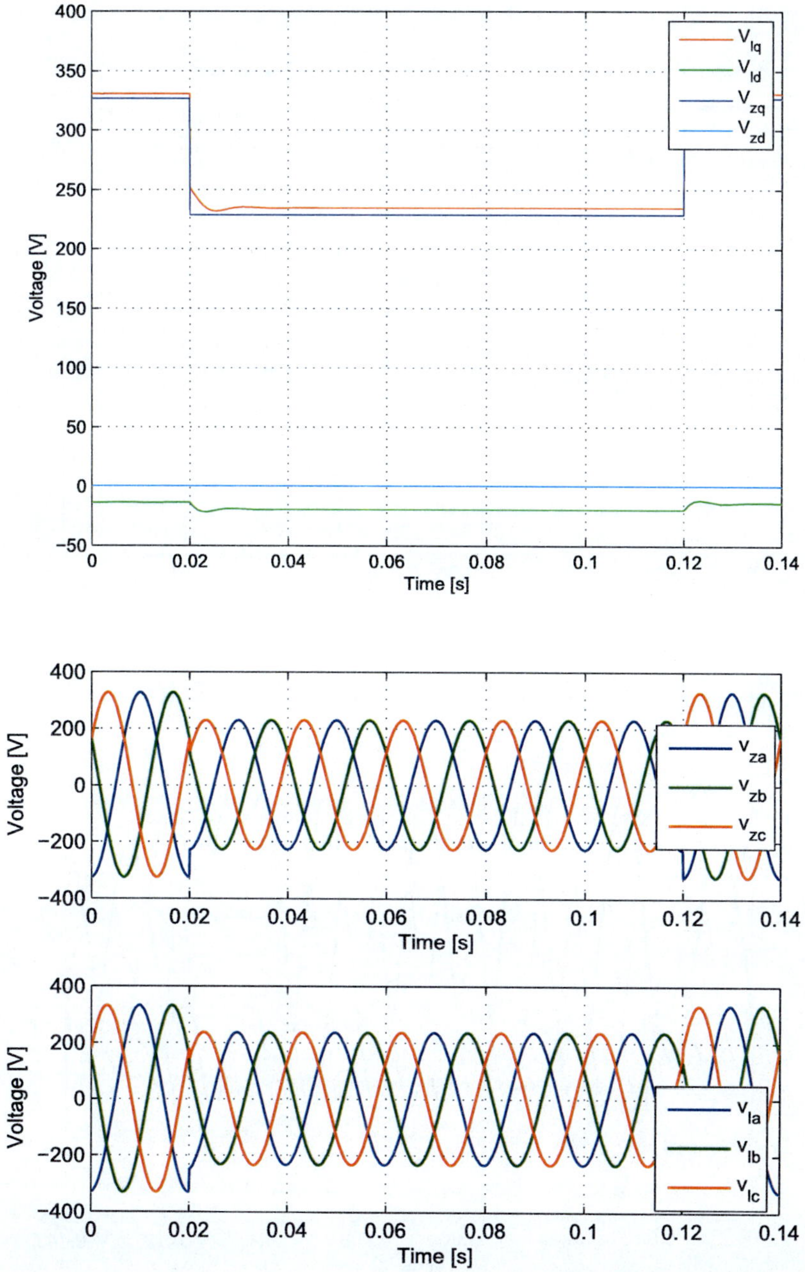

Fig. 26 Grid and converter voltages in the *abc* and *qd* frames

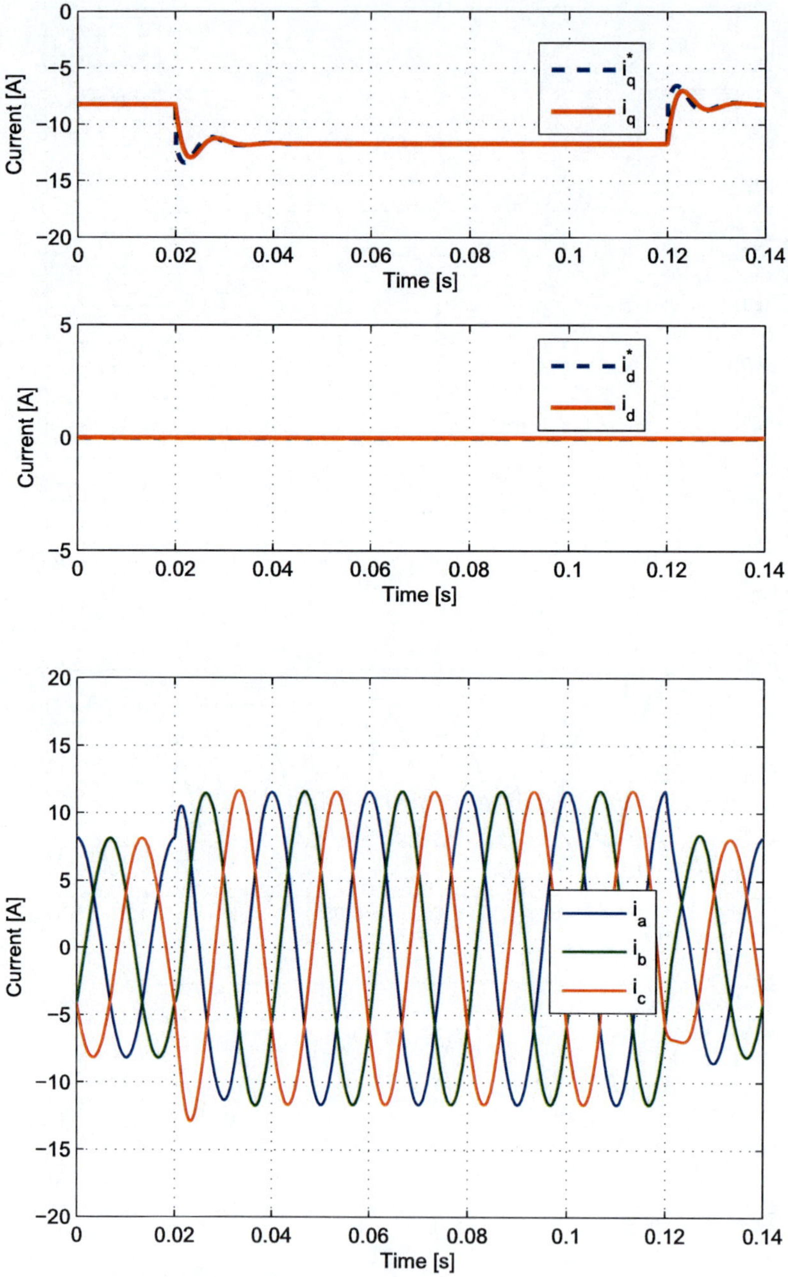

Fig. 27 Current and reference currents in the *abc* and *qd* frames

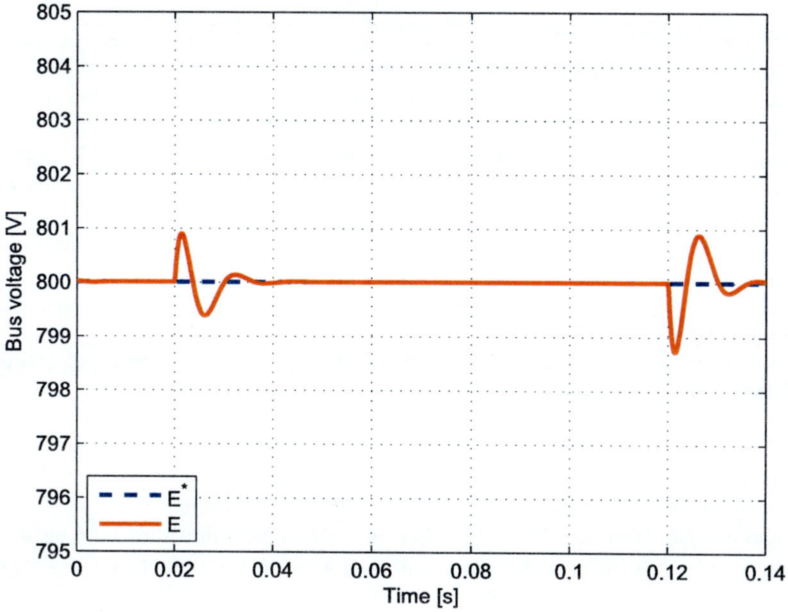

Fig. 28 DC bus voltage E_{DC} response

response during the AC fault. It can be seen that the control adapts rapidly when the sag occurs by adjusting new current references and keeping the DC bus voltage stable.

5 Conclusions

Active and reactive power control for distributed generation and storage systems connected to the grid by means of voltage source converters has been described. The mentioned generation and storage systems are connected to a three-phase three-wire grid. The system components have been described and modelled.

The overall active and reactive power control scheme based on the instantaneous power theory has been described including discussion on the different relevant control blocks required. Simulations results have been included to exemplify the described control approach with storage and generation systems.

Acknowledgements. This work was supported by the *Ministerio de Ciencia e Innovación* under the project ENE2009-08555.

References

1. Farhangi, H.: The path of the smart grid. IEEE Power and Energy Magazine 8(1), 18–28 (2010)
2. Venkataramanan, G., Marnay, C.: A larger role for microgrids. IEEE Power and Energy Magazine 6(3), 78–82 (2008)
3. Kroposki, B., Lasseter, R., Ise, T., Morozumi, S., Papatlianassiou, S., Hatziargyriou, N.: Making microgrids work. IEEE Power and Energy Magazine 6(3), 40–53 (2008)
4. Mohan, N., Undeland, T., Robbins, W.P.: Power Electronics: Converters, Applications, and Design. Wiley, Chichester (1995)
5. Bose, B.K.: Modern Power Electronics and AC Drives. Prentice Hall PTR, Englewood Cliffs (2001)
6. Kazmierkowski, M.P., Krishnan, R., Blaabjerg, F.: Control in power electronics. Elsevier, Amsterdam (2002)
7. Carrasco, J.M., Franquelo, L.G., Bialasiewicz, J.T., Galvan, E., Guisado, R.C.P., Prats, M.A.M., Leon, J.I., Moreno-Alfonso, N.: Power-electronic systems for the grid integration of renewable energy sources: A survey. IEEE Transactions on Industrial Electronics 53(4), 1002–1016 (2006)
8. Malesani, L., Tomasin, P.: Pwm current control techniques of voltage source converters-a survey. In: International Conference on Industrial Electronics, Control, and Instrumentation. Proceedings of the IECON 1993, November 1993, vol. 2, pp. 670–675 (1993)
9. Kazmierkowski, M.P., Malesani, L.: Current control techniques for three-phase voltage-source pwm converters: a survey. IEEE Transactions on Industrial Electronics 45(5), 691–703 (1998)
10. Holtz, J.: Pulsewidth modulation for electronic power conversion. Proceedings of the IEEE 82(8), 1194–1214 (1994)
11. Song, H.-S., Nam, K.: Dual current control scheme for pwm converter under unbalanced input voltage conditions. IEEE Transactions on Industrial Electronics 46(5), 953–959 (1999)
12. Hu, J., He, Y.: Modeling and control of grid-connected voltage-sourced converters under generalized unbalanced operation conditions. IEEE Transactions on Energy Conversion 23(3), 903–913 (2008)
13. Noguchi, T., Tomiki, H., Kondo, S., Takahashi, I.: Direct power control of pwm converter without power-source voltage sensors. IEEE Transactions on Industry Applications 34(3), 473–479 (1998)
14. Rodriguez, J., Pontt, J., Silva, C.A., Correa, P., Lezana, P., Cortes, P., Ammann, U.: Predictive current control of a voltage source inverter. IEEE Transactions on Industrial Electronics 54(1), 495–503 (2007)
15. Akagi, H., Kanazawa, Y., Nabae, A.: Generalized theory of the instantaneous reactive power in three-phase circuits. In: International Power Electronics Conference, pp. 1375–1386 (1983)
16. Akagi, H., Watanabe, E., Aredes, M.: Instantaneous power theory and Applications to power conditioning. Wiley, Chichester (2007)
17. Clarke, E.: Circuit Analysis of AC Power. John Wiley and Sons, Chichester (1941)
18. Park, R.H.: Two-reaction theory of synchronous machines. AIEE Transactions 48, 716–730 (1929)

19. Harnefors, L., Nee, H.-P.: Model-based current control of ac machines using the internal model control method. IEEE Transactions on Industry Applications 34(1), 133–141 (1998)
20. Chung, S.-K.: A phase tracking system for three phase utility interface inverters. IEEE Transactions on Power Electronics 15, 431–438 (2000)
21. Bergas, J.: Control del motor d'inducció considerant els límits del convertidor i del motor. PhD thesis, Technical University of Catalonia, UPC (2000)

Modeling and Control for Islanding Operation of Active Distribution Systems

Seung Tae Cha, Qiuwei Wu, Arshad Saleem, Jacob Østergaard, and Yi Ding

Abstract. Along with the increasing penetration of distributed generation (DG) in distribution systems, there are more resources for system operators to improve the operation and control of the whole system and enhance the reliability of electricity supply to customers. The distribution systems with DG are able to operate in islanding operation mode intentionally or unintentionally. In order to smooth the transition from grid connected operation to islanding operation for distribution systems with DG, a multi-agent based controller is proposed to utilize different resources in the distribution systems to stabilize the frequency. Different agents are defined to represent different resources in the distribution systems. A test platform with a real time digital simulator (RTDS), an OPen Connectivity (OPC) protocol server and the multi-agent based intelligent controller is established to test the proposed multi-agent based frequency controller. The modeling of different DG is discussed in details. Two distribution systems with DG are used to carry out case studies to illustrate the proposed multi-agent controller.

1 Introduction

Focus on the smart grid technologies has certainly brought renewed interests in enhancement of all parts of power systems. Traditionally, electric power has been produced in bulk, transmitted over long distances and distributed at load centers or to end users via a primary distribution system. For many years, power distribution systems have been fed through just one source – substations. This means that the existing distribution system infrastructure, designed in a passive, limited capability and less intelligent manner, is insufficient to adapt technology advances with some non-traditional generation technologies such as wind, solar, micro-turbine,

Seung Tae Cha · Qiuwei Wu · Jacob Østergaard · Yi Ding
Center for Electric Technology (CET), Department of Electrical Engineering, Technical University of Denmark (DTU)

Arshad Saleem
Industrial Information and Control Systems (ICS), School of Electrical Engineering, Royal Institute of Technology (KTH)

L. Wang (Ed.): *Modeling and Control,* Green Energy and Technology, pp. 83–125.
springerlink.com

fuel cells, biomass, etc. A great deal of importance is given to the renewable distributed energy generation (RDEG).

Typically, RDEG or distributed generation (DG) refers to an emerging evolution of the electric power generation systems, in which all the generating technologies available in a given centralized or decentralized region are integrated in the power supply system according to the availability of their respective resources. These resources are known as distributed energy resources (DERs). DER comprises DG, the storage of electrical and thermal energy and/or flexible loads. DER units are operated either independently of the electrical grid or connected to the low or medium voltage distribution level of the main grid. They are located at or close to the point of consumption nodes. This trend has the potential to play a much larger role in the future and has spurred a lot of interest in and growth of such DG around the globe.

Penetration of DG across the globe has not yet reached significant levels. However, that situation is changing rapidly and requires attention to issues related to high penetration of DG within the distribution system. The described paradigm is also strongly motivated by the increasing concern for governmental policy commitments to harness cleaner and greener power on a large scale deployment. Therefore, the interest towards small and decentralized generation is an emerging mode of operations that is a growing alternative to the traditional centralized power generation infrastructure as shown in Fig. 1.

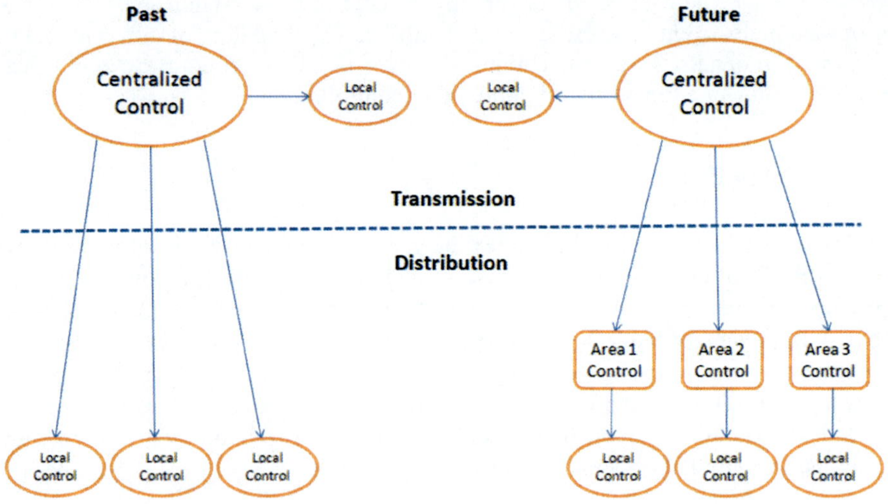

Fig. 1 Transition of power system control

Some analysts have reported that RDEG market will experience strong growth over the next several years, with total system revenues increasing from \$50.8 billion in 2009 to \$154.7 billion by 2015. During this period, the annual RDEG capacity additions will increase from 5.9 GW in 2009 to 15.1 GW in 2015 [1]. Many types of these new technologies are based on power electronic converters for grid coupling unlike conventional synchronous generators. Power electronics-based distribution controllers are the key elements for transforming distribution system from being passive to active (that is, flexible, controllable, and securely utilized up to its capacity). The system will differ substantially from the present grid, which is far more active and has much capability to absorb intermittent generation. The term active is used to express the shift from the conventional and more passive operation approach towards more advanced smart grid concepts. As renewable is gaining a greater share of local energy production and becomes more mainstreamed, how will it impact our existing distribution systems? What are the challenges as the penetration of DG increases over the next few years? Operators and researchers will have to be prepared to face the control and operational challenges that necessarily follow any new trend. Are we there yet?

These systems are interconnected to the medium voltage distribution network, but they can also be isolated from the main grid when a fault occurs in the main grid. Islanding operation can occur when a DG or a group of DG units continue to energize a portion of the distribution system that has been separated from the main grid. A typical power distribution system is shown in Fig 2. An island situation occurs, for example, when re-closer A opens. DG1 will feed into the resultant island in this case. Large DG units are typically connected to the primary feeders (DG1 and DG 2). Most of them are synchronous or induction generators at present. Small DG units such as inverter based Photovoltaic (PV), and fuel cell systems are connected to the low voltage secondary feeders (DG3). Many customers could be supplied from decentralized sources. Hence, the adoption of DG sources can considerably increase reliability of the power system in case of utility outage or when a portion of grid needs to be independent, and is becoming a viable option for technical and economical reasons.

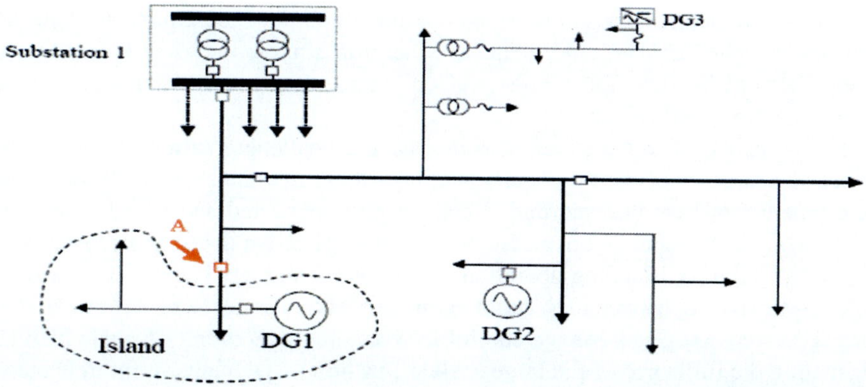

Fig. 2 Typical distribution system with DGs

However, the arising problem of islanding operation of active distribution systems is the possible instability in terms of voltage and frequency due to the weakness of such a small electric network operated in islanding mode. Regarding the frequency, the problems are caused mainly because of the rather low inertia (*H*) of the system and frequency controlled reserve. As it is shown in the network electromechanical equation 1, a mismatch between produced power (*Pmech*) and consumed power (*Pelec*) generates variations in the angular frequency ω of the generators and thereby the frequency of the system according to the equation. These variations are dumped by the inertia of the system, and since the inertia, in this case, is much smaller than in non-isolated operation, the frequency will be more affected (i.e. the higher the inertia of the system is, the smaller the frequency deviations are, and vice versa).

$$\frac{2H}{\omega}\frac{\partial\omega}{\partial t} = P_{mech} - P_{elec} \tag{1}$$

In the case of voltages, the reactive power variations of the demand and DG have a more significant impact on the total reactive power balance of the system compared with the interconnected operation. If the excitation systems of the synchronous generators are not fast enough to compensate such variations, the voltage will be affected.

In this regard, the main challenge of operating DG based systems is the coordination of DGs in order to control the system frequency and voltages within the islanded systems. In order to ensuring stable operation during network disturbances and maintain stability and power quality in the islanding operation in the presence of arbitrary varying loads or under a variety of different network operating conditions, it is necessary to develop sophisticated and comprehensive control strategies. Realization of these concepts further requires that power systems should be of distributed nature, consisting of autonomous components, which are able to coordinate, communicate, and adapt to emerging situations. Hence, intelligent control should be used for control and operation of active distribution systems consisting of several DG units.

The objective of this chapter is to discuss the challenges raised by the emergence of a large number of DG units, and how these interconnected DG units can be efficiently operated as microgrids both in grid-connected and islanded modes. Moreover, the transient and dynamic performance of active distribution systems is investigated during islanding operation and a control strategy is proposed for reliable power sharing between DG units in the network. The focus on control strategies is on frequency and voltage control for islanding operation. This can help in evaluating the influence of the large scale utilization of DG units in future's electrical power networks in order to realize smooth and secure islanding transition without sacrificing the system stability.

The rest of the chapter is arranged as follows. Control concepts for frequency control and multi-agent based controller for islanding operation of active distribution systems are described in Sect. 2. The test platform with a real time digital simulator (RTDS), a MatrikonOPC server for SCADA DNP 3 and a multi-agent based controller is described in Sect. 3 and the communication test for the test plat form is presented in the same section as well. In Sect. 3.4, two active distribution systems with DGs are described. Case study results of the two distribution systems are presented in Sect. 5 in order to illustrate the proposed control concept. In the end, a brief summary of the chapter is given.

2 Control Concepts for Islanding Operation of Active Distribution Systems

2.1 Frequency Control Concepts for Islanding Operation of Active Distribution Systems

The connection of DG units will drastically change the operation of the electrical power system. Currently, the system frequency is controlled by the conventional power plants. Hence, if the consumers increase their consumptions, the power will be delivered directly from the conventional power plants. Conventional power plants generally use synchronous generators. The goal of the frequency control is to maintain the power balance at all time and keep the system frequency within the pre-defined limits.

In this section, frequency control concept will be shortly discussed. If, for instance, consumption is larger than production, the rotational energy stored in large synchronous machines is utilized to keep the balance between production and consumption. The rotational speed of the generators decrease which in return decrease the system frequency. After a decrease of the system frequency, power plants will immediately release energy from their rotating mass. The energy stored in this rotating mass is given by Equation 2.

$$E = \frac{1}{2} * J * \omega_m^2 \tag{2}$$

where J is the total inertia of the machine and *Wm* the rotational speed of the machine. In electrical engineering, the inertia constant, H, is often used which is defined as:

$$H = \frac{J\omega_m^2}{2H} \tag{3}$$

where S the nominal apparent power of the generator.

The inertia constant has the dimension time and gives an indication of the time that the generator can provide nominal power by only using the energy stored in its rotating mass. Typical inertia constants for the generators of the large power plants are in the range of 2 – 9s, depending on the type of power plant in which they are used and on the nominal rotational speed [2]. The inertia of conventional generators plays a significant role in today's power systems in order to stabilize the system frequency during a transient situation. The response is determined by the dynamics of the system and is called 'inertial response'. In the period immediately following a disturbance, the frequency deviation is dependent on the inertia of the system. When the frequency deviation exceeds a pre-defined threshold value, the primary frequency controllers of the conventional power plants will react. These units usually have frequency-sensitive equipment and are called primary control units. Controllers will be activated to change the power input to the prime movers until the balance between production and consumption is restored. This is the second phase. Even the power balance is restored. There will still be a steady-state frequency deviation. The further frequency deviation depends on the droop characteristics, the primary control reserve and the deployment time. This is called the primary frequency control. The droop constant gives the additional power that is supplied as a function of the frequency deviation. The droop constant K_D is a ratio without dimension and generally expressed as a percentage:

$$K_D = \frac{-\Delta f / f_n}{\Delta P_G / P_{Gn}} \times 100\% \tag{4}$$

Two different droop characteristics are shown in Fig. 3. The following figure shows a diagram of variation in the generating output of two generators, a and b, of different droop under equilibrium conditions, but with identical primary control reserve. From this figure, the definition of the primary control reserve can be seen. It is the range of a generator from the working point prior to the disturbance to the maximum power, at which the generator can provide primary control. The primary control reserve is denoted by ΔK. The deployment time is the time that the generator needs to increase its output power. The frequency will stabilize at a frequency different from the nominal frequency [3].

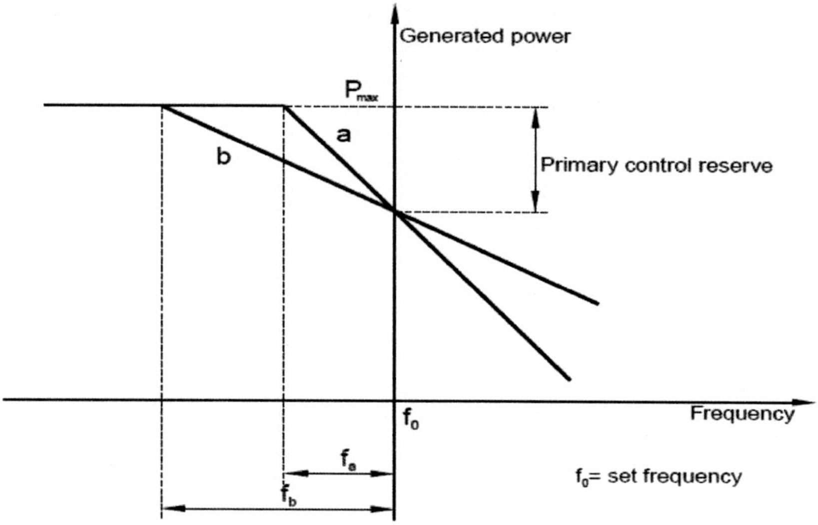

Fig. 3 Different droop characteristics

As mentioned earlier, the integration of DGs gives new challenges from the frequency control perspective and a significant part of these DG units will be connected to the grid with power electronic converters, as shown in Fig. 4.

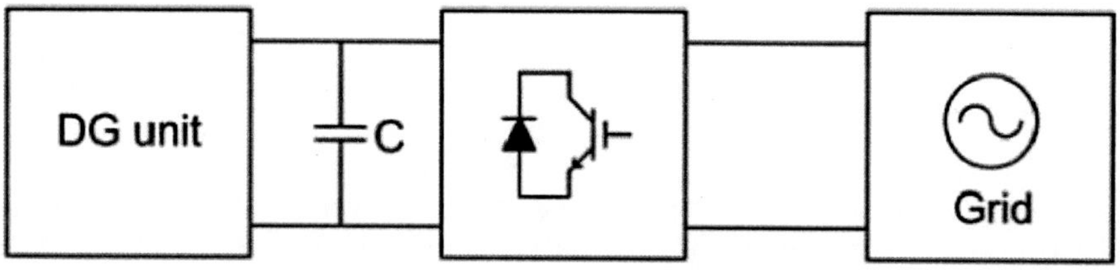

Fig. 4 Power electronic based converters

This gives these generators a behavior that is fundamentally different from the conventional generators particularly in terms of their impact on electromechanical stability. Some types of DG units supply a dc current, which is converted to AC by an inverter. They are inherently 'inertia-less' and typically have no direct relation between power and frequency. The inertia contribution of DG units, for example, is much less than that of conventional generators. Other types of DG units are based on machines, but the converter decouples their rotational speed from the grid frequency to make variable speed operation possible, thus preventing the generator from responding to system frequency changes. And, they also don't have the direct relation between their inertia and the system frequency. As they use other generation technologies than conventional power plants, they have a limited capability of participating in primary frequency control in the same way as conventional generators do. With an increasing penetration level of DG units, assuming that they replace conventional generation, frequency control will become more difficult because of the decreasing level of inertia that is seen by the grid [4]. In order to avoid large frequency deviations, DG units will have to contribute to the frequency control in the islanding operation mode. The islanding operation is defined as a condition in which a portion of the grid that contains both generation and load is isolated from the grid and continues to be energized by DGs connected to the isolated subsystem. Islanding may occur in many conditions. For example, below is a three-phase balanced network consisting of a DG unit, local load and power grid identified that can be simplified as shown in Fig. 5.

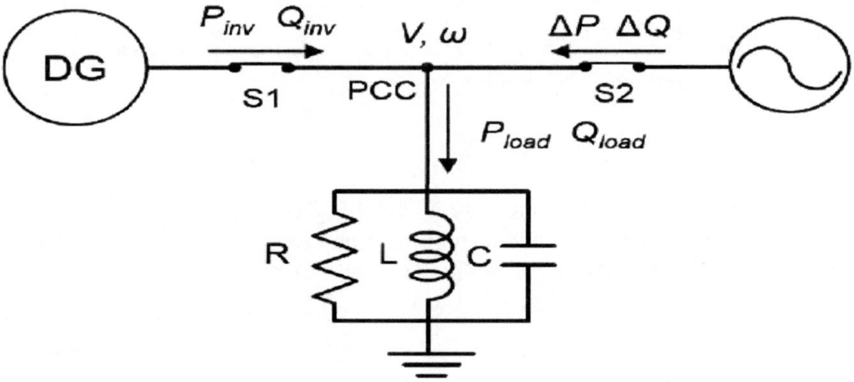

Fig. 5 Simple network with a DG, RLC and Grid

Power injections from the grid to the local distribution system, ΔP and $\Delta \mathbf{Q}$, are defined as

$$\Delta P = P_{load} - P_{inv} \tag{5}$$

$$\Delta Q = Q_{load} - Q_{inv} \tag{6}$$

where P_{load} and Q_{load} are active and reactive power demand of the RLC load, and P_{inv} and Q_{inv} are active and reactive power output of the DG.

When the grid-side breaker S2 opens owing to network faults, an islanding condition occurs. The islanding condition is an unregulated power system if there is not a controller to coordinate the DGs' responses. Its behavior is unpredictable due to the power mismatch between the load and generation and the lack of voltage and frequency control.

In order to be able to contribute to the frequency control for the islanding operation of active distribution systems, DG units must be able to adjust their output power. The first considered type of DG unit is the variable speed wind turbine. The supplied power depends on the wind, which is not controllable. Therefore, wind turbines can't participate in primary frequency control in the classical way. The large blades of the turbine gives a significant inertia and the kinetic energy stored in this rotating mass can contribute, for a short time frame, to the inertial response. In general, the variable speed wind turbines have a speed controller, which has the task to keep the optimal tip speed ratio λ over different wind speeds, a feature not possible with fixed speed induction generators, by adapting the steady state generator speed to its reference value. This reference value is normally obtained from a predefined power-speed curve. For low wind speeds, the generator speed is kept at a fixed low speed and for wind speeds above the rated speed the speed control loop prevents the rotor/generator speed from becoming too large by progressively pitching the blades in order to limit the aerodynamic power.

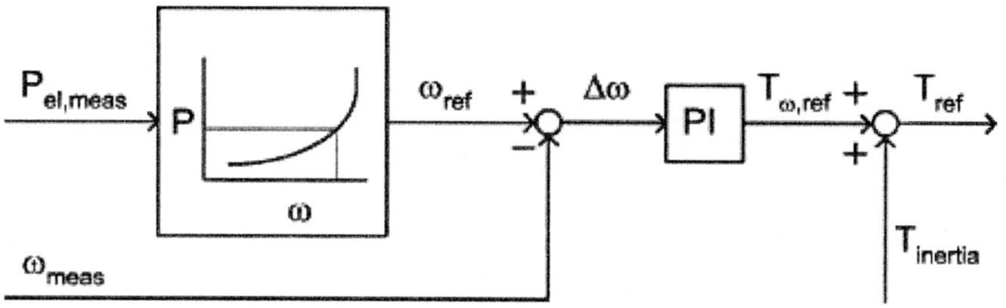

Fig. 6 Speed controller

Fig. 6 is a typical speed control loop and the generator speed reference is obtained from the predefined static *P*-*ω* characteristic. It corresponds to the generator speed which is optimal for a certain power. The error between the actual and the reference speed is sent to a PI controller, which gives a set point for the current controller of the turbine.

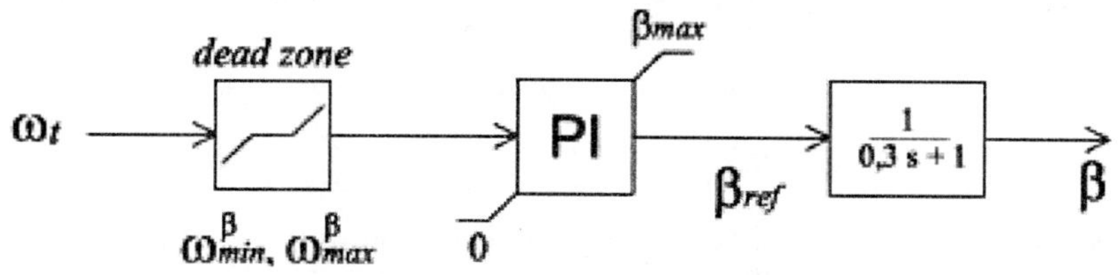

Fig. 7 Simplified pitch control

Besides the speed control, the pitch angle of the turbine will be controlled in another loop as shown in Fig. 7. For high wind speeds, the generated power is kept constant and the rotational speed is limited by the simplified pitch control. The output of the speed controller is the reference torque for the rotor current controller of the doubly-fed induction generator (DFIG) [5].

In order to realize the inertial response from the DFIG, a control loop is required as the rotational speed of the turbine is decoupled from the grid. A brief description is given below. Wind turbine generator (WTG) frequency response means a WTG's active power output as a function of the grid frequency. In [6], DFIG frequency control is introduced through a functional block that generates a power component signal ΔPe as a function of the grid frequency which is shown in Equation (7).

$$\Delta P_e = \Delta P_{IE}(f) + \Delta P_{DROOP}(f) \tag{7}$$

This signal is added to the normal power reference of the WTG as shown in Fig. 8.

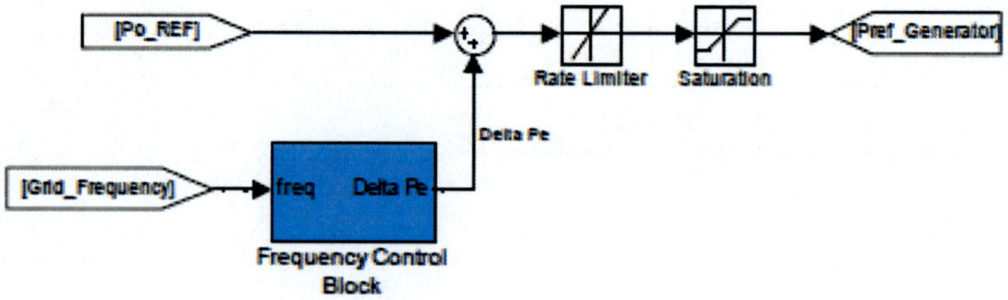

Fig. 8 Frequency control signal in the wind turbine

The frequency response signal ΔPe has a component that emulates the inertia response of a synchronous machine, called inertia emulation power *ΔPie*, and a component that depends on the grid frequency deviation from the nominal value, called primary control power *ΔP droop*. The inertia emulation power component is generated using Equation 8.:

$$\Delta P_{IE}(f) = -2H_t * f * \frac{df}{dt}[pu] \tag{8}$$

where *Ht* is the wind turbine inertia calculated according to Equation (3).

The primary control power component is generated using Equation 9.

$$\Delta P_{DROOP}(f) = 100 * \frac{1-f}{R}[pu] \tag{9}$$

where *R* is the wind farm droop factor in %.

Like the above example, a control loop should be implemented to give these DG units a 'virtual inertia' or to let the DG units contribute to primary frequency control. It might be required that DG units contribute to or participate in secondary frequency control in the near future. In this chapter, only the contribution to primary frequency control will be considered. Contribution to inertial response and secondary frequency control is not taken into account. In addition, the possible governor configurations can also be used to control the frequency of the islanded system.

The Danish distribution systems are characterized by a significant penetration of small gas turbine generators and fixed speed wind turbine generators. Many generators use speed droop governor to change the governor reference speed as loads change. However, depending on the power mismatch on the islanded system, the system frequency may settle outside the power quality limit. The problem of frequency settling outside the power quality limit can be solved by an isochronous speed controller that has the ability to bring the gas turbine speed to the reference speed after islanding. In such a situation, speed governors can be implemented to control the frequency of the system. The GAST model for the gas turbine generator is shown in Fig. 9, which is one of the most commonly used dynamic models.

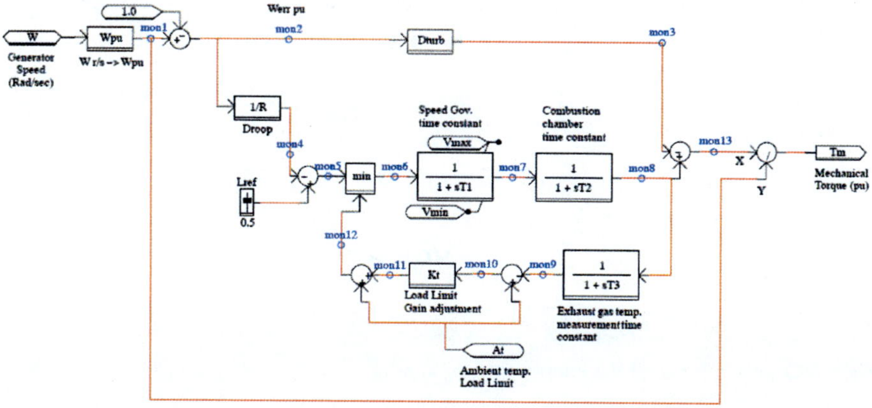

Fig. 9 GAST governor

The GAST model above represents a gas turbine and its associated speed governor. A permanent droop loop is included as part of the speed governor so that system load will be shared among multiple generators. The load reference may be adjusted by a system frequency controller to maintain rated frequency with changing load. Maximum turbine output may be limited due to temperature sensitive feedback loop. The gas turbine has to have the ability to operate in different modes when there is a change in states. In grid connected mode, it operates in a constant power factor with speed droop control where as in islanding mode, it operates in voltage control and isochronous control. The switching between the two control strategies has to be optimally implemented [7].

Moreover, the energy storage system should play an important role in maintaining the frequency and voltage of the islanded system by ensuring the balance between energy generation and consumption. The inverter controller of energy storage systems responds in milliseconds. In general, the power output of energy storage systems may be fixed when the active distribution system is operated in grid connected mode, and the fixed power control of energy storage system supplies constant power. It can't provide proper frequency and voltage control in islanding operation. Therefore, the control scheme of the energy storage system must be switched from fixed power control to droop or constant frequency/voltage control during islanding operation. Fig. 10 shows the detailed control block of a typical energy storage system [8].

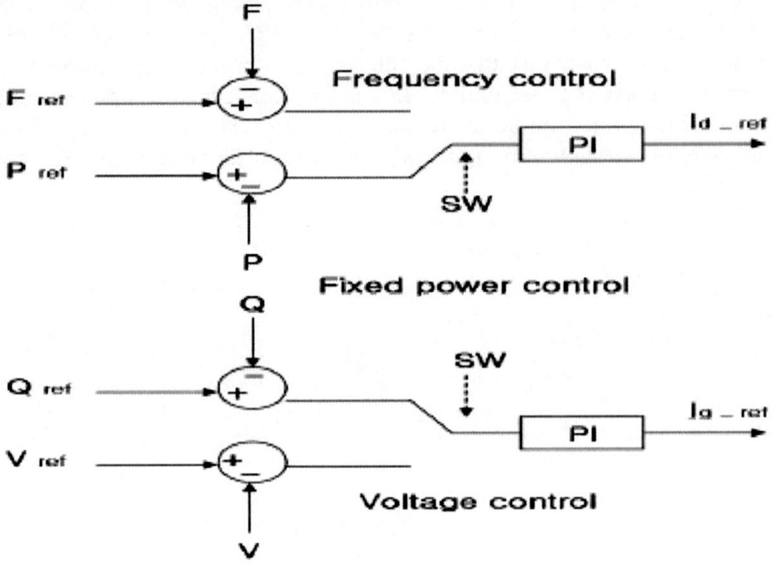

Fig. 10 Control scheme of the energy storage system

The frequency can be effectively controlled by applying a droop control scheme in the energy storage system, but the control capability of the energy storage system may be limited by its available energy capacity. Therefore, the power output of the energy storage system should be brought back to zero as soon as possible. The secondary regulation of Microgrid Management System (MMS) is in charge of returning the current power output of the energy storage system to the pre-planned value, which is usually set at zero. When operating in islanding mode, the energy storage system regulates the frequency and the voltage in local control actions, and the MMS calculates the proper power outputs of each microsource to make the power output of the energy storage system equal zero, as shown in Fig. 11. The calculation procedure of MMS is not shown here.

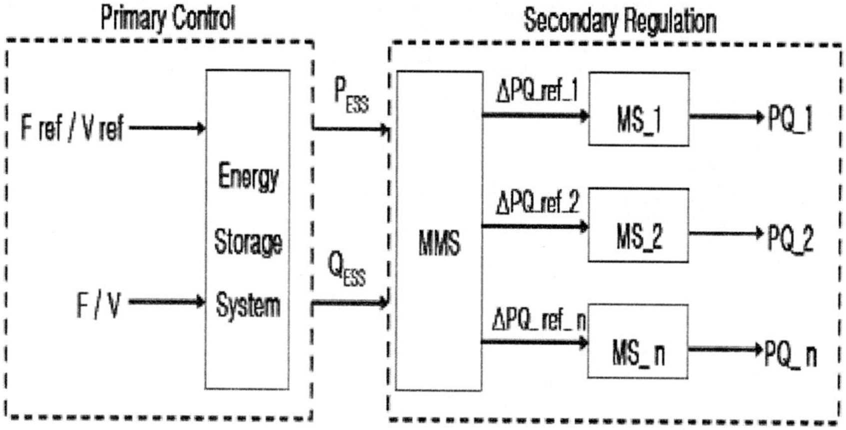

Fig. 11 Coordinated control structure

2.2 *Multi-Agent Based Controller for Islanding Operation of Active Distribution Systems*

In this chapter, we present our work in devising a novel multi-agent system for balancing and voltage control for islanding operation of distribution systems with DGs. In the attributed mechanism, intelligent agents represent different components in electric power distribution grid such as DGs and electric power loads. Our approach is based upon a dynamic service oriented mechanism where DG and Load agents offer regulation and load shedding services respectively for systems balancing and voltage control. Decision of assigning specific services provision to agents is taken dynamically through explicit communication. This is done in a distributed manner through an auction mechanism. DG and Load agents calculate their local cost functions based upon their current state and capabilities. Based upon the value of this cost function, DG and Load agents send a bid to the DF agent for provision of regulation services. Whenever there is a new situation, e.g. voltage drop at nodes of the loads, the load agents contact DF agent and ask for available regulation services. The DF agent uses an accumulation function for comparing and assigning service provision to specific DG and load agents.

A. Multi-agent Based Hierarchical Control

In electric power systems control, agents are applied at different levels of the control. Starting from a low level control of devices, it goes to higher level of planning and optimization. This is shown in Fig. 12 which presents an overview of agents at different levels of the control in electric power systems.

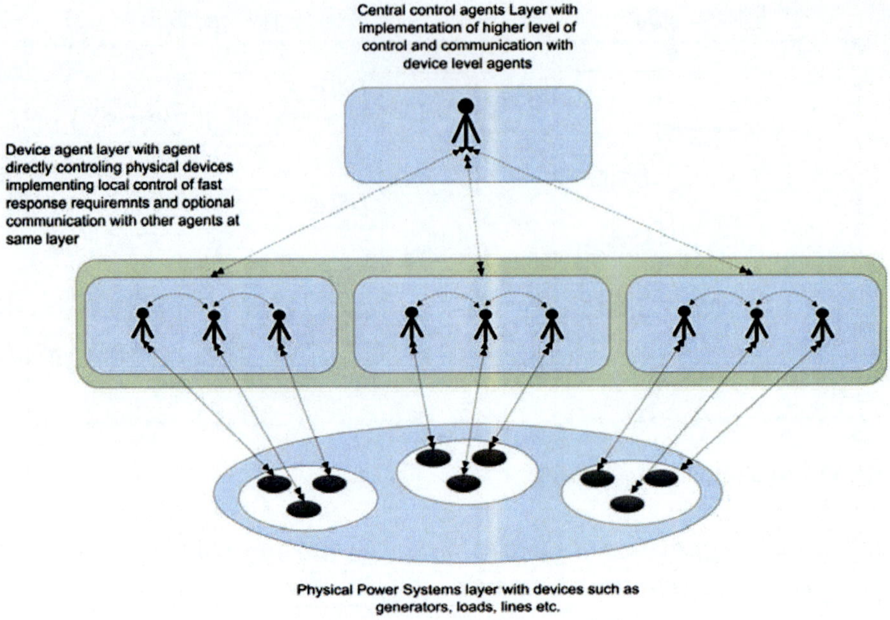

Fig. 12 Multi-agent based hierarchical control Scheme

The agents at the device layer interact directly with devices at physical system layer. In most cases, an agent acts as the controller of a physical device and performs control functions, e.g. a generator agent control active and reactive power set points of a generator, and a breaker agent would perform functions of opening and closing a breaker. Agents at this level of control have a higher requirement for timely execution of actions and thereby usually do not implement high level mechanism for decision making. Agents at this level may communicate with other agents at same level and to the agents at higher level of control. The so called local or distributed control is implemented at this level. Though being able to communicate with other agents at same level as well as higher level, agents at this level should be able to make local decision independently.

The agents at the control coordination level usually do not directly interact with physical electric power system devices. Instead, they communicate with agents at lower level of control, i.e. device level agents. Status information is communicated from lower level to the high level whereas control commands are sent from higher level to the lower. Agents at this level implement less time critical requirements and more sophisticated decision making mechanisms for system planning and control.

DG Agent Control Logic

DG agents represent distributed power generators in electrical networks. Every DG agent calculates its cost function. The cost function of DG agent is based upon its current state, e.g. active power set point capabilities, ability to control frequency, etc. The cost function of DG agents is

$$\delta_{c,DG}(S_{cur}, C_{cur}) \Rightarrow U_{role} \tag{10}$$

It is a function that maps current state of DG agent and current capabilities into a role utility for provision of regulation service.

DG agent sends a bid based on value of this cost function. DF agent cumulates bids from all DG agents and sends a message to Load agents with information such as selection for service provision and amount of regulation power to be delivered.

Load Agent Control Logic

Every Load agent calculates its cost function. The cost function of load agents is based upon its current state, e.g. connected/disconnected and capabilities, auto-connect (and re-connect). The cost function of Load agents is

$$\delta_{c,Load}(S_{cur}, C_{cur}) \Rightarrow U_{role} \tag{11}$$

DF Agent Control Logic

The DF agent implements a mapping function f_{tr} that bid of each DG and Load agents and maps into a role utility for service provision.

$$f_{tr} = \sum A_i(\beta) \Rightarrow R_{srv} \tag{12}$$

where β is the bid value for each agent A_i and R_{srv} is the role utility for service provision.

This function decides which DG agent shall provide a regulation service and how much regulation power should be provided through the service provision. It is important to note that the role utility allocation is performed based upon the bid value submitted by DG and Load agents and this bid value is calculated considering current state and capabilities of those agents. This ensures the optimal decision on selection of appropriate agent and the amount of regulation value according to a current situation.

B. Control Agents

In our approach, software agents act as smart controllers of different components of electric power systems, e.g. DGs (DG agents) and electric power loads (Load agents). Such agents reside in a dynamic multi-agent software platform as shown in Fig. 13.

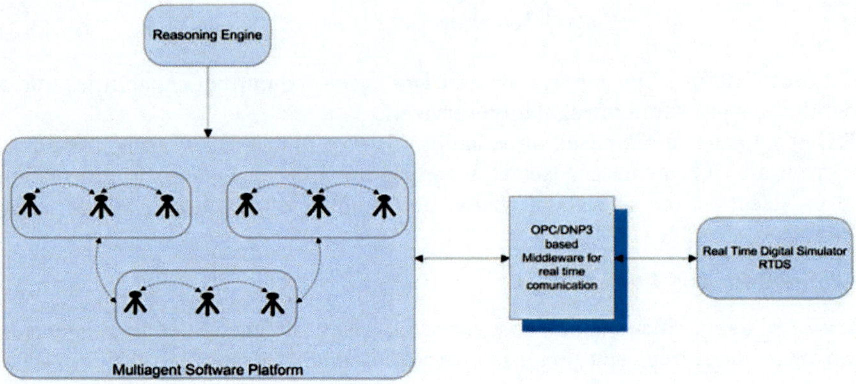

Fig. 13 Overall structure of agent and service

The platform consists of one main container, and several sub containers. Each sub container represents an island in a distribution network consisting of DG and load agents. Both DG and load agents can join or leave the network dynamically according to the changes in the network. New islands can also be created following any situation in the network. The software platform also includes some utility agents and services. Some of the important utilities and agents are:

i) DF (Directory Facilitator) agent works as a market agent. DG and load agents interact with this agent to register and discover agent services.
ii) AMS (Agent Management Services) agent provides white page services. This agent is responsible for creating, destroying and managing agents and containers in the software platform.
iii) MTS (Message Transport Service) is responsible for message transportation between agents within a container and across containers. This service also enables synchronization of messages when several messages are sent and received from different agents in parallel. In order to take full advantage of agent capabilities such as autonomy, local control, scalability and high level communication, the software platform is implemented as fully compliant with Foundation of Intelligent Physical Agents (FIPA) standards on JADE platform.

C. Agent Communication

Real-time communication between the software platform and RTDS is achieved by implementing a middleware based upon OPC and DNP3 protocols [9]. This middleware is implemented using Java Native Interface (JNI) and fully conforms to the mentioned standard. Through an OPC server, software agents can connect to respective device models in RTDS and perform control actions, e.g. changing the set point of a generator or opening and closing of a breaker during simulations.

Each agent in the software platform creates its own instance of connection and has an individual channel of control commands, which ensures that decentralized nature and robustness of the control mechanism is not compromised [10].

3 Test Platform with RTDS and MatrikonOPC Server for SCADA DNP3

In order to validate the developed control concepts and test the dynamic performance of active distribution systems for islanding operation, a test platform has been established in Center for Electric Technology (CET), Technical University of Denmark. The setup of the test platform and the communication between different parts of the test platform are presented in Sect. 3.3.1 and Sect. 3.3.2, respectively.

3.1 Test Platform Setup

The test platform is comprised of a real time digital simulator (RTDS) and a MatrikonOPC server for SCADA DNP3, and is illustrated in Fig. 14.

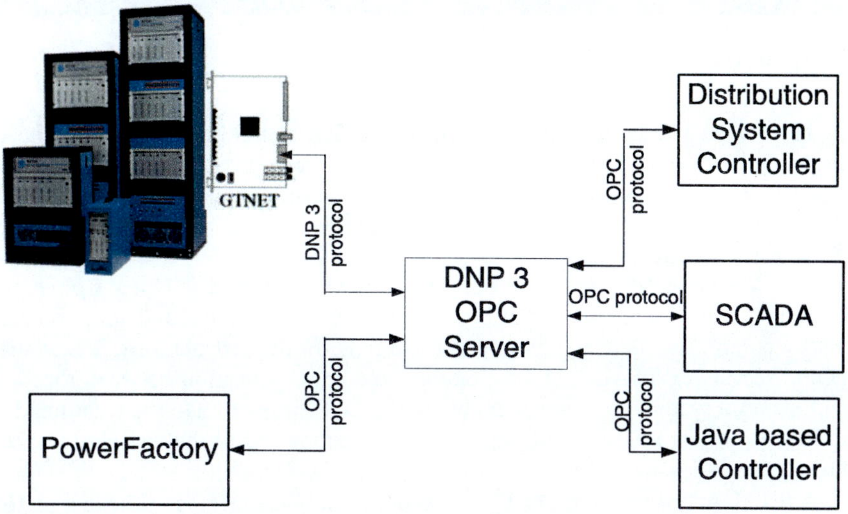

Fig. 14 Test System Setup

The RTDS in DTU has 5 racks and each rack has 4 RISC processor cards (GPC) and can model a system with 110 buses. The RTDS is shown in Fig. 15.

Fig. 15 RTDS in DTU

The RTDS is equipped with analogue and digital I/O cards to enable the data exchange between the RTDS and external hardware. In addition, the RTDS has one Gibabit Transceiver Network Interface System (GTNET) card which enable the RTDS communicate with external systems with Distributed Network Protocol (DNP) and IEC 61850 protocol.

The MatrikonOPC server for SCADA DNP 3 provides Open Connectivity (OPC) access to devices compatible with the DNP 3 protocol and supports supervisory control and data acquisition applications. In the test platform, the MatrikonOPC server for SCADA DNP 3 enables the communication between the RTDS and external systems or controllers using the OPC protocol. On the right hand side in Fig. 14, a few possible options that can be connected to the test platform to do testing or studies are shown.

A SCADA system using the OPC protocol can be connected to the test platform and illustrate the operation and control of power systems. The status of the simulated power system can be transported to the SCADA system and shown on the human machine interface (HMI) of the SCADA system. The control action can be realized by sending control signals from the SCADA system to the specific power components in the simulated power system. Such a system can be a training system for power systems engineers in the control room and students.

Besides using the platform as an education and training system, the platform can be used to test the developed controller for active distribution systems. Since the OPC protocol is widely used in the industry, it is quite possible that the controller will use the OPC protocol. A typical active distribution system with different kinds of RES can be modeled in the RTDS and different operating conditions can be simulated to validate the functionality of the developed controller for active distribution systems.

3.2 Communication Testing between the RTDS and the MatrikonOPC Server for SCADA DNP 3

In order to verify the data exchange between the RTDS and the MatrikonOPC server for SCADA DNP 3, the communication test has been done. A small test case has been built to verify the data exchange for analogue and binary variables between the RTDS and the MatrikonOPC server for SCADA DNP 3. The test case is shown in Fig. 16.

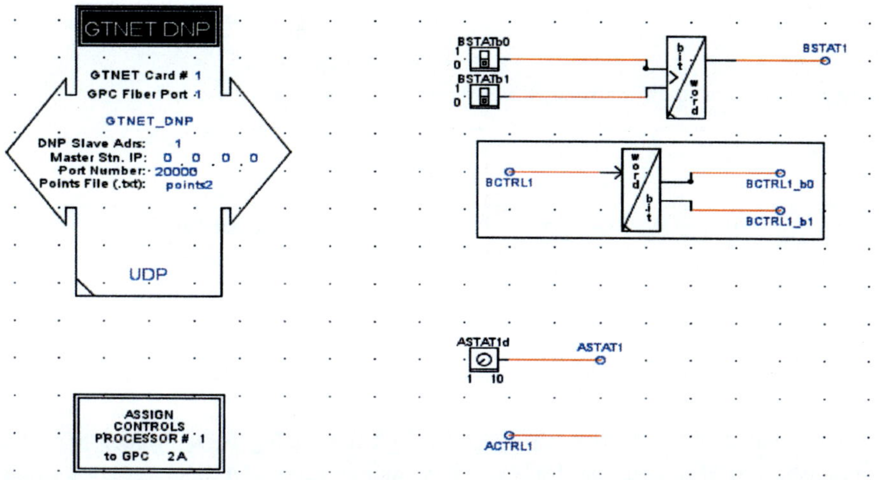

Fig. 16 Test Case for Communication between the RTDS and the MatrikonOPC Server for SCADA DNP 3

In order to avoid the ambiguity, terms of 'status' and 'control' are used to represent the variables from the RTDS and to the RTDS, respectively.

In the test case, one analogue status variable and one analogue control variable are defined to test the analogue data exchange; two binary status variables and two binary control variables are defined to test the binary data exchange. The defined variables are listed in Table 1.

Table 1 Variables in the Test Case in RTDS

Variable Name	Variable Type
BSTATb0	Binary Status Variable
BSTATb1	Binary Status Variable
BCTRL1_b0	Binary Control Variable
BCTRL1_b1	Binary Control Variable
ASTAT1	Analog Status Variable
ACTRL1	Analog Control Variable

The values of binary and analog status variables will be transported from the RTDS to the MatrikonOPC server and the values of binary and analog control variables will be transported from the MatrikonOPC server to the RTDS.

The MatrikonOPC Explorer is used to illustrate the values of status variables and to change the values of control variables. In the MatrikonOPC Explorer, tags for all variables have to be defined. The used tags are listed in Table 2.

Table 2 Tags used in the MatrikonOPC Explorer

Tag Name	Value
AnalogInput 030.0.x	Analog status value
AnalogOutput 040.0.x	Analog control value
AnalogOutputBlockShortFP 041.3.x	Analog control setting
BinaryInput 001.0.x	Binary status value
BinaryOutput 010.0.x	Binary control value
ControlBlockRelay 012.1.x	Binary control setting

The AnalogInput 030.0.x tag is used to get the analog status values from the simulated power system in RTDS. The AnalogOutput 040.0.x tage is used to change the analog control values to the simulated power system which will be the set points for the control variables. The AnalogOutputBlockShotFP 041.3.x is used to change the value of AnalogOutput 040.0.x.

The BinaryInput 001.0.x tag is used to get the binary status values from the simulated power system in RTDS. The BinaryOutput 010.0.x is used to change the binary control values in the simulated power system. The ControlBlockRelay 012.1.x is used to change the value of BinaryOutput 010.0.x.

The status and control values in the RTDS and the MatrikonOPC Explorer are shown in Fig. 17 and Fig. 18.

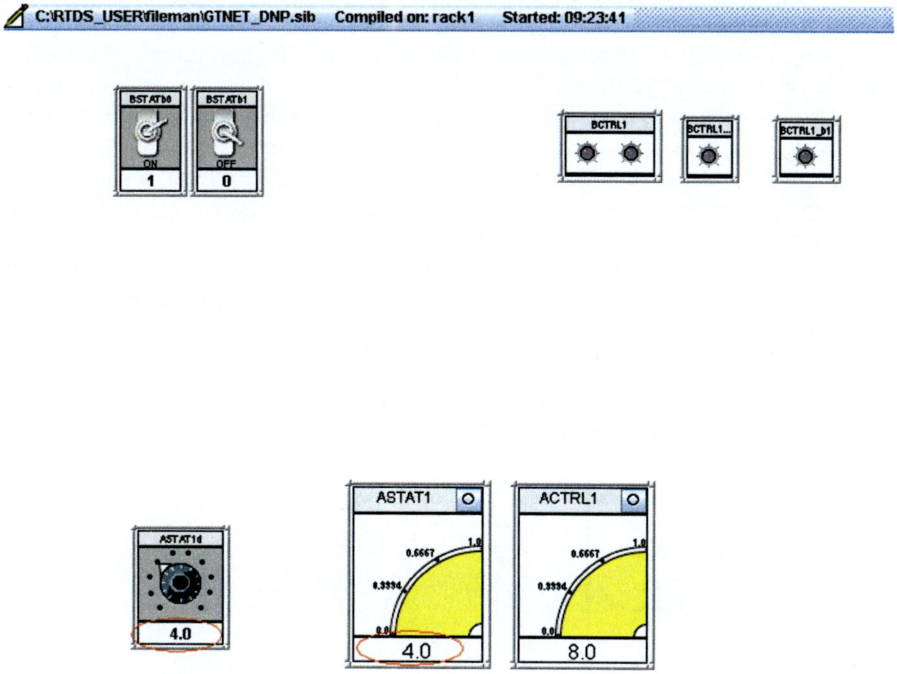

Fig. 17 Status and Control Values in RTDS

Contents of 'Group0'

Item ID	Access Path	Value	Quality
Rtds.Rtdshost.Rtdsdnp3.AnalogInput 030.0.0		7	Good, non-specific
Rtds.Rtdshost.Rtdsdnp3.AnalogOutput 040.0.0		3	Good, non-specific
Rtds.Rtdshost.Rtdsdnp3.AnalogOutputBlockShortFP 041.3.0		0	Good, non-specific
Rtds.Rtdshost.Rtdsdnp3.BinaryInput 001.0.0		True	Good, non-specific
Rtds.Rtdshost.Rtdsdnp3.BinaryInput 001.0.1		True	Good, non-specific
Rtds.Rtdshost.Rtdsdnp3.BinaryOutput 010.0.0		True	Good, non-specific
Rtds.Rtdshost.Rtdsdnp3.BinaryOutput 010.0.1		True	Good, non-specific
Rtds.Rtdshost.Rtdsdnp3.ControlBlockRelay 012.1.0		3,1,0,0,0	Good, non-specific
Rtds.Rtdshost.Rtdsdnp3.ControlBlockRelay 012.1.1		3,1,0,0,0	Good, non-specific

Fig. 18 Status and Control Values in MatrikonOPC Explorer

The test results show that both analog and binary status values can be obtained from the simulated system in RTDS and the values of control variables can be sent back to the simulated system in RTDS. Therefore, with the system setup as such, the functionality of the distribution system controller and the Java based controller can be tested using the real time performance of the simulated distribution system.

4 Modeling of Active Distribution Systems

4.1 Distributed Generation System

The penetration of DGs causes changes in frequency control philosophy for active distribution systems. There is a strong need in development of power system model for study of frequency control features in considered areas. In this regard, modeling is essential for development of the control. Inappropriate selection of models can lead to inaccurate control. Therefore, precise modeling of the distribution system elements is necessary. At the same time, models must have a reasonable order. This is usually the case especially for DG units, some of which have quite complicated dynamic behavior. For the study of DG systems, a sample distribution system which includes several different DG units has been developed. Also, investigated various DG configurations and chosen several specific units for use in this study. The DG system simulation model is shown in Fig. 19 and Fig. 20, respectively and a description of each DG unit, along with associated controls is provided.

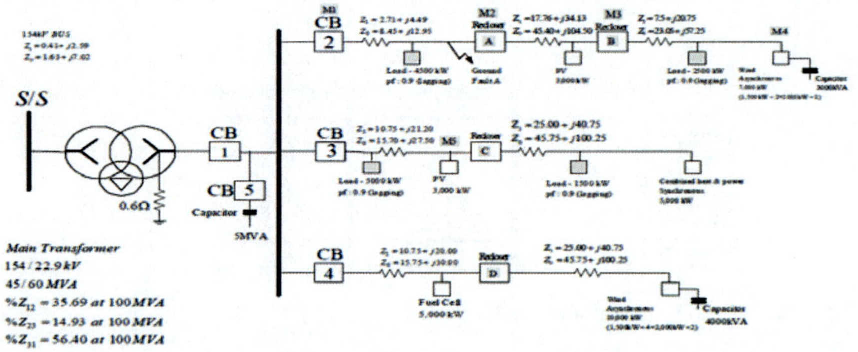

Fig. 19 Distributed generation system model

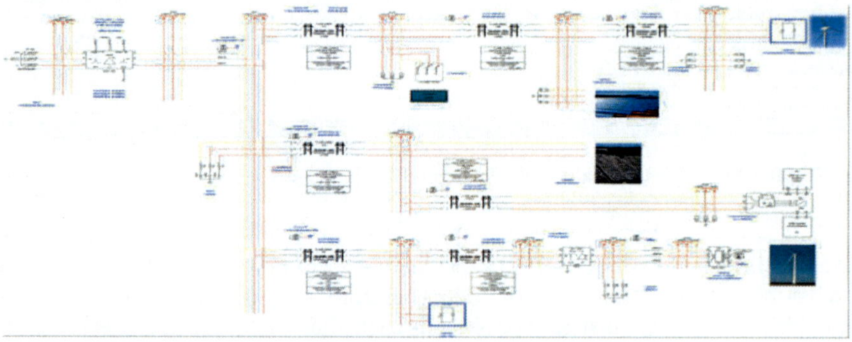

Fig. 20 RTDS implementation of distributed generation system model

4.1.1 Variable Speed Wind Turbine

The variable speed wind turbine model used in the study system is based on the work of [11]. The principles and mathematical models of DFIG are well documented in literature and hence are only briefly outlined. An interface transformer, a high pass filter and two Voltage Source Converters (VSCs), as well as the DFIG are included in the small time step portion of this model. The controls and the wind turbine model are run in the large time step portion of the model. The DFIG has become the choice of many utilities for harvesting wind energy due to its ability to provide reactive power support to the power system during low voltage conditions. The DFIG can also extract the optimum wind energy over a wide range of wind speeds, a feature not possible with fixed speed induction generators. The parameters of the DFIG are scaled up from a laboratory sized machine to a 7 MW, 2.5 kV machine. The actual DFIG parameters used are typical of a 2 MW, 690 V machine. The high pass filter is rated at 10 percent of the interface transformer MVA and tuned to remove the switching voltage and current components generated by the VSCs (21 x 60 Hz). The grid side VSC operates to control voltage on the capacitor connected between it and the rotor side VSC, while the rotor side VSC controls voltages applied to the rotor to maintain reactive power flow into the DFIG and to control torque. The grid side VSC can also regulate reactive power flow into the system. The wind turbine allows the selection of either of two commonly used power coefficient curves along with user entered parameters. Both wind speed and rotor speed are inputs into the model and mechanical torque is the output. Rotor speed is determined by the DFIG based on inertia and on the difference between mechanical and electrical torques. An optimum torque reference is included to produce maximum power at any rotor speed. Also, a simple pitch control was used to protect the rotor from over speed in the event of strong wind.

4.1.2 Photovoltaic Generators and Fuel Cell Generator

The two Photovoltaic (PV) generators are connected to the ac system through standard voltage source converters, interface transformers and high pass filters all running in the small time step portion of this model. Control references for capacitor voltage and reactive current are included. No outer loop controls such as capacitor voltage control for power transfer optimization are included due to insufficient time to study and investigate such control. A Norton current source in parallel with the VSC capacitor is used to simulate current injections from PV cells. This injected current results in real power production from the VSC.

4.1.3 Fixed Speed Wind Turbine

The fixed speed wind turbine uses the same wind turbine model as mentioned above and is connected in this case, to a typical Danish squirrel cage induction generator. No controls are implemented except the beta control for the turbine blades to prevent excessive slip.

4.1.4 Synchronous Generator

The machine is based on the default synchronous generator provided in the standard RSCAD library. It is connected to the network with a step up transformer and is controlled with a standard static exciter (EXST1), turbine and governor (IEESGO) components.

4.1.5 Control System

Each of the VSCs in the study circuit has its own control system. In the case of the two PV systems, the Fuel Cell system and the grid side VSC of the variable speed wind turbine, the controls are very similar. Three phase current into the VSC is measured and decomposed into direct and quadrature components. These are regulated by measuring errors with respect to references and then passing them to proportional plus integral gain blocks. Cross coupling terms are then removed resulting in a computed direct and quadrature voltage. The controls then use the two computed voltages to produce a three phase set of sinusoidal modulation references with the correct magnitude and phase. Finally, the three phase reference set is compared against a triangle wave to generate the square wave switching sequence required by each leg of the VSC. The magnitude and phase of the modulation reference controls the magnitude and phase of the VSC internal node voltages. Voltage magnitude differences with respect to the external AC voltage give rise to reactive current just as in synchronous condenser. Phase differences give rise to real current. Real current is regulated to control the capacitor voltage. The controls for the variable speed wind turbine rotor VSC are similar to the others, but there are major differences. These controls define an internal synchronous direct and quadrature reference frame where the stator flux is aligned with the direct axis. The rotor flux is then located on that reference and the rotor slip currents controlled to generate a quadrature and direct rotor flux components. The direct axis rotor current then controls reactive current flowing into the stator and the quadrature axis rotor current controls the electric torque. The cross coupling terms are removed to find the required voltages as in the other VSC controls, but computing them is somewhat more complicated. The final control structure is shown below.

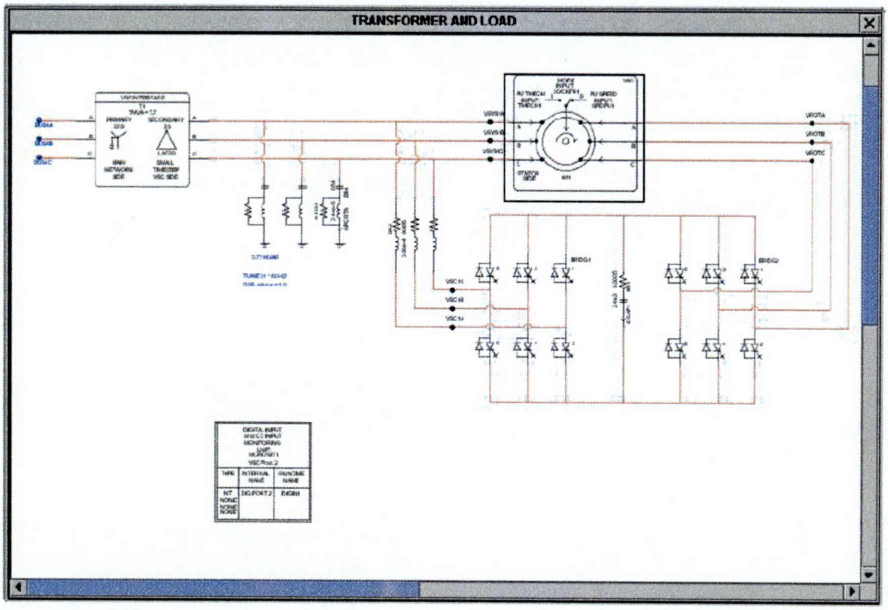

Fig. 21 DFIG with its VSC

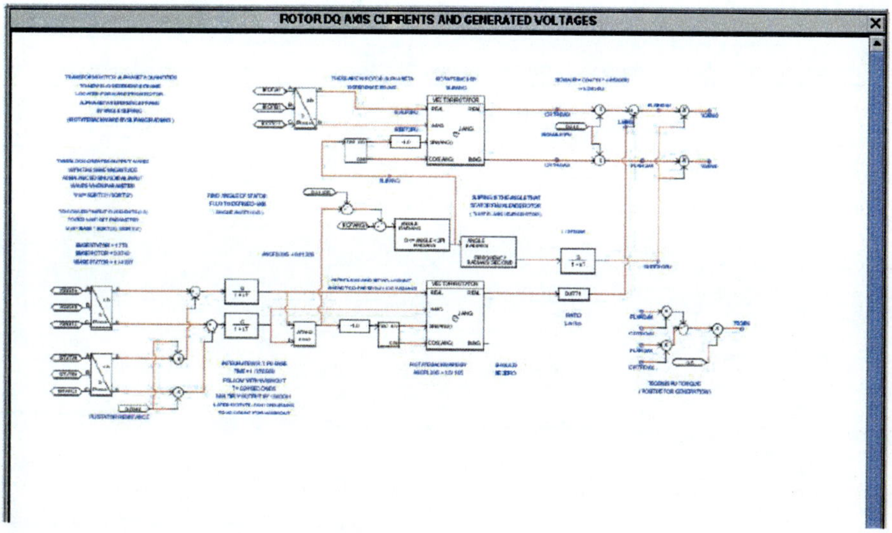

Fig. 22 VSC control system - rotor DQ axis currents and generated voltage control

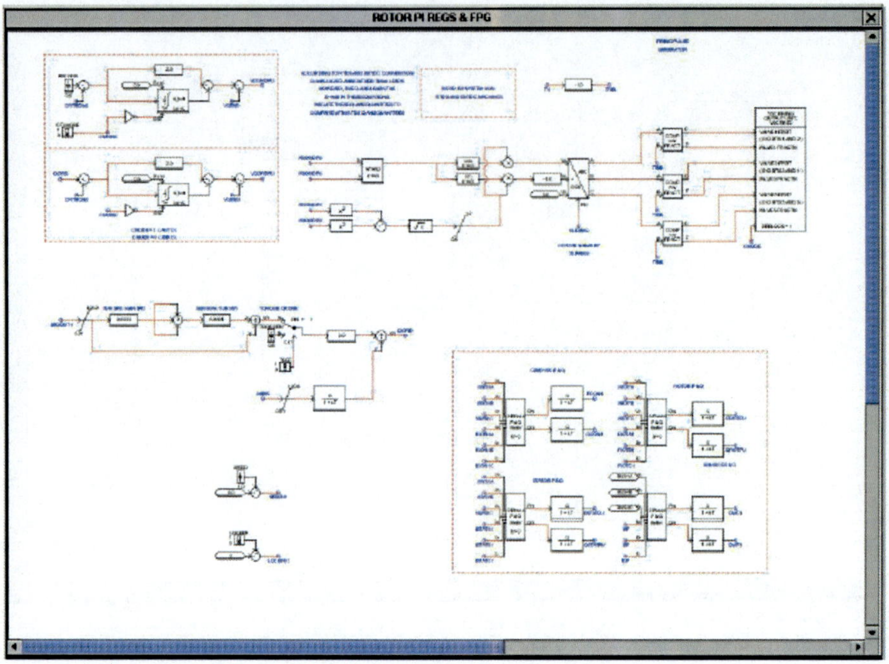

Fig. 23 VSC control system - rotor PI regulation and FP generation

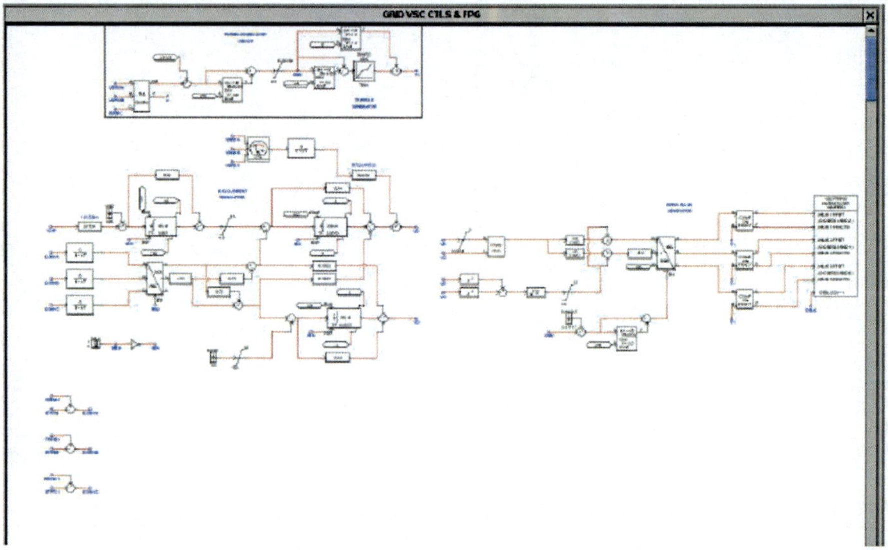

Fig. 24 VSC control system - Grid VSC controls and FP generation

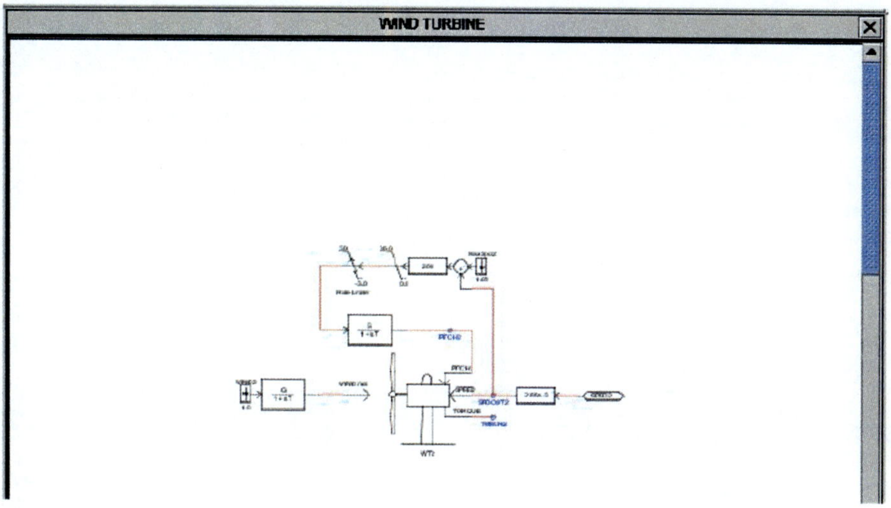

Fig. 25 Wind and wind turbine model

4.2 *IEEE 9-Bus System*

The modified IEEE 9-bus test system is used to illustrate the islanding operation of an active distribution networks, and to validate the proposed multi-agent based controller for the islanding operation of active distribution systems. The single-line diagram of the modified IEEE 9-bus test system in DIgSILENT PowerFactory and RSCAD are shown in Fig. 26 and Fig. 27, respectively. The modified IEEE 9-bus system comprises a 60kV, 50 Hz grid which feeds an 11 kV network through a 60/11kV transformer. The system consists of three DG units, eight transmission lines, one transformer and four loads. Three DG units with nominal power of 4.85 MVA are connected to bus 1, 3 and bus 4, respectively, and simulated with 1.5 MW power output before islanding operation. The loads, totaling 6.0 MW, are distributed along the bus 5,6,7,8 and are modeled as constant impedance static and ZIP dynamic loads [12].

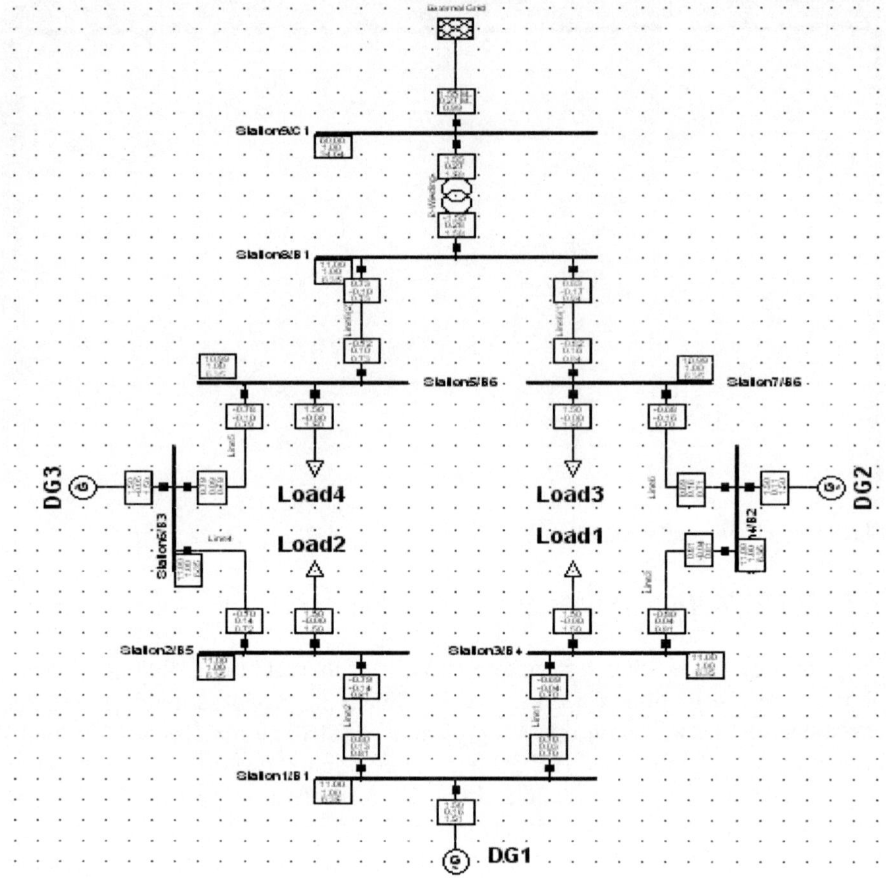

Fig. 26 Modified IEEE 9-bus test system

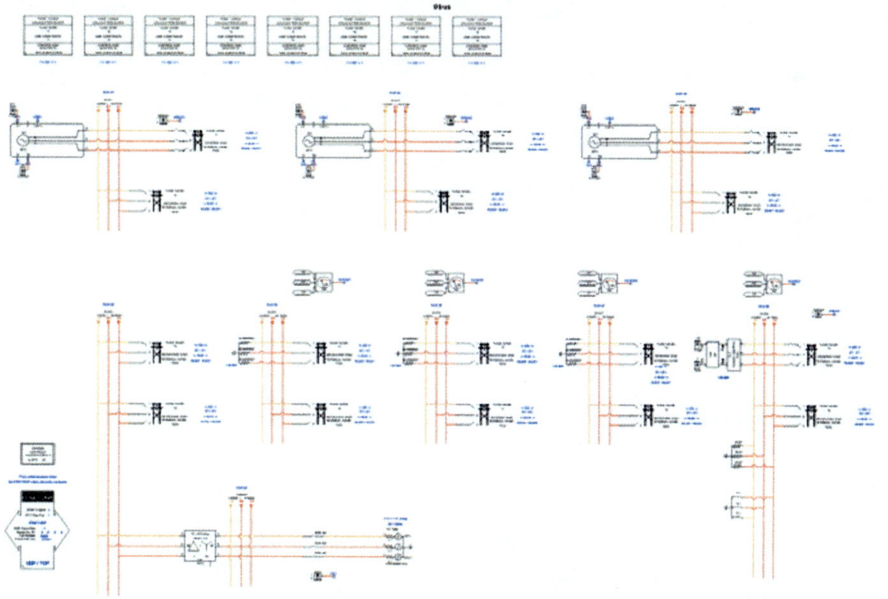

Fig. 27 RTDS implementation of IEEE 9-bus test system

Table 3 Characteristics of the IEEE 9-bus test system

System characteristic	Value
Number of buses	8
Number of DGs	3
Number of loads	4
Number of transmission lines	8
Total generation	6.0 MW / 0.379 MVAr
Total load	6.0 MW / 0 MVAr

The commercial cogeneration systems use the natural gas, coal or oil to drive the steam turbine. The steam turbine couples with the rotor of the synchronous generator to generate electric power, and they are operated according to the heat demand. The transient and sub-transient impedances of the synchronous generator are considered so that more accurate transient stability analysis can be obtained [13]. To regulate the bus voltage of the islanded system, the excitation system as depicted in Fig. 28, the IEEE standard type ST1 model, is used for the steam turbine cogeneration units. Also, in order to achieve the quick response of the steam turbine mechanical power output to the frequency variation, the IEEEG1 governor model shown in Fig. 29 is applied in the speed governing system. Table 4 gives the parameters of the governor model.

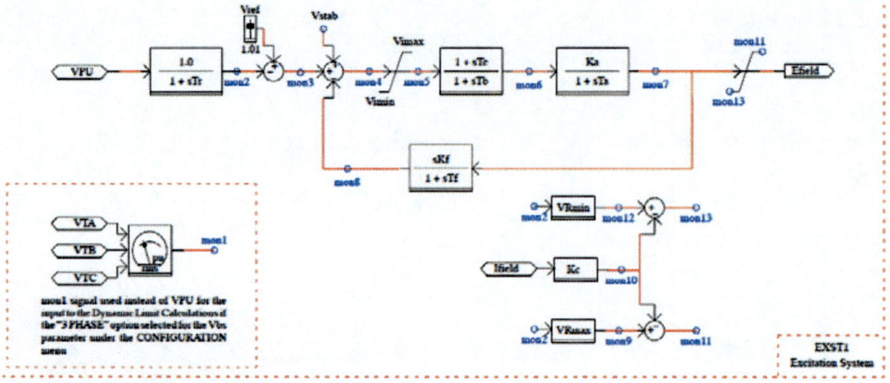

Fig. 28 RTDS exciter model

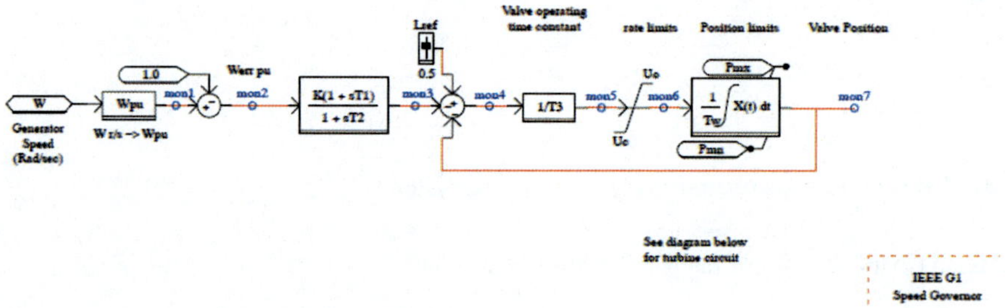

Fig. 29 RTDS speed governor model

Table 4 Parameters of the governor model

Parameter	Value	Parameter	Value
K	20	K2	0
T1	0.001	T5	5
T2	0.001	K3	0.7
T3	0.25	K4	0
U0	0.1	T6	0
Uc	-0.2	K5	0
Pmx	1	K6	0
Pmn	0	T7	0
T4	0.2	K7	0
K1	0.3	K8	0

For all the simulation cases, when the grid is connected to the 11 kV bus, the DGs are operating in the PQ mode and the voltage and frequency at the 11 kV bus are regulated by the grid. Under the islanding operation mode, the DG units are switched from PQ mode to V-f mode, and the 11 kV bus voltages and the system frequency are then regulated by the DG units. In our approach, agents act as smart controllers of different components of electric power systems e.g. DG agents and Load agents. Such agents reside in a dynamic multi-agent platform as previously shown in Sect. 3.3. JADE is a popular platform for application of the multi-agent technology in power system applications. Therefore, the secondary load-frequency control scheme was implemented in a JAVA application which calls the RTDS dynamic simulation computation module. This computation module runs dynamic simulations for a given controller sample time intervals. At the end of each dynamic simulation, the integral and absolute value of frequency deviation is read from RTDS's output file. Using this information, new power set points are computed and stored in the RTDS's information file for the next dynamic simulation. This process is repeated up to frequency stabilization or a maximum simulation time.

5 Case Studies for Frequency Control of Active Distribution Systems

5.1 Case Studies with a Sample Distributed Generation System

One base distribution network model was used for all three test cases. The three test cases considered are listed in the Table 5 below.

Table 5 Test cases

Test Number	Test Description
Case I	Opening re-closer A
Case II	Opening re-closer B
Case III	Ground fault at point A

5.1.1 Case I: Simulation of Islanding 1

This test involved the opening of re-closer A between bus #2 and bus #3. The main result seen from this test shows the voltage between M3 and M4 rises as the current increases to feed the local load of 2.5 MW. The controls for the variable speed wind turbine do not change the power reference and the power output stays constant. This requires the voltage to rise till all the power is consumed in the local load. If the controls had contained a voltage regulator, they would likely have become unstable. If the controls had contained a feature to run back the power level to match the local load (load frequency control) and voltage regulator, they would have had little impact on the local system.

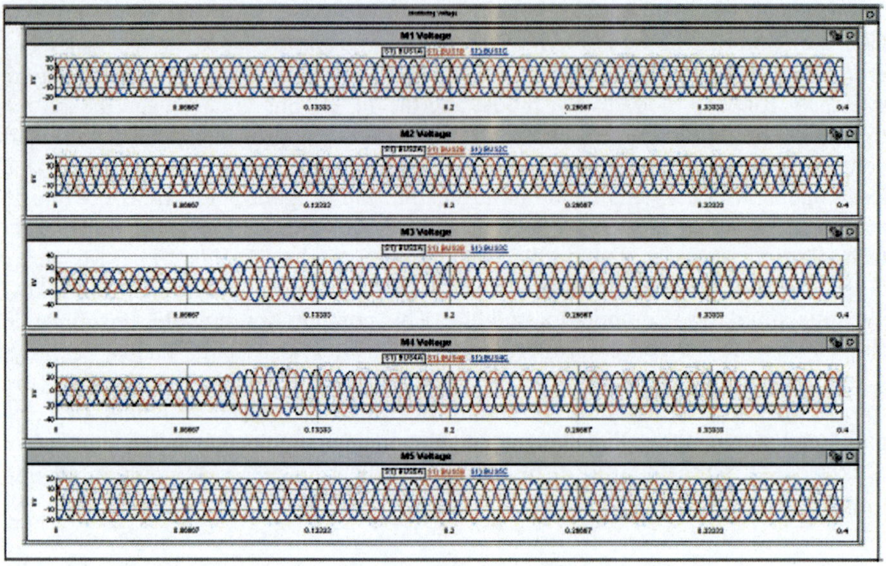

Fig. 30 Measured voltage at every measuring point (M1-M5) in Case 1

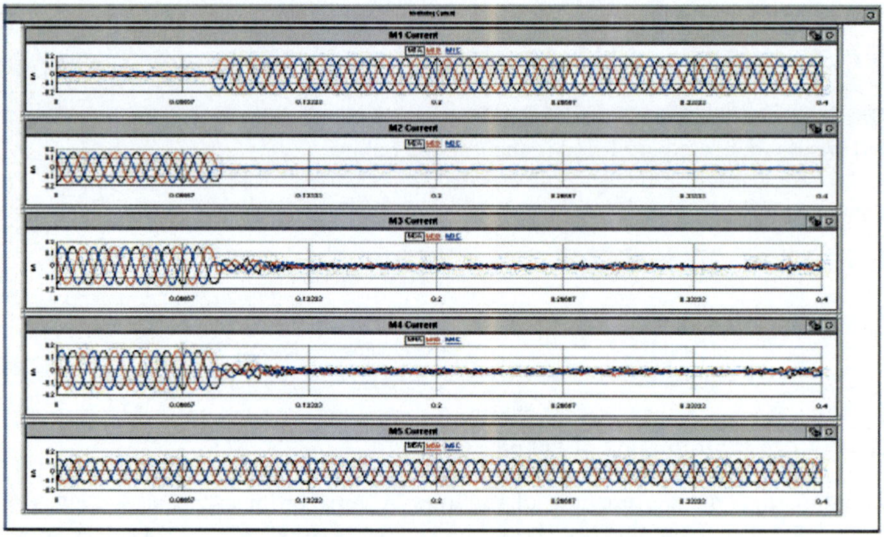

Fig. 31 Measured current at every measuring point (M1-M5) in Case 1

5.1.2 Case II: Simulation of Islanding 2

This test required the opening of recloser B. The main results from this test show the voltage at M4 rises as does the current to feed the local load of 2.5 MW. The controls for the variable speed wind turbine do not change the power reference and the power output stays at 7MW constant. This requires the voltage to rise until all the power is consumed in the local load. If the controls had contained a voltage regulator, they would likely have become unstable. If the controls had contained a feature to run back the power level to match the local load (load frequency control) and voltage regulator, they would have had little impact on the local system. The result is similar to that of test 1.

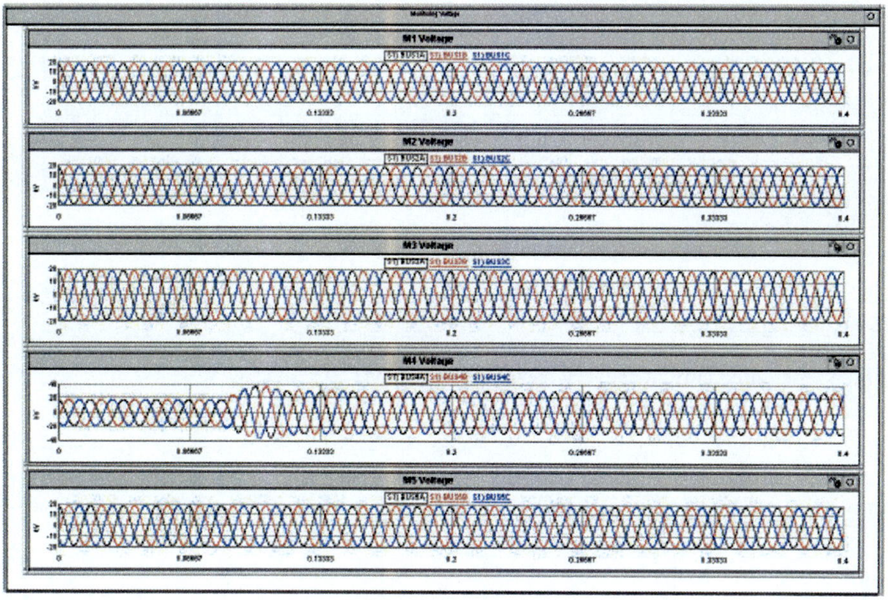

Fig. 32 Measured voltage at every measuring point (M1-M5) in Case 2

5.1.3 Case III: Single Phase to Ground Fault at Point A

In this case, a single line to ground fault was applied to the bus #2 at t=0.03s. Fig. 34 illustrates the timing sequence for test case 3. During the fault, local voltages and currents from the VSCs had larger harmonics than normal operation and there were significant negative sequence components as expected. The network and the controls experienced very little disturbance following the clearing of the fault after 0.3s and seem to have recovered well.

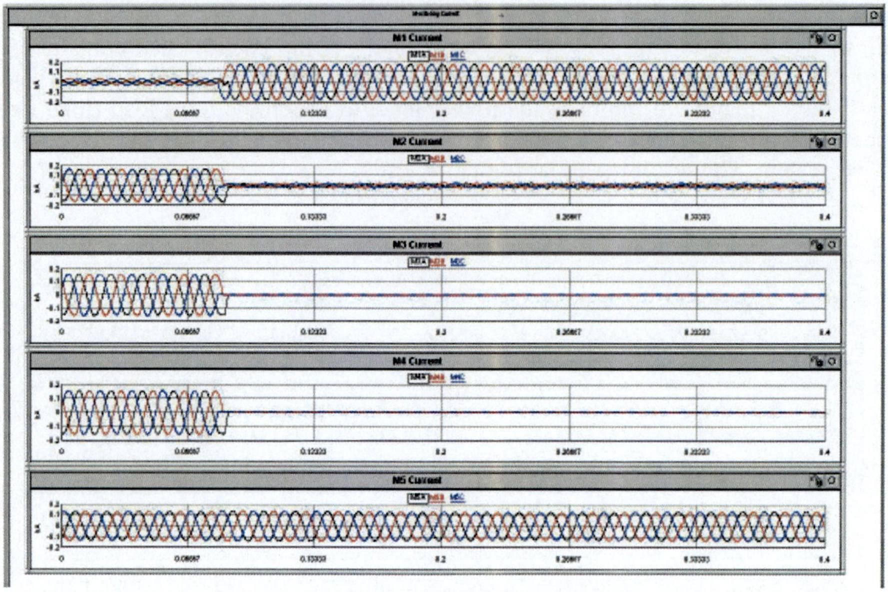

Fig. 33 Measured current at every measuring point (M1-M5) in Case 2

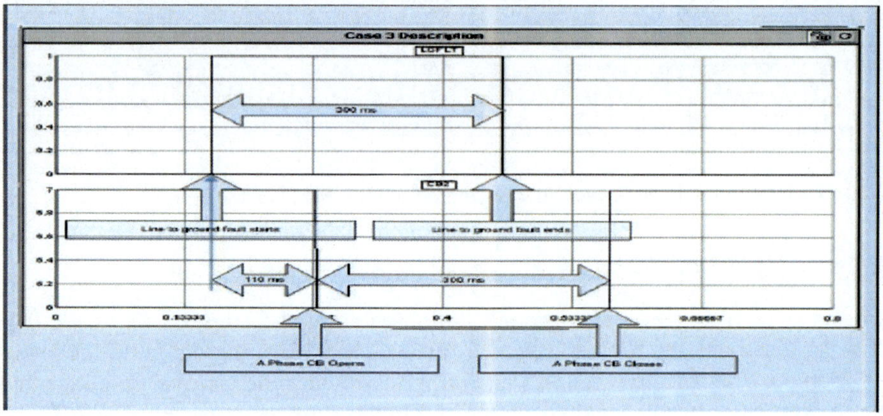

Fig. 34 Fault sequence and CB operation in Case 3

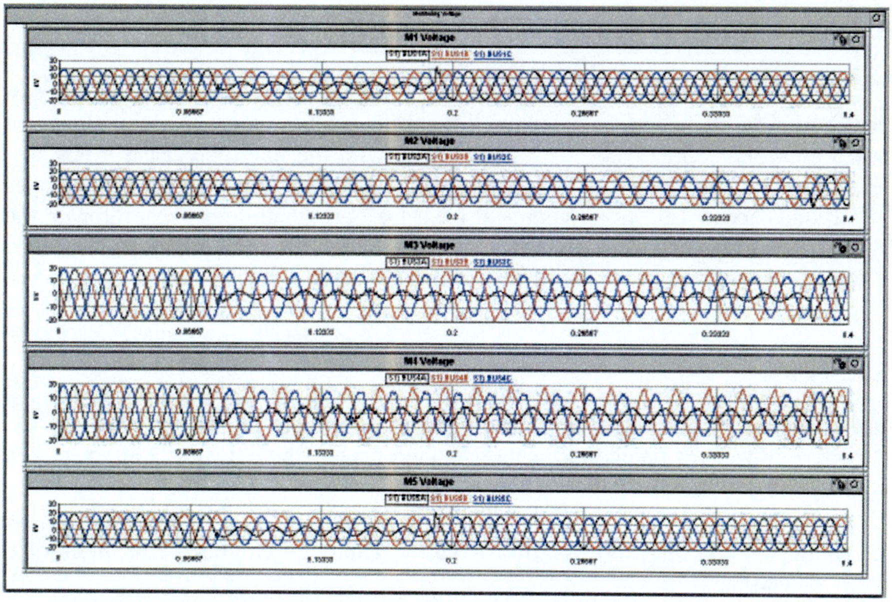

Fig. 35 Measured voltage at every measuring point(M1-M5) in Case 3

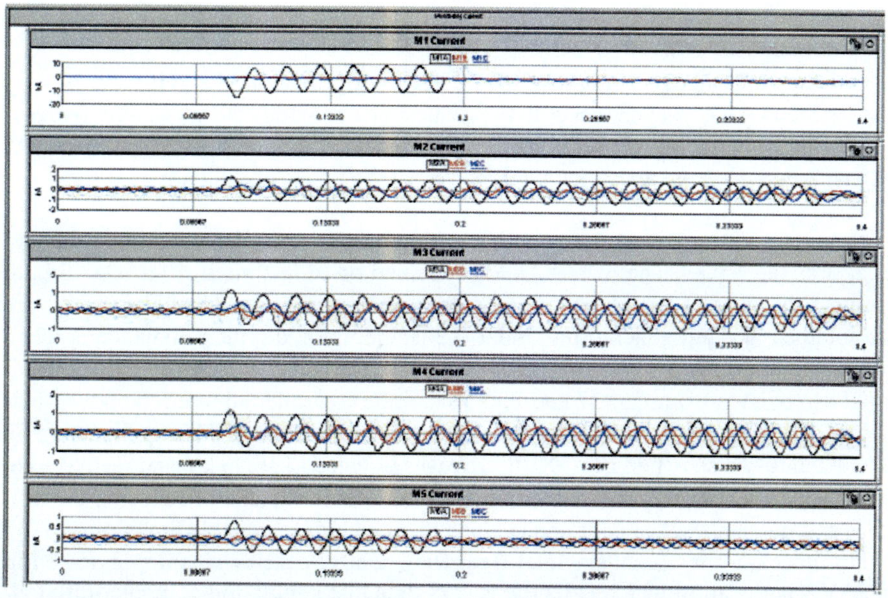

Fig. 36 Measured current at every measuring point (M1-M5) in Case 3

5.2 *Case Studies with the IEEE 9-Bus System*

In order to further investigate the steady state and dynamic performance of active distribution systems under islanding operation with the proposed multi-agent based controller, three case studies are set up with the modified IEEE 9-bus system. Initially, the test distribution system is balanced and all voltages at the nodes of all loads are at 1pu. Load agents at each load are continuously monitoring the voltage and frequency to response to any disturbance. The contingency considered is a three phase fault at the grid connection. Each study case has a test description as follows;

5.2.1 Case I

In this case, only one DG is providing regulation service. The DG provides an extra active power for balancing in the islanded part of the network. The execution sequence is described as follows:

- Step1. Loss of 1.5 MW power from the grid due to an outage or intentional islanding
- Step2. Created an imbalance in the islanded part of the network
- Step3. Load agents observe voltage and frequency drop
- Step4. Load agents contact DF agent for any available regulation service
- Step5. DF agents informs the current service availability and provides its reference
- Step6. Load agents request DG #2 agent for provision of service
- Step7. DG #2 agent accepts the request and provides the service by increasing its active power set point
- Step8. Voltage and frequency recover at the nodes of all loads

The control strategy is to use one DG as the master controller while the other remains as the slaves. The master DG is switched on to V-f mode after islanding to control the overall island voltage and frequency while the other slave DGs are maintained at fixed generation. Under the islanding mode, the imbalance is taken up by the master DG.

It can be seen from Fig. 37 and Fig. 38 that the controller adjusted the power output of DG #2 to the desired value and the required power needed is supplied only by DG #2 unit. As shown in Fig. 38, the voltages dropped to 10.96 kV and recovered to the operational range quickly. The master controller is capable of reducing the voltage and frequency excursions and keeping them within permissible limits.

In this case, the secondary load-frequency control scheme was implemented in a JAVA application which calls the RTDS dynamic simulation computation module. This computation module runs dynamic simulations for a given controller sample time intervals. At the end of each dynamic simulation, the integral and absolute value of frequency deviation is read from RTDS's output file. Using this information, new power set points are computed and stored in the RTDS's information file for the next dynamic simulation. This process is repeated up to frequency stabilization or a maximum simulation time.

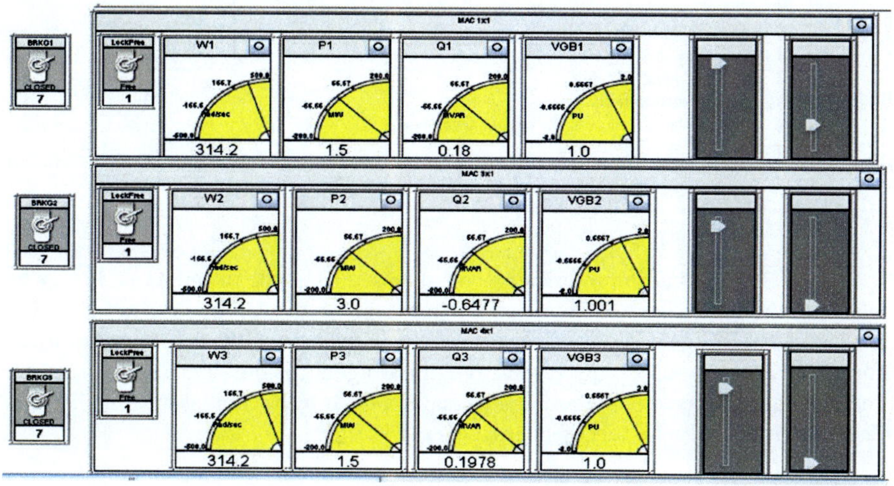

Fig. 37 System frequency and DG output in case I

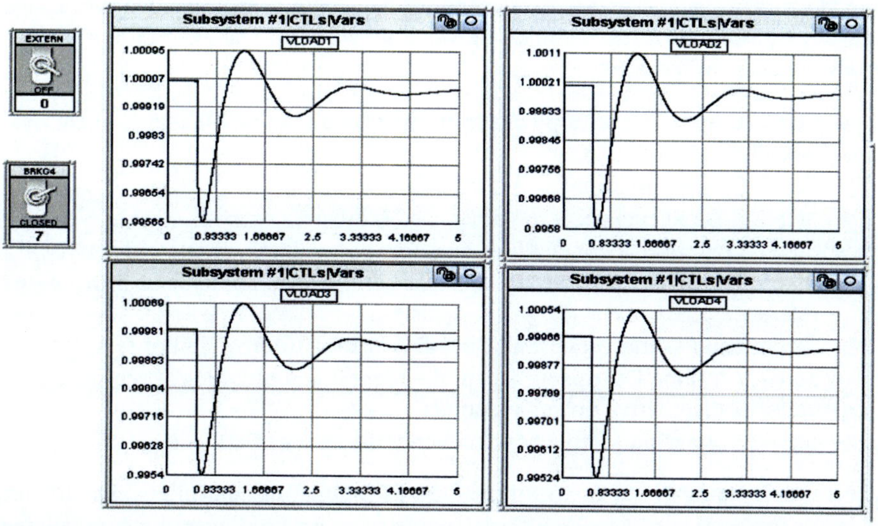

Fig. 38 Bus voltages in case I

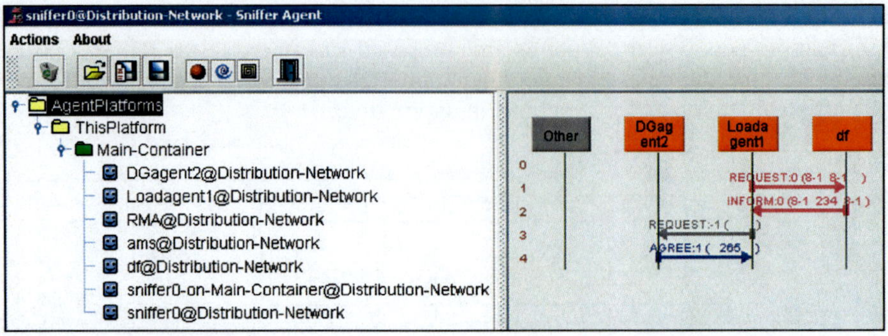

Fig. 39 Agent communication in case I

Fig. 39 shows the communication of agents during the simulation. It can be seen that the load agent sends a request message to the DF agent for regulation service. The DF agent sends back a message with the available service information. Load agent sends a request message to DG agent for provision of this service. The content of request is the amount of active power to be increased. Finally, DG #2 agent replies with an 'agree' message.

5.2.2 Case II

In this case, all three DG units are providing regulation services. The DG units provide an extra active power if required for balancing in the islanded part of the network by sharing or rescheduling between three units. The execution sequence is described as follows;

- Step1. Loss of 1.5 MW power from the grid due to an outage or intentional islanding
- Step2. Created an imbalance in the islanded part of the network
- Step3. Load agents observe voltage and frequency drop
- Step4. Load agents contact DF agent for any available regulation service
- Step5. DF agents informs the current service availability and provides its reference
- Step6. Load agents request all three DG agents for provision of services
- Step7. All three DG agents accept the request and provide the service by increasing their active power set points
- Step8. Voltage and frequency recover at the nodes of all loads

The control strategy is to use all three DGs as the master controllers. The master DGs are switched to V-f mode after islanding to control the overall island voltage and frequency. Under the islanding mode, the imbalance is taken up by all three master DGs. It can be seen from Fig. 40 and Fig. 41 that the controllers adjusted the power output of DG #1 – DG #3 to desired values and the required power needed is supplied by all three DG units. As shown in Fig. 41, the voltage dropped to 10.96 kV and recovered to the operational range quickly. All three master controllers are capable of reducing the voltage and frequency excursions, and keeping them within permissible limits.

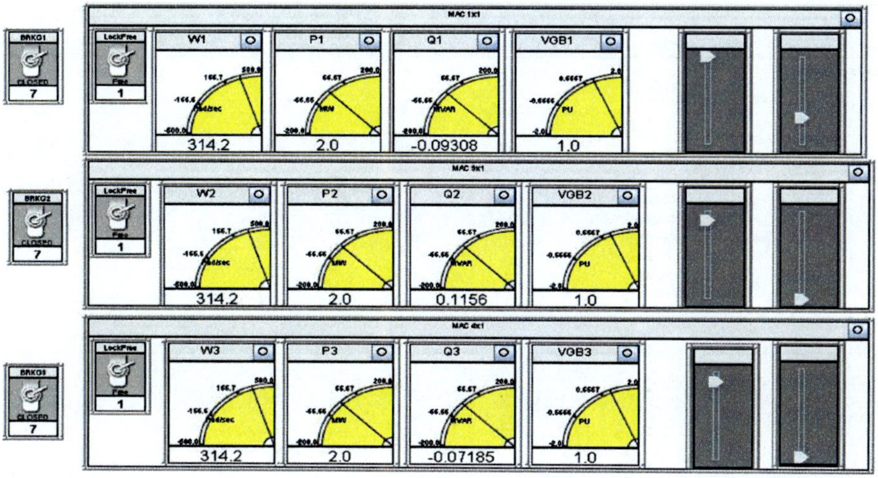

Fig. 40 System frequency and DG output in case II

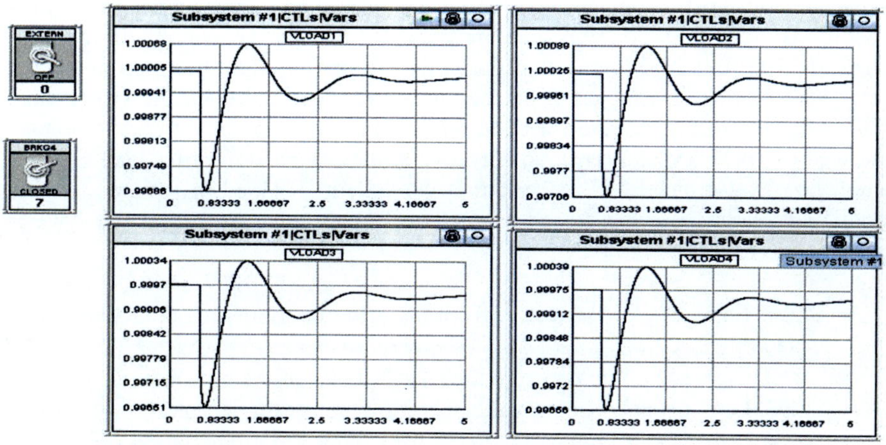

Fig. 41 Bus voltages in case II

Fig. 42 shows communication of agents during the simulation. It can be seen that the load agent sends a request message to DF agent for available regulation service. The DF agent sends back a message with available service information. Load agent sends a request message to all units of DG agents for provision of service. The content of request is the amount of active power to be increased by all three DG units. Finally, DG #1- DG #3 agent replied with 'agree' messages.

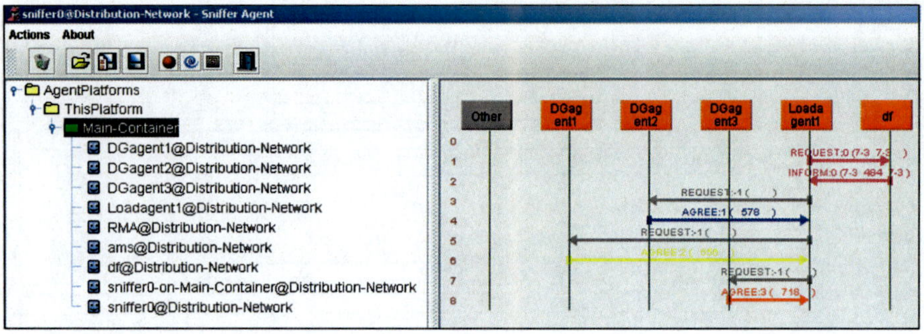

Fig. 42 Agent communication in case II

5.2.3 Case III: Load Shedding

In this case, it was investigated to use load shedding under islanding operation. These results show that the power balance in the island created by disconnection from the main grid can be maintained by load shedding. Under normal operating conditions, the generation and load are balanced and the generator terminal voltages are 1pu. Initially, the system is running in normal condition, but at t = 0.5 sec, power source is isolated from the system due to a fault on the utility side. This isolation creates an island with 25% imbalance. As Fig. 44 shows, the voltage dropped to 10.57 kV and the load shedding strategies shed the first load encountered, i.e., load 4 at bus #8. After the system sheds the load, it restores the frequency and voltage as the power balance is restored. The execution sequence is described as follows:

- Step1. Loss of 1.5 MW power from the grid due to an outage or intentional islanding
- Step2. Created an imbalance in the islanded part of the network
- Step3. Load agents observe voltage and frequency drop
- Step4. Load agents contact DF agent for any available regulation service
- Step5. DF agents informs the unavailability of any regulation service
- Step6. Load agent informs this situation to Load Shedding agent
- Step7. A load shedding of 1.5 MW at bus #8
- Step8. Voltage and frequency recover at the nodes of all loads

This is case where none of DG units are providing regulation service when a part of distribution network is islanded from the main grid. Load agents are informed about unavailability of any regulation service by DF agent. Load agent informs this situation to the Load Shedding agent. Load shedding agent to shed 1.5 MW in order to achieve balance.

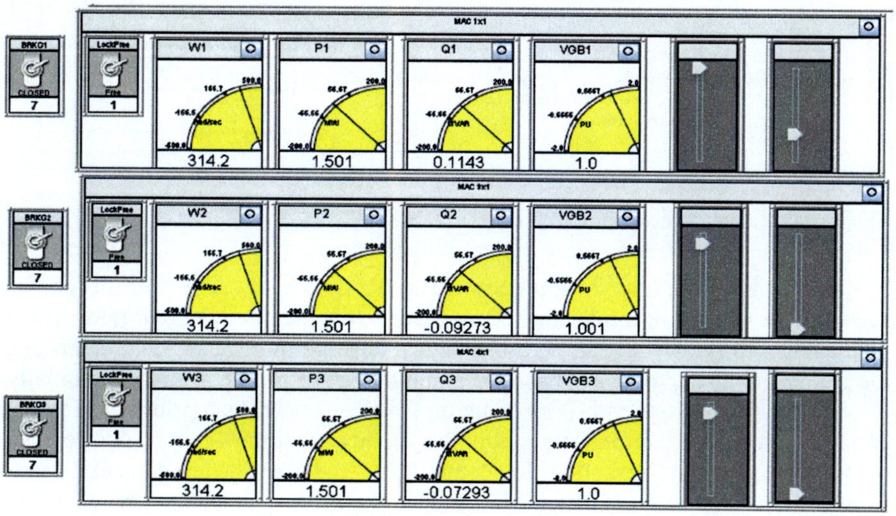

Fig. 43 System frequency and DG output in case II

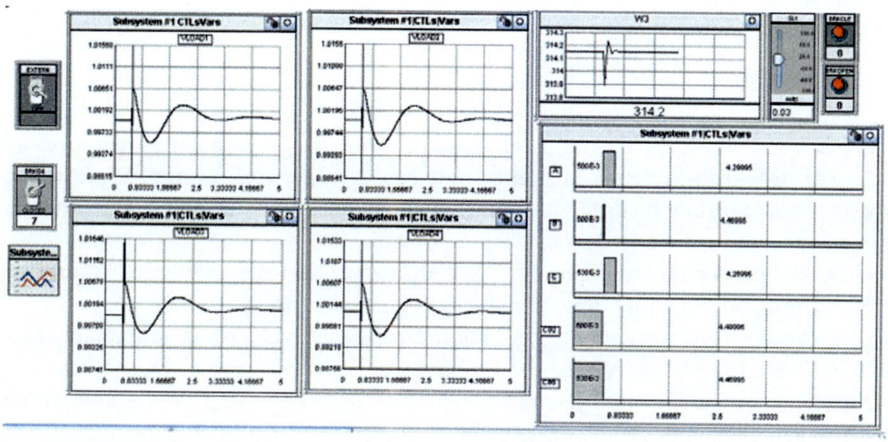

Fig. 44 Bus voltages and fault time sequence in case III

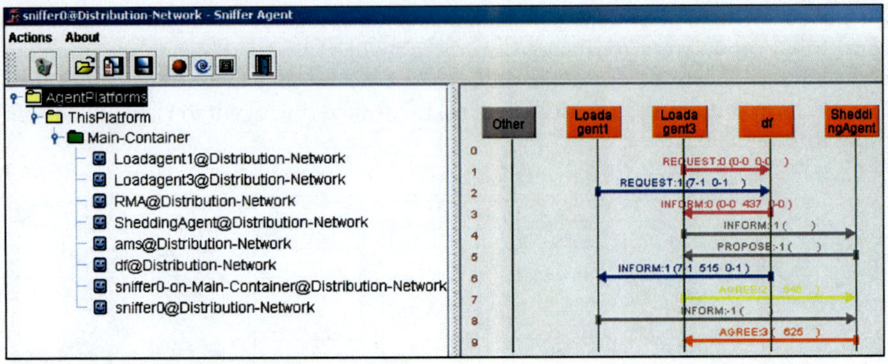

Fig. 45 Agent communication in case III

Fig. 45 shows communication between agents during the simulation. It can be seen that the control agent detects the fault at t = 0.5, the control agent informs the load agent and the DF agent, both of which exchange information and determine the amount of loads to be shed and the amount of power to be produced internally in order to stabilize the grid. Depending on the pre-defined load priority set by user and the available internal generation, the loading shedding agent can disconnect non-critical loads and certain critical loads. Thus, the load shedding agent disconnects the non-critical loads of 1.5 MW and leaves the critical load intact. All agent actions from detecting the fault, disconnecting the main circuit breaker, disconnecting the non-critical loads to stabilize the grid can be accomplished within half an electrical cycle, i.e. less than 0.008 second.

6 Summary

This chapter describes the frequency control concept for islanding operation of active distribution systems. A multi-agent based framework is used to design a frequency controller in order to utilize different resources in active distribution systems to stabilize the frequency.

The test platform comprising RTDS, OPC server and the java based controller is described and the communication test is presented as well.

An example DG system and the modified IEEE 9 bus system are depicted in details. The example DG system is comprised of different types of DG such as variable speed wind turbine generator, photovoltaic, fuel cell, fixed speed wind turbine generator and small synchronous machine. The modified 9 bus IEEE bus system has three DGs.

Case studies were implemented with the example DG system and the IEEE 9 bus system to illustrate the islanding operation of active distribution systems. The results show that the multi-agent based controller can efficiently stabilize the frequency of the distribution systems using different distributed resources.

References

[1] Pike Research Report, http://www.pikeresearch.com
[2] Grainger, J.J., Stevenson, W.D.: Power System Analysis. McGraw-Hill, New York (1994)
[3] UCTE operational Handbook, Appendix 1, A1-3
[4] Morren, J., Pierik, J., de Haan, S.W.H.: Inertia response of variable speed wind turbines. Electric Power Systems Research 76(11), 980–987 (2006)
[5] Morren, J., de Haan, S.W.H., Fereira, J.A.: Contribution of DG units to primary frequency control. In: Proc. 2005 International Conference on Future Power Systems, pp. 1–6 (2005)
[6] German, C.T., Phillip, C.K., Poul, E.S., Østergaard, J.: Study on variable speed wind turbines capability for frequency response. In: Proc. 2009 European Wind Energy Conference and Exhibition, pp. 190–193 (2009)
[7] Mahat, P., Chen, Z., Birgitte, B.J.: Control and operation of distributed generation in distribution systems. Electric Power Systems Research 81(2), 495–502 (2010)
[8] Jeon, J.H., Kim, J.Y., Kim, S.K., Ahn, O.B., Park, J.H.: Development of HILS(Hardware in-loop simulation) system for MMS(Microgrid management system) by using RTDS. In: Proc. 2008 Power Electronics and Motion Control Conf., pp. 2492–2497 (2008)
[9] Saleem, A., Heussen, K., Lind, M.: Agent services for situation aware control of power systems with distributed generation. In: Proc. 2009 IEEE Power Energy Society General Meeting, pp. 1–8 (2009)
[10] Solanki, J.M., Schulz, N.N.: Multi-agent system for islanded operation of distribution systems. In: Proc. 2006 IEEE Power Energy Society Power Systems Conference and Exhibition, pp. 1735–1740 (2006)
[11] Pena, R., Clare, J.C., Asher, G.M.: Doubly fed induction generator using back-to-back PWM converters and its application to variable speed wind energy generation. IEE Proceedings on Electrical Power Applications 143(3), 231–241 (1996)
[12] Cha, S.T., Wu, Q., Østergaard, J., Saleem, A.: Multi-agent based controller for islanding operation of active distribution networks with distributed generation. In: Electric Utility Deregulation and Restructuring and Power Technologies (2011)
[13] Anderson, P.M., Fouad, A.A.: Power System Control and Stability. Iowa State University Press, Iowa (1977)

Control Methods for Grid Side Converters under Unbalanced Operating Conditions in Wind Power Applications

Ana Vladan Stankovic, Dejan Schreiber, and Shuang Wu

Abstract. The chapter presents two control methods for complete harmonic elimination of a grid side inverter under unbalanced operating conditions. The first control method is suitable for unbalanced grid voltages and balanced line impedances whereas the second method is general and can be used for extreme unbalanced operating conditions. Under severe fault conditions in the distribution system, not only grid voltages, but also line impedances must be considered as unbalanced. An analytical approach for complete harmonic elimination shows that a grid side inverter can operate under extreme unbalanced operating conditions with adjustable power factor resulting in a smooth (constant) power flow from dc to ac side. Both methods can be used for unbalanced falt-ride through control of a wind turbine inverter. Simulation results show excellent behavior of the wind power system under sever fault conditions.

1 Introduction

In this chapter two novel control methods for grid side converters under unbalanced operating conditions in wind power applications are presented. The grid

Ana Vladan Stankovic
Electrical and Computer Engineering Department
Cleveland State University, Cleveland, Ohio 44115, U.S.A.
e-mail: a.stankovic@csuohio.edu

Dejan Schreiber
Semikron Elektronik GmbH & Co KG
Sigmundstr. 200, 90431 Nürnberg, Germany
e-mail: dejan.schreiber@semikron.com

Shuang Wu
Electrical and Computer Engineering Department
Cleveland State University, Cleveland, Ohio 44115, U.S.A.
e-mail: sharon_shuangwu@hotmail.com

L. Wang (Ed.): *Modeling and Control*, Green Energy and Technology, pp. 127–153.
springerlink.com

side converter controller is responsible for maintaining the DC link voltage at the reference value by exporting the active power to the grid. It is also designed to exchange the reactive power between the converter and the grid when required. Unfortunately, the features that PWM converter offers are fully realized only when the grid voltages are balanced. Under unbalanced grid voltages there is a deterioration of the converter input and output characteristics. The imbalance in grid voltages may occur frequently especially in weak systems. In order to provide a smooth operation of the grid side converter under unbalanced operating conditions with low harmonic content in line currents and the DC link capacitor, two control methods are proposed. First control method is suitable for unbalanced grid voltages and balanced line impedances whereas the other one is applicable to extreme unbalanced operating conditions when both grid voltages and line impedances have to be considered unbalanced. To illustrate the operation of the grid side inverter under unbalanced operating conditions when proposed control methods are implemented, simulation data of a variable speed wind turbine with PM synchronous generator and two controlled PWM back to back converters are shown.

2 Recent Studies on Unbalanced Grid-Side Inverter Operation in Wind Power Applications

Causes and effects of unbalanced grid voltages have been explained in detail. In a weak power system network, an unbalanced load at the distribution lines can cause unbalanced voltage condition. This is particularly true for rural power systems where the wind turbines are normally connected. Unbalanced condition can also be caused by unsymmetrical transformer windings, unbalanced transmission-line impedances or faults. Regardless of the cause, unbalanced input voltages have a severe impact on the performance of the grid-side inverter used in variable-speed wind power systems. Unbalanced grid voltages result in the appearance of large low order harmonics at the inverter input as well as low order harmonics in line currents that pollute the utility [1-7].

Muljadi *et al* [8] summarized the problems caused by unbalanced voltages in a wind power system connected directly to an induction generator. The unbalanced grid will cause the stator currents of the induction generator to be unbalanced. The unbalanced currents create unequal heating on the stator winding which will degrade the insulation of the winding and shorten its life expectancy. Unbalanced stator currents also create torque pulsation on the shaft, resulting in audible noise and extra mechanical stress.

For the variable-speed wind power system with two back-to-back PWM converters [9-17], the impacts on the generator might decrease because the converter decouples the generator from the grid. But due to the inherent drawbacks of the PWM VSI, the unbalanced grid brings other problems to the system. It has been shown in references [18] that unbalanced voltages contain a significant negative sequence component, causing the derived current reference to vary in time, which will cause a significant second order harmonic in the inverter DC voltage.

This will in turn cause third-order harmonics in ac line currents. The third order harmonics in ac currents will reflect back causing the forth order harmonics in the DC link voltage and so on. This will result in appearance of even harmonics in the DC link voltage and odd harmonics in the line currents. More importantly, the grid faults cause an overvoltage in the DC link capacitor which yields to power imbalance causing the acceleration of the turbine's rotor and increase damaging risks of the PWM converter as well as the DC link capacitor [19].

Standard control strategy used during grid faults consist of an additional chopper connected in parallel to the DC link capacitor. During faults, the surplus power is dissipated in the chopper resistance [20]. Even though the standard control strategy protects the system during grid faults, it does not prevent the grid pollution by harmonic currents.

Hansen *et al.* [21] proposed the control method for the wind power system under unbalanced operating conditions. However, the fault ride-through control is based on the control of the machine side converter. A damping controller, which is to use the DC capacitor as short-term energy storage is implemented to counteract the torque and speed oscillations and ensure a stable operation of the wind turbine under the grid fault. This method only minimizes the power pulsations and harmonic currents.

Abedini *et al.* [22] proposed a control method by adding a limiter to limit the grid currents and decrease the power generated from the machine during grid faults. The focus in this paper is on the capacitor overvoltage protection during grid faults but the harmonic distortion in line currents has not been considered.

Y. Zhang *et al.* [23] proposed the inverter control strategy for the wind power system with a permanent-magnet generator under unbalanced three-phase voltages. The negative sequence current is decomposed and added to the current template which is calculated based on the phase lock loop (PLL). However, the proposed method does not completely eliminate harmonics in line currents. In addition, the control of the reactive power under unbalanced input voltages has not been considered.

Lazarov *et al.* [24] applied grid inverter control method based on [25], to control the positive and negative sequence currents in d-q rotating frame. Under unbalanced operating conditions the active and reactive power flowing into the grid are given below,

$$P = P_o + P_{c2}\cos(2wt) + P_{s2}\sin(2wt) \tag{1}$$

$$Q = Q_o + Q_{c2}\cos(2wt) + Q_{s2}\sin(2wt) \tag{2}$$

By nullifying pulsating terms of the active power shown in (1) and reactive power shown in (2) harmonics can be minimized under unity power factor operation. Equation (3) represents the relationship between the active and reactive power as a function of grid voltages and currents under the condition of second harmonic elimination.

$$\frac{2}{3}\begin{bmatrix} P_0 \\ Q_0 \\ P_{s2} \\ P_{c2} \end{bmatrix} = \begin{bmatrix} P_0 \\ 0 \\ 0 \\ 0 \end{bmatrix} = \begin{bmatrix} E_d^p & E_q^p & E_d^n & E_q^n \\ E_q^p & -E_d^p & E_q^n & E_d^n \\ E_q^n & -E_d^n & -E_q^p & E_d^p \\ E_d^n & E_q^n & E_d^p & E_q^p \end{bmatrix} \begin{bmatrix} I_d^p \\ I_q^p \\ I_d^n \\ I_q^p \end{bmatrix} \tag{3}$$

where, $E_d^p, E_d^n, E_q^p, E_q^n, I_d^p, I_d^n, I_q^p, I_q^n$ represent positive and negative sequence grid voltages and currents in d-q reference frame. The reference values for positive and negative sequence d-q components of the line currents are obtained from (4) and given below,

$$\left[I_{dq}\right] = \left[E_{dq}\right]^{-1}\left[S\right] \tag{4}$$

where $\left[S\right]$ is a vector given by $\frac{2}{3}\begin{bmatrix} P_0 \\ Q_0 \\ P_{s2} \\ P_{c2} \end{bmatrix}$

Chong *et al.* [26] applied the control method based on [24] and [25]. The control scheme is shown in Figure 1.

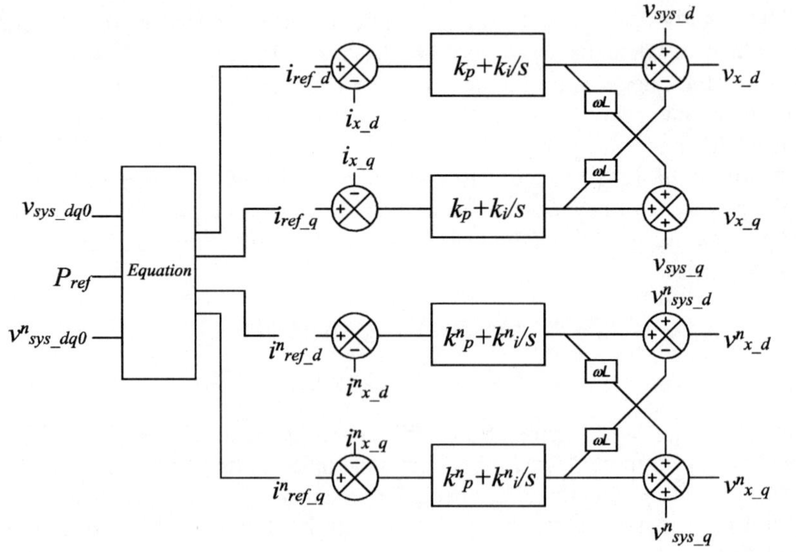

Fig. 1 Grid-side inverter control scheme in ref. [26]

However, the method proposed in [24], [25] and [26] ignores pulsating power on line inductors and cannot be effective under extreme unbalanced operating conditions. In addition the reactive power cannot be controlled.

Hu *et al.* [27] proposed the control scheme based on [28], [29] and [30]. Figure 2 shows the proposed control scheme.

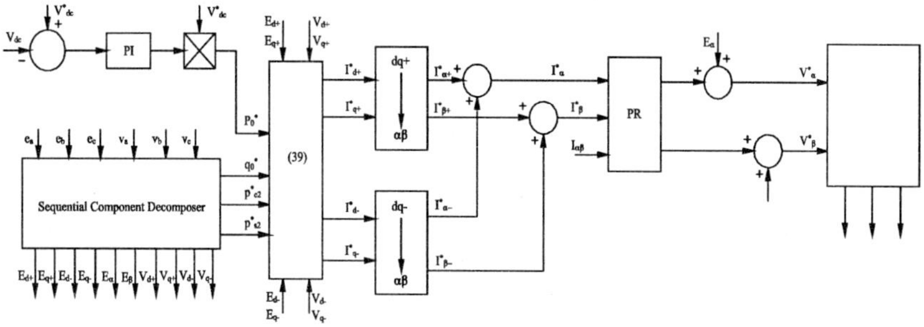

Fig. 2 Grid-side inverter control scheme [27]

Rodriguez *et al.* [31] [32] presented and compared five reactive and active power control strategies developed for the grid side inverter operating under unbalance operating conditions.

Unlike the majority of control methods presented in literature that only minimize the harmonics of a grid side inverter under unbalanced operating conditions, two control methods presented in this chapter provide complete harmonic elimination and smooth power flow from the wind turbine to the grid. The second control method developed in the abc reference frame does not require complicated transformations and is applicable for extreme simultaneous unbalance in grid voltages and line impedances. The proposed method is general and can be used for all levels of unbalance in grid voltages and line impedances. The power factor can be adjusted in addition to the harmonic elimination. Based on the measurements of the grid voltage and impedance unbalance, both magnitudes as well as phase angles of line currents are adjusted. In spite of the level of unbalance in grid voltages and line impedances, high quality of line currents and dc link voltage are obtained. Analytical and simulation results are in excellent agreement

3 Grid-Side Converters under Unbalanced Operating Conditions

Variable speed wind power generation enables the operation of the turbine at its optimal power over a range of wind speeds. Figure 3 shows the variable-speed wind power system.

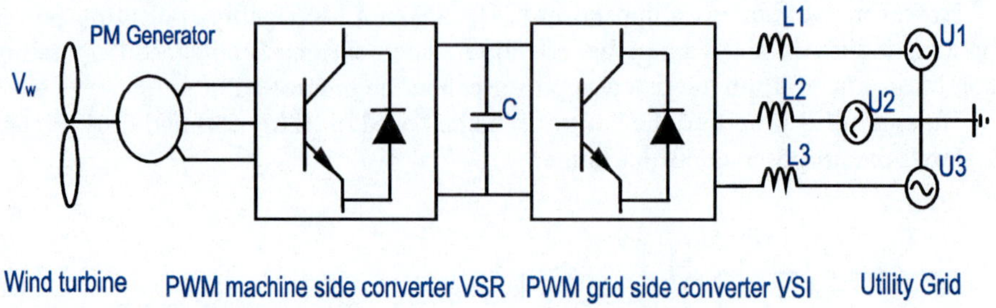

Fig. 3 The variable-speed wind power system with two PWM converters and a permanent magnet (PM) generator.

The machine side converter sets the torque demand according to the wind speed to obtain the optimal power and transfers the power to the DC link. The grid side inverter, transfers the DC link power to the power grid with constant frequency and magnitude. At the same time, the grid side converter is controlled to maintain the DC link voltage at a constant value [33], [34], [35].

Under unbalanced grid voltages there is a deterioration of the converter input and output characteristics [36], [37]. The imbalance in grid voltages may occur frequently especially in weak systems. Nonuniformly distributed single phase loads, faults or unsymmetrical transformer windings could cause imbalance in the three phase voltages both in magnitude and in phase. Regardless of the cause, unbalanced voltages have a severe impact on the performance of the PWM grid-side converter. Actually, the huge harmonics of lower frequencies, not present in the PWM switching functions appear at both the input and output ports of the converter. The problems include a significant distortion in line current waveforms and increase in the dc capacitor ripple current and voltage. These additional low frequency components cause additional losses and should be considered in filter design of these converters.

3.1 Analysis of the Grid-Side Converter under Unbalanced Operating Conditions

The unbalanced input voltages cause an abnormal second harmonic at the dc link voltage which reflects back to the ac side causing the third-order harmonic current to flow. Next, the third harmonic current causes the fourth-order harmonic voltage at the output, and so on. This results in the appearance of even harmonics at the dc link voltage and odd harmonics in the line currents. The grid–side converter is shown in Fig.4.

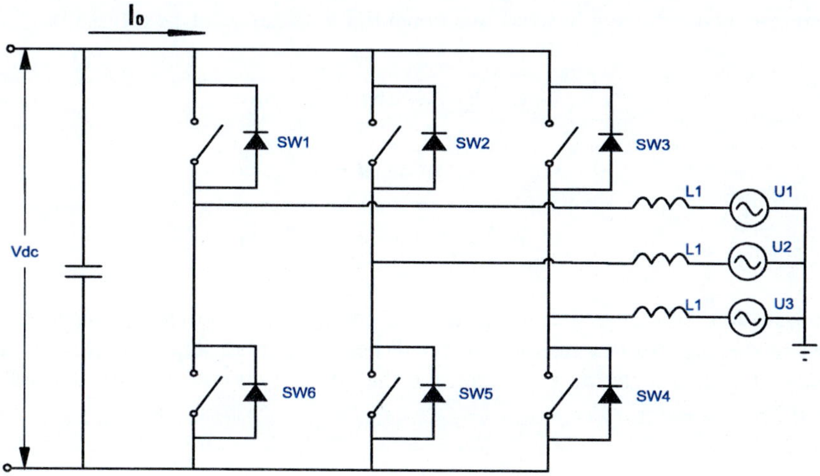

Fig. 4 Grid-side converter

The DC link current I_0 of the matrix converters is a function of the converter transfer function vector T and the line current vector i and is given by,

$$I_0 = Ti \tag{5}$$

The converter transfer function vector T is composed of three independent line to neutral switching functions.

$$T = \begin{bmatrix} SW_1 & SW_2 & SW_3 \end{bmatrix} \tag{6}$$

The line current vector is given by,

$$i = \begin{bmatrix} i_1 \\ i_2 \\ i_3 \end{bmatrix} \tag{7}$$

The line to neutral switching functions are balanced and can be represented only by their fundamental components.

$$SW_1(t) = S_1 \sin(wt - \Theta)$$

$$\begin{aligned} SW_2(t) &= S_1 \sin(wt - \Theta - 120^0) \\ SW_3(t) &= S_1 \sin(wt + 120^0 - \Theta) \end{aligned} \tag{8}$$

Therefore, converter synthesized line to neutral voltages can be expressed as,

$$V_{s1} = \frac{1}{2} V_{dc} S_1 \sin(wt - \Theta)$$
$$V_{s2} = \frac{1}{2} V_{dc} S_1 \sin(wt - 120^0 - \Theta) \quad (9)$$
$$V_{s3} = \frac{1}{2} V_{dc} S_1 \sin(wt + 120^0 - \Theta)$$

The equation (9) shows that the grid- side converter synthesized voltages are always balanced. For this reason, there will be no negative sequence voltage component present at its terminals. It follows that the line currents are unbalanced and given by,

$$\begin{bmatrix} I_1 \\ I_2 \\ I_3 \end{bmatrix} = \begin{bmatrix} 1 & 1 & 1 \\ 1 & a^2 & a \\ 1 & a & a^2 \end{bmatrix} \begin{bmatrix} I^0 \\ I^+ \\ I^- \end{bmatrix} \quad (10)$$

where I^0, I^+, I^- are zero, positive and negative sequence currents. Since $I_1 + I_2 + I_3 = 0$, the zero sequence current never flows in this circuit. $I^0 = 0$

Line to neutral voltages are unbalanced and given by,

$$\begin{bmatrix} U_1 \\ U_2 \\ U_3 \end{bmatrix} = \begin{bmatrix} 1 & 1 & 1 \\ 1 & a^2 & a \\ 1 & a & a^2 \end{bmatrix} \begin{bmatrix} U^0 \\ U^+ \\ U^- \end{bmatrix} \quad (11)$$

where U^0, U^+ and U^- are zero, positive and negative sequence line voltages.

In time domain, the fundamental components of the three phase currents are given by,

$$i_1(t) = I_1 \sin(wt - \phi_1)$$
$$i_2(t) = I_2 \sin(wt - 120^0 - \phi_2) \quad (12)$$
$$i_3(t) = I_3 \sin(wt + 120^0 - \phi_3)$$

According to equation (5), the output current $I_0(t)$ is given by,

$$I_0(t) = I_1 \sin(wt - \phi_1) S_1 \sin(wt - \Theta) +$$
$$I_2 \sin(wt - 120^0 - \phi_2) S_1 \sin(wt - 120 - \Theta)$$
$$+ I_3 \sin(wt + 120^0 - \phi_3) S_1 \sin(wt + 120^0 - \Theta) \quad (13)$$

By using a trigonometric identity, $I_0(t)$ becomes,

$$\begin{aligned} I_0(t) &= \frac{1}{2}I_1S_1[\cos(\Theta-\phi_1)-\cos(2wt-\Theta-\phi_1)] \\ &+\frac{1}{2}I_2S_1[\cos(\Theta-\phi_2)-\cos(2wt-240^0-\Theta-\phi_2)] \\ &+\frac{1}{2}I_3S_1[\cos(\Theta-\phi_3)-\cos(2wt+240^0-\Theta-\phi3)] \end{aligned} \tag{14}$$

The DC link current consists of a dc and a harmonic current.

$$I_0(t) = I_{dc} + I_{sh}(2wt) \tag{15}$$

where $I_{sh}(2wt)$ is the second-order harmonic current and is given by,

$$\begin{aligned} I_{sh}(2wt) &= -\frac{S_1I_1}{2}\cos(2wt-\Theta-\phi_1)-\frac{I_2S_1}{2}\cos(2wt-240^0-\Theta-\phi_2) \\ &-\frac{I_3S_1}{2}[\cos(2wt+240^0-\Theta-\phi_3)] \end{aligned} \tag{16}$$

Therefore, the dc link voltage will also contain the second-order harmonic, which will reflect back to the output causing the third-order harmonic current to flow. The third harmonic current will reflect back to the input causing the fourth-order harmonic to flow. As the literature indicates, even harmonics will appear at the input and odd harmonics at the output of the converter under unbalanced voltages. The second and third-order harmonics are of the primary concern.

4 A Proposed Method for Complete Harmonic Elimination of a Grid-Side Converter under Unbalanced Grid Voltages and Balanced Line Impedances

The circuit shown in Figure 4 is analyzed under unbalanced input voltages and balanced input impedances. This case will later be extended to a more general situation which will also consist of unbalanced input impedances. *The harmonic elimination is achieved by generating unbalanced switching functions* [38]. The assumptions used in the derivation are:

- The system is lossless.
- The switching functions are unbalanced and contain no zero sequence
- Only fundamental components of switching functions and input currents are taken into account.

By using symmetrical component theory, line to neutral switching functions SW_1, SW_2 and SW_3 are obtained and given by,

$$\begin{bmatrix} SW_1 \\ SW_2 \\ SW_3 \end{bmatrix} = \begin{bmatrix} 1 & 1 & 1 \\ 1 & a^2 & a \\ 1 & a & a^2 \end{bmatrix} \cdot \begin{bmatrix} 0 \\ S^+ \\ S^- \end{bmatrix} \tag{17}$$

Since the zero sequence current is never present in this circuit, currents I_1, I_2 and I_3 are given by,

$$\begin{bmatrix} I_1 \\ I_2 \\ I_3 \end{bmatrix} = \begin{bmatrix} 1 & 1 & 1 \\ 1 & a^2 & a \\ 1 & a & a^2 \end{bmatrix} \cdot \begin{bmatrix} 0 \\ I^+ \\ I^- \end{bmatrix} \tag{18}$$

where S^+ and S^- are positive and negative sequence switching functions, I^+ and I^- are positive and negative sequence currents and $a = 1\angle 120^0$.

As it was shown earlier, output current I_0 is given by,

$$I_0 = SW_1 I_1 + SW_2 I_2 + SW_3 I_3 = (S^+ + S^-)(I^+ + I^-) + (a^2 S^+ + a S^-)$$
$$(a^2 I^+ + a I^-) + (a S^+ + a^2 S^-)(a I^+ + a I^-) = 3 S^+ I^- + 3 S^- I^+ \tag{19}$$

In the time domain, the pulsating component of the output current is expressed as:

$$\begin{aligned} I_0(t) &= 3Real(\left|S^+\right| e^{j*w*t} e^{j*\theta_s^+})\, Real(\left|I^-\right| e^{j*w*t} e^{j*\theta_I^-}) \\ &+ 3Real(\left|S^-\right| e^{j*w*t} e^{j*\theta_s^-}) Real(\left|I^+\right| e^{j*w*t} e^{j*\theta_i^+}) \\ &= 3\left|S^+\right|\left|I^-\right| cos(wt + \Theta_s^+) cos(wt + \Theta_i^-) \ldots\ldots \\ &+ 3\left|S^-\right|\left|I^+\right| cos(wt + \Theta_s^-) cos(wt + \Theta_i^+) \end{aligned} \tag{20}$$

By using trigonometric identity cos(α)cos(β) = 1/2[cos($\alpha + \beta$) + cos($\alpha - \beta$)],one arrives to this result:

$$I_0(2wt) = 3/2 \left|S^+\right| \left|I^-\right| cos(2wt + \Theta_s^+ + \Theta_i^-) + 3/2 \left|S^-\right| \left|I^+\right| cos(2wt + \Theta_s^- + \Theta_i^+) \tag{21}$$

Equation (21) shows the presence of the second-order harmonic current at the input of the inverter. The above equation indicates that the second-order harmonic can be eliminated under the following conditions.

$$\begin{aligned} &\textbf{1. } \left|S^+\right|\left|I^-\right| = \left|S^-\right|\left|I^+\right| \\ &\textbf{2. } \Theta_s^- + \Theta_i^+ = \pi + \Theta_s^+ + \Theta_i^- \end{aligned} \tag{22}$$

where $|S^+|$, $|S^-|$, Θ_s^+ and Θ_s^- are the magnitudes and phase angles of the positive and negative sequence switching functions, respectively. $|I^+|$,$|I^-|$, Θ_i^+ and Θ_i^- are the magnitudes and phase angles of the positive and negative sequence line currents.

The per-phase positive and negative sequence equivalent circuit under unbalanced input voltages and unbalanced synthesized voltages at the inverter terminals are shown in Figure 5.

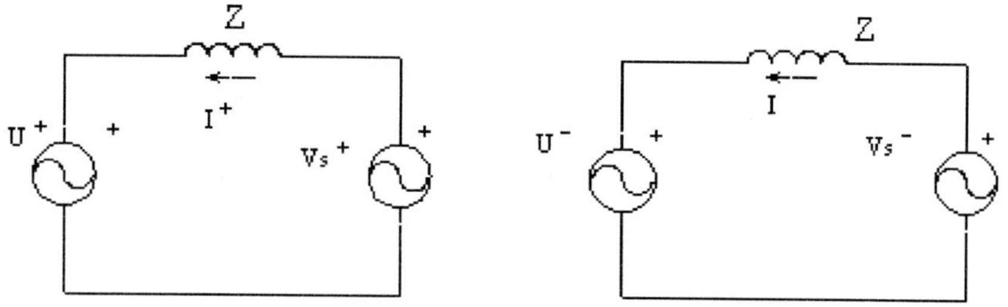

Fig. 5 Per-phase positive and negative sequence equivalent circuit

From Figure 5, two additional equations are obtained and given by,

$$ZI^+ = V_s^+ - U^+ \tag{23}$$

$$ZI^- = V_s^- - U^- \tag{24}$$

where V_s^+ and V_s^- are the positive and negative sequence voltages at the inverter terminals and U^+ and U^- are the positive and negative sequence grid voltages.

$$V_s^+ = V_{dc}\, S^+ /(2\sqrt{2}\,),\, V_s^- = V_{dc}\, S^- /(2\sqrt{2}) \tag{25}$$

where V_{dc} is the dc link voltage.

From the power equation, the following relationship is obtained,

$$P_{dc} = \text{-}Real\,(3U^+ I^+ + 3U^- I^-) \tag{26}$$

where P_{dc} is average DC power at the inverter input.

The solution for harmonic elimination is obtained by combining equations (22), (23), (24), (25) and (26). The solution is given by the following set of equations:

$$\frac{|S^+|}{|S^-|} = \frac{|U^+|}{|U^-|} \tag{27}$$

$$\Theta_u^{\,-} - \Theta_s^{\,-} = \Theta_s^{\,+} - \Theta_u^{\,+} \tag{28}$$

$$\left|S^{+}\right| = \frac{2\sqrt{2}\left|U^{+}\right|}{V_{dc}} \cos\left(\Theta_u^{\,+} - \Theta_s^{\,+}\right) \tag{29}$$

$$sin2(\Theta_u^{\,+} - \Theta_s^{\,+}) = -\frac{2P_{dc}Z}{3(\left|U^{+2}\right| - \left|U^{-2}\right|)} \tag{30}$$

The set of four equations shown above represents the open loop steady-state solution for input-output harmonic elimination of the grid side converter under unbalanced line voltages and balanced line impedances.

4.1 Control Method for Complete Input – Output Harmonic Elimination of a Grid-Side Converter under Unbalanced Line Voltages and Balanced Line Impedances

Based on the analysis of the open loop configuration presented above, a feed-forward control method is proposed. In order to control the dc link voltage and eliminate harmonics at the input and output of the PWM grid-side inverter under unbalanced line voltages, not only current magnitudes but also their phase angles have to be controlled. The DC bus error is used to synthesize magnitudes and the phase angle of the positive and negative sequence reference currents. The sequence components are then transformed into three phase (abc) quantities which become reference signals for the hysteresis controller.

The steady-state solution for harmonic elimination in the open loop configuration was derived and given by equations (27), (28), (29) and (30). Based on the steady state solution, the relationship between positive and negative sequence line voltages and currents is given by,

$$1.\ \frac{\left|U^{+}\right|}{\left|U^{-}\right|} = \frac{\left|I^{+}\right|}{\left|I^{-}\right|} \tag{31}$$

$$2.\ \Theta_u^{+} - \Theta_i^{+} = \Theta_i^{-} - \Theta_u^{-} - \pi \tag{32}$$

where $\left|U^{+}\right|, \left|U^{-}\right|, \Theta_u^{+}, \Theta_u^{-}$ are the magnitudes and phase angles of the positive and negative sequence line voltages and $\left|I^{+}\right|, \left|I^{-}\right|, \Theta_i^{+}, \Theta_i^{-}$ are the magnitudes and phase angles of the positive and negative sequence line currents.

However, it is always true that average DC power is equal to average AC power and given by the following equation,

$$P_{dc} = V_{dc}I_0 = -[3\left|U^{+}\right|\left|I^{+}\right|\cos(\Theta_u^{+} - \Theta_i^{+}) + 3\left|U^{-}\right|\left|I^{-}\right|\cos(\Theta_u^{-} - \Theta_i^{-})] \tag{33}$$

The dc link voltage is proportional to the positive and negative sequence currents (magnitudes), so is the error signal $(V_{ref} - V_{dc})$. In order to satisfy equation (26), the condition for second-harmonic elimination, the positive and negative sequence commands for current magnitudes should satisfy equations (34) and (35).

$$\left|I^+\right| = K_p \left|U^+\right|(V_{ref} - V_{dc}) \tag{34}$$

$$\left|I^-\right| = K_p \left|U^-\right|(V_{ref} - V_{dc}) \tag{35}$$

Equations (34) and (35) represent positive and negative sequence magnitude commands.

The positive and negative sequence commands for phase angles are derived from equation (36) and equation (37). The line current phase angle commands are given by,

$$\Theta_i^+ = \Theta_u^+ - K_{p1} a\cos(V_{ref} - V_{dc}) \tag{36}$$

$$\Theta_i^- = \Theta_u^- + \pi + K_{p1} a\cos(V_{ref} - V_{dc}) \tag{37}$$

where *acos* is an inverse cosine function.

Equations (36) and (37) are approximated by the following two equations:

$$\Theta_i^+ = \Theta_u^+ - K_{p1}(V_{ref} - V_{dc}) \tag{38}$$

$$\Theta_i^- = \Theta_u^- + \pi + K_{p1}(V_{ref} - V_{dc}) \tag{39}$$

Two proportional controllers are proposed (one for magnitude, the other for phase angle control). An integral controller could be added to reduce the steady state error.

Positive and negative sequence current commands are transformed to abc quantities and used as references for hysteresis controller.

The transformation is given by,

$$i_{1ref}(t) = \left|I^+\right| \sin(wt + \Theta_i^+) + \left|I^-\right| \sin(wt + \Theta_i^-) \tag{40}$$

$$i_{2ref}(t) = \left|I^+\right| \sin(wt + \Theta_i^+ - \frac{2\pi}{3}) + \left|I^-\right| \sin(wt + \Theta_i^- + \frac{2\pi}{3}) \tag{41}$$

$$i_{3_{ref}}(t) = \left|I^+\right| \sin(wt + \Theta_i^+ + \frac{2\pi}{3}) + \left|I^-\right| \sin(wt + \Theta_i^- - \frac{2\pi}{3}) \tag{42}$$

The proposed control method is shown in Figure 6.

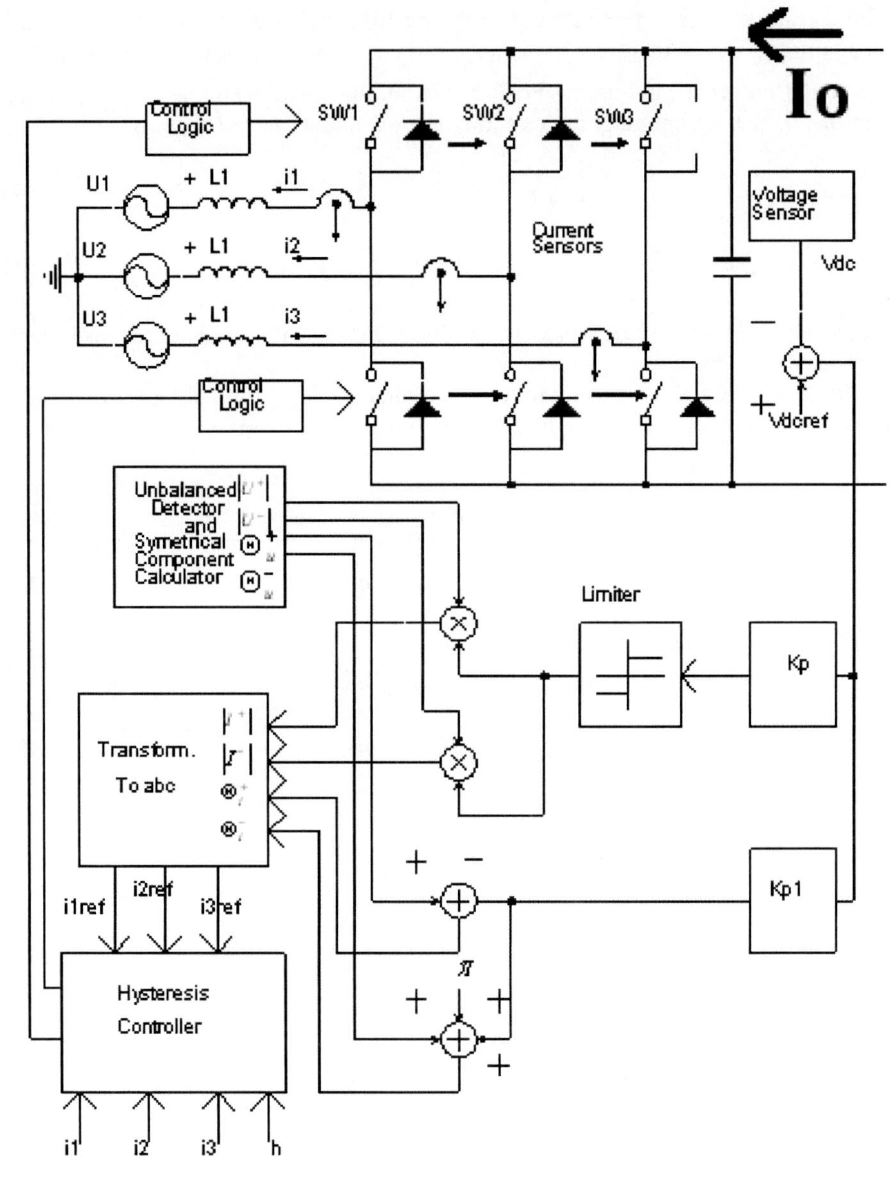

Fig. 6 Control method for complete harmonic elimination under unbalanced line voltages and balanced line impedances

5 Generalized Method for Harmonic Elimination of a Grid Side Converter under Simultaneous Unbalanced Grid Voltages and Line Impedances

Under severe fault conditions in the distribution system, not only line voltages, but also line impedances must be considered unbalanced. Theoretical approach for complete harmonic elimination of the grid side inverter under unbalanced line voltages and impedances is presented. Based on the analysis in the open loop, closed loop solution is proposed.

5.1 Theoretical Approach

The circuit shown in Fig.4 is analyzed under the following assumptions:

- The line voltages are unbalanced
- The line impedances are unbalanced
- The converter is lossless

Harmonic elimination can be achieved by generating unbalanced reference commands for three line currents under unbalanced voltages and impedances [39].

5.1.1 Derivation

From the circuit shown on Fig. 4 three equivalent per phase circuits are obtained and shown in Fig. 7.

$$V_{s1} = U_1 + z_1 I_1 \quad (43)$$

$$V_{s2} = U_2 + z_2 I_2 \quad (44)$$

$$V_{s3} = U_3 + z_3 I_3 \quad (45)$$

$$I_1 = -I_2 - I_3 \quad (46)$$

$$S^* = -(U^* I_1 + U_2^* I_2^* + U_3^* I_3) \quad (47)$$

$$SW_1 I_1 + SW_2 I_2 + SW_3 I_3 = 0 \quad (48)$$

where $U_1, U_2, U_3, I_1, I_2, I_3, z_1, z_2, z_3, V_{s1}, V_{s2}, V_{s3}, S, SW_1, SW_2$ and SW_3 are grid voltages, line currents, line impedances, synthesized voltages at the converter terminals, apparent power and switching functions, respectively, represented as *phasors*.

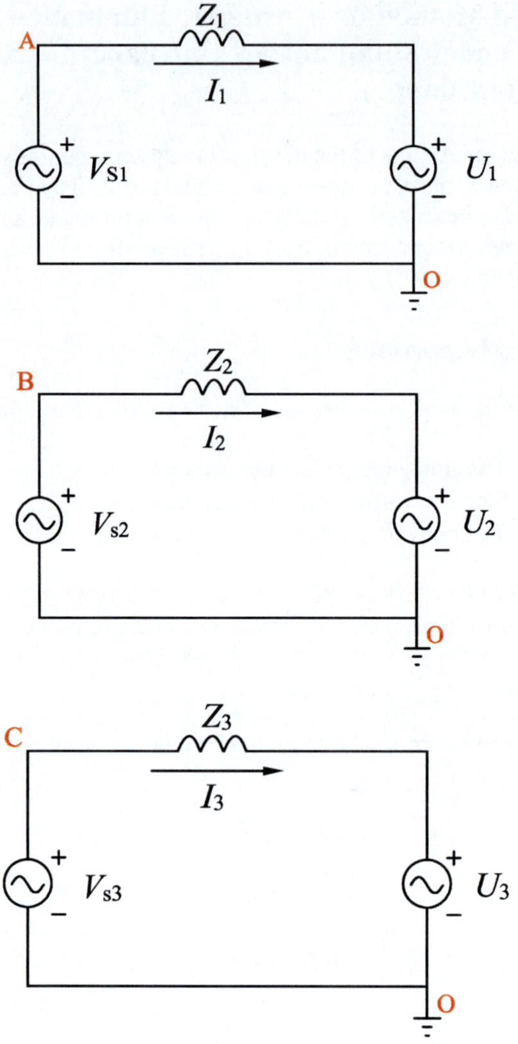

Fig. 7 Per Phase equivalent circuits under unbalanced line voltages and impedances

Equation (48) represents the condition for the second harmonic elimination. Synthesized voltages V_{s1}, V_{s2} and V_{s3} can be expressed as,

$$V_{s1} = SW_1 \frac{V_{dc}}{2\sqrt{2}} \tag{49}$$

$$V_{s2} = SW_2 \frac{V_{dc}}{2\sqrt{2}} \tag{50}$$

$$V_{s3} = SW_3 \frac{V_{dc}}{2\sqrt{2}} \tag{51}$$

where V_{dc} is the DC link voltage.

By substituting equations (49), (50) and (51) into (43), (44) and (45) the following set of equations is obtained,

$$U_1 = SW_1 \frac{V_{dc}}{2\sqrt{2}} - z_1 I_1 \tag{52}$$

$$U_2 = SW_2 \frac{V_{dc}}{2\sqrt{2}} - z_2 I_2 \tag{53}$$

$$U_3 = SW_3 \frac{V_{dc}}{2\sqrt{2}} - z_3 I_3 \tag{54}$$

$$I_1 = -I_2 - I_3 \tag{55}$$

$$S^* = -(U_1^* I_1 + U_2^* I_2 + U_3^* I_3) \tag{56}$$

$$SW_1 I_1 + SW_2 I_2 + SW_3 I_3 = 0 \tag{57}$$

By multiplying equations (52), (53) and (54) by I_1, I_2 and I_3, respectively, and adding them up the following equation is obtained,

$$U_1 I_1 + U_2 I_2 + U_3 I_3 = -z_1 I_1^2 - z_2 I_2^2 - z_3 I_3^2 + \frac{V_{dc}}{2\sqrt{2}}(SW_1 I_1 + SW_2 I_2 + SW_3 I_3) \tag{58}$$

By substituting equation (57) into (58) the following equation is obtained,

$$U_1 I_1 + U_2 I_2 + U_3 I_3 = -z_1 I_1^2 - z_2 I_2^2 - z_3 I_3^2 \tag{59}$$

The set of six equations with six unknowns, (52) to (57), reduces to three equations with three unknowns and are given by,

$$I_1 = -I_2 - I_3 \tag{60}$$

$$S^* = -(U_1^* I_1 + U_2^* I_2 + U_3^* I_3) \tag{61}$$

Equations (59), (61) and (61) represent a set of three equations with three unknowns.

By substituting equation (60) into equations (59) and (61), the following set of equations is obtained and given by,

$$U_1(-I_2 - I_3) + U_2 I_2 + U_3 I_3 = -z_1(-I_2 - I_3)^2 - z_2 I_2^2 - z_3 I_3^2 \tag{62}$$

$$S^* = -(-U_1^* I_2 - U_1^* I_3 + U_2^* I_2 + U_3^* I_3) \tag{63}$$

Equation (62) can be simplified as,

$$I_2(U_2 - U_1) + I_3(U_3 - U_1) = -(z_1 + z_2)I_2^2 - (z_1 + z_3)I_3^2 - 2z_1 I_2 I_3 \tag{64}$$

From equation (63) current, I_2, can be expressed as,

$$I_2 = \frac{-S^* - I_3(U_3^* - U_1^*)}{U_2^* - U_1^*} \tag{65}$$

Finally by substituting equation (65) into equation (64),

$$\begin{aligned} &\frac{-S^* - I_3(U_3^* - U_1^*)}{U_2^* - U_1^*}(U_2 - U_1) + I_3(U_3 - U_1) = \\ &= -(z_1 + z_2)\frac{S^{*2} + 2S^* I_3(U_3^* - U_1^*) + I_3^2(U_3^* - U_1^*)^2}{(U_2^* - U_1^*)^2} + \\ &-(z_1 + z_2)I_3^2 - 2z_1 \frac{-S^* - I_3(U_3^* - U_1^*)}{U_2^* - U_1^*} I_3 \end{aligned} \tag{66}$$

$$\begin{aligned} &[-\frac{2z_1(U_3^* - U_1^*)}{U_2^* - U_1^*} + \frac{(z_1 + z_2)(U_3^* - U_1^*)^2}{(U_2^* - U_1^*)^2} + (z_1 + z_3)\ ]I_3^2 + \\ &+[\ (U_3 - U_1) - \frac{(U_3^* - U_1^*)(U_2 - U_1)}{U_2^* - U_1^*} - \frac{2z_1 S^*}{U_2^* - U_1^*} + \frac{2S^*(z_1 + z_2)(U_3^* - U_1^*)}{(U_2^* - U_1^*)^2}\ ]I_3 + \\ &-\frac{S^*(U_2 - U_1)}{U_2^* - U_1^*} + \frac{(z_1 + z_2)S^{*2}}{(U_2^* - U_1^*)^2} = 0 \end{aligned} \tag{67}$$

Currents I_2 and I_1 can be obtained from equations (60) and (65).

Equations (60), (65) and (67) represent the steady state solution for line currents under both unbalanced grid voltages and unbalanced line impedances. An analytical solution represented by equation (67) always exists unless all the coefficients of the quadratic equations are equal to zero. *For given power, S, grid voltages, U_1, U_2, U_3 and line impedances z_1, z_2 and z_3, line currents, I_1, I_2 and I_3, can be obtained from the above set of equations.*

5.1.2 Critical Evaluation

The analytical solution that has been obtained is general. The only constraint that exists, as far as the level of unbalance is concerned, is governed by constraints of the operation of the PWM converter itself. The proposed generalized method for input-output harmonic elimination is valid if and only if $U_i, z_i \neq 0$, where $i = 1,2,3$. In other words the solution exists for all levels of unbalance in line voltages and impedances, except for cases where both voltage and impedance in the same phase are equal to zero. Therefore, the maximum level of voltage imbalance with balanced line impedances, for which the proposed solution is still valid, is given by,

$$\begin{aligned} &U_1 \neq 0 \\ &U_2 = U_3 = 0 \\ &z_1 = z_2 = z_3 \neq 0 \end{aligned}$$

The maximum level of imbalance in both line voltages and impedances for which the proposed solution is still valid is given by,

$$\begin{aligned} &U_1 \neq 0 \\ &U_2 = U_3 = 0 \\ &z_1 = 0 \\ &z_2 \neq z_3 \neq 0 \end{aligned}$$

5.2 *Control Method for Harmonic Elimination of a Grid Side Converter under Simultaneous Unbalanced Grid Voltages and Line Impedances*

Based on the analysis of the open loop configuration presented above, a feed forward control method is proposed. The line voltages as well as line impedances have to be measured. Based on this information and a DC bus error, reference currents are calculated according to equations (60), (65) and (67) which become reference signals for the hysteresis controller [7]. Only one PI controller is utilized, which has been shown to be sufficient for good regulation. The proposed control method is shown in more detail in Fig.8.

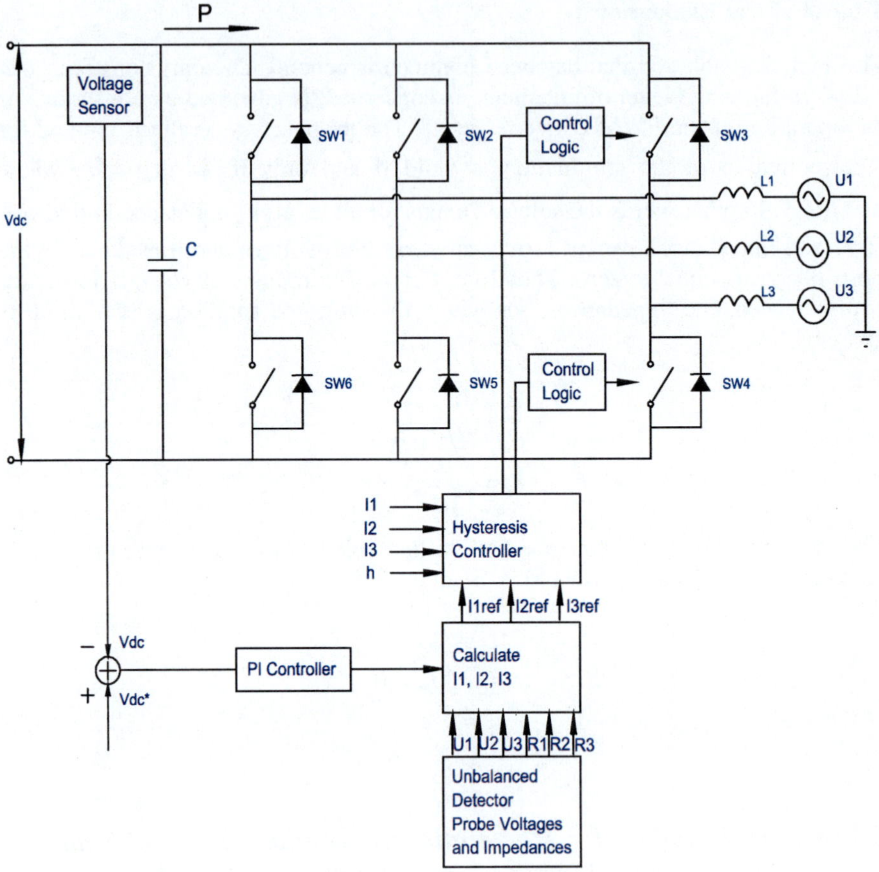

Fig. 8 Generalized control method for grid side inverter under unbalanced operating conditions

6 Example

In this example a variable speed wind power system shown in Fig.3 is simulated in Simulink. Table 1 lists parameters used in simulation. Grid-side converter operates under extreme unbalanced operating conditions

Figures 9,11, 13 and 15 show simulation results obtained when generalized control method shown in Fig.8 is used. Figures 10, 12, 14 and 16 show simulation results when the traditional control method is used. It is obvious that huge harmonic currents flow in line currents and DC link voltage under extreme unbalanced operating conditions when traditional control is utilized. In addition pulsating power appears at the output of the generator cousing torque pulsations.

Table 1 Parameters used in simulation

Stator phase Rs	2.8750 ohm	**f** $_{grid}$	60 Hz	**Hysteresis band**	0.02 A
Stator Inductances	L_d(H)=51 mH L_q(H)=51 mH	**C** $_{dclink}$	300μF	**Sampling time**	0.02 ms
Numberof Poles	16	**Line impedance**	L_1=5 mH	L_2=0 mH	L_3=5 mH
Power Rating	3 kW	**Grid voltage**	$U_a=0\angle 0$	$U_b=110\angle -120$	$U_c=220\angle 120$
Rated Voltage	480 V	**PF**	0.7 leading		

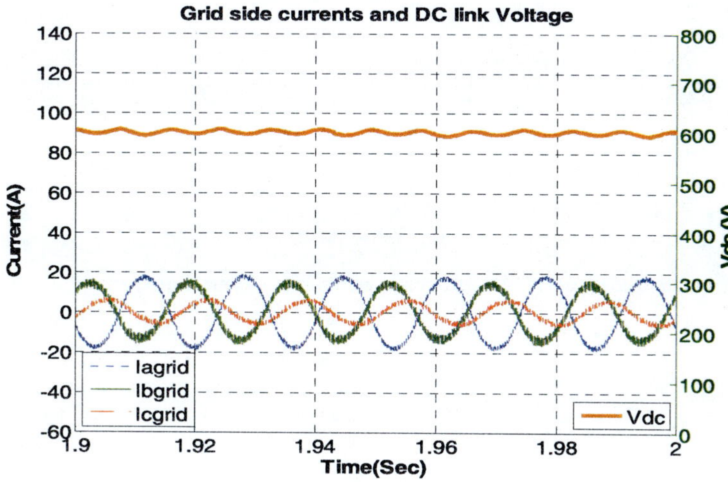

Fig. 9 Three-phase grid currents and DC link voltage with generalized control method

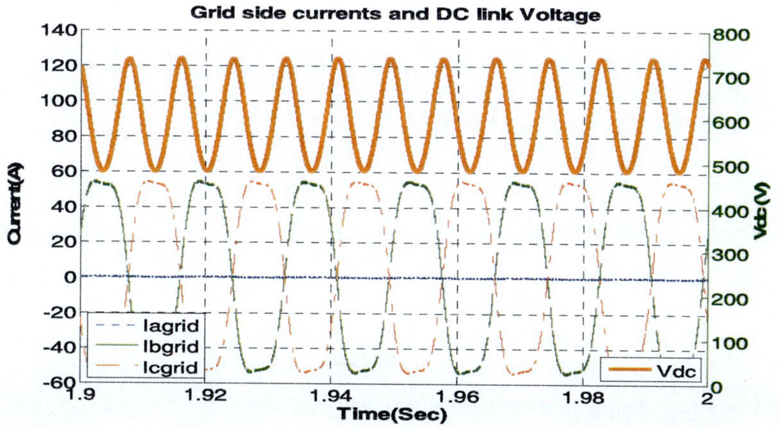

Fig. 10 Three-phase grid currents and DC link voltage with the traditional control method

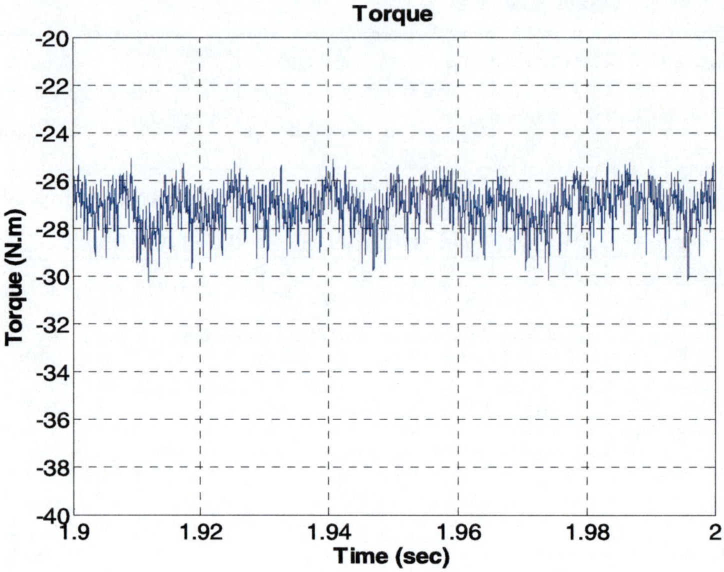

Fig. 11 Generator torque with the proposed generalized control method.

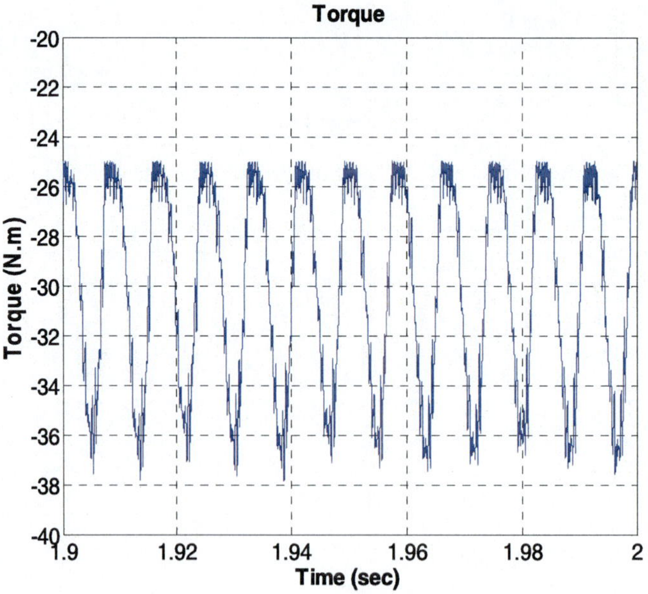

Fig. 12 Generator torque with the traditional control method

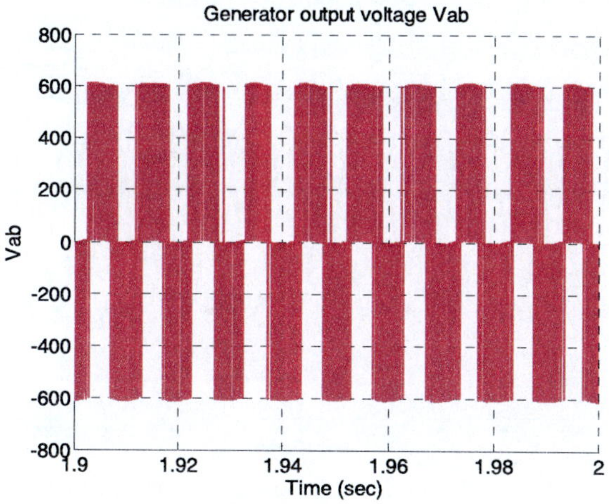

Fig. 13 Generator line-line voltage with the proposed generalized control method

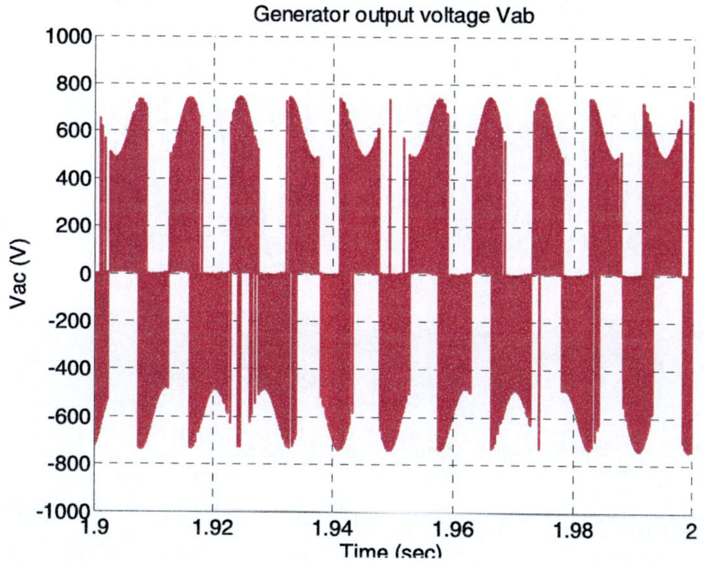

Fig. 14 Generator line-line voltage with the traditional control

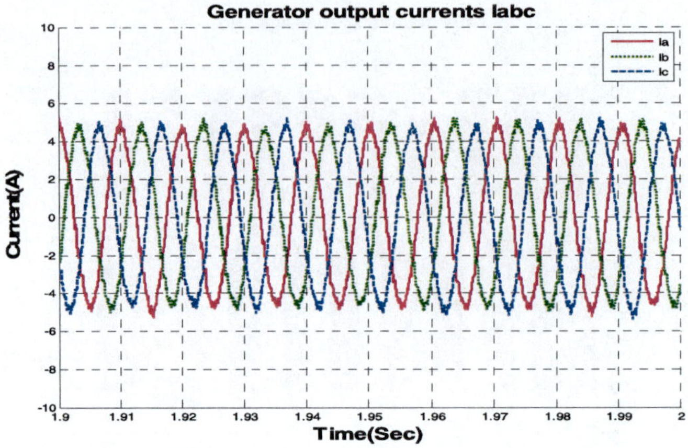

Fig. 15 Generator currents with the proposed generalized control method

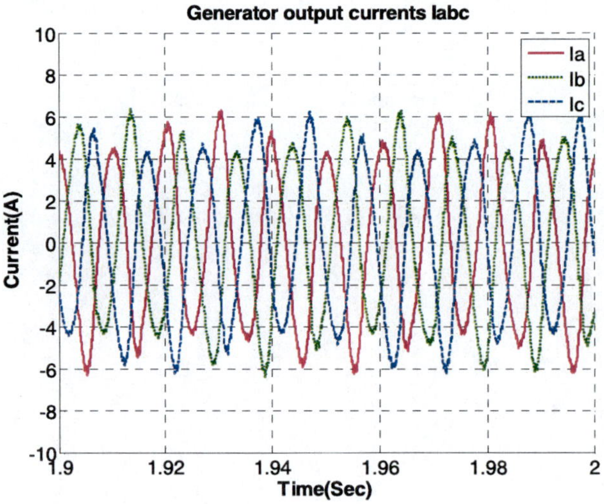

Fig. 16 Generator currents with the traditional control method

7 Conclusion

In this chapter two control methods for complete input-output elimination of a grid side convert under unbalanced operating conditions in variable-speed wind power applications is presented. The first control methods is suitable for cases where only line voltages are considered unbalanced. The other method presented in this chapter is general and can be used for all levels of unbalance in line voltages and impedances with complete harmonic elimination in line currents and DC link voltage. The analytical method for inverter control under unbalanced operating conditions with complete harmonic elimination in line currents and DC link voltage and adjustable power factor is presented. Based on the analytical solution, a closed loop control method is proposed and verified on a SIMULINK model of a variable speed wind power system. The simulation results obtained by applying the generalized control method of harmonic elimination are compared with the simulation results obtained by using a traditional grid-side inverter control method. The results show that under severe fault conditions in the power system, high quality line currents and DC link voltage are obtained by using the proposed control technique. The same wind power system was simulated by using a traditional method. Results show that huge low order harmonics flow in line currents and DC link voltage causing torque pulsations.

Acknowledgment. I would like to thank my graduate students, Shuang Wu and Xiangpeng Zheng for their help and support.

References

1. Zhang, C.: Analysis of the interaction between voltage unbalance and wind power plant operation. Wind Engineering 20(5), 307–318 (1996)
2. Wilson, J.W.: The Forced-Commutated Inverter As a Regenerative Rectifier. IEEE Transactions on Industry Applications IA-14(4), 335–340 (1978)
3. Stankovic, A.V., Lipo, T.A.: A Novel Control Method for Input-Output Harmonic Elimination of the PWM Boost Type Rectifier Under Unbalanced Operating Conditions. IEEE Transactions on Power Electronics 16(5), 603–611 (2001)
4. Dixon, J.W., Ooi, B.T.: Indirect Current Control of a Unity Power Factor Sinusoidal Current Boost Type Three-phase Rectifier. IEEE Trans. Ind. Electron 35(4), 508–515 (1988)
5. Brod, D.M., Novotny, D.W.: Current Control of VSI-PWM Inverters. IEEE Transactions on Industry Applications IA-21(4), 769–775 (1984)
6. Lipo, T.A.: Recent Progress and Development of Solid State AC Motor Drives. IEEE Trans. On Power Electronics 3(7.2), 105–117 (1988)
7. Zhang, C.: Analysis of the interaction between voltages unbalanced and wind power plant operation. Wind Engineering 20(5), 307–318 (1996)
8. Muljadi, E., Yildirim, D., Batan, T., Butterfield, C.P.: Understanding the unbalanced-voltage problems in wind turbine generation. In: Proc. of 34th Annual Meeting of the IEEE Industry Applications, Phoenix, USA, vol. 2, pp. 1359–1365 (1999)
9. Carrasco, J.M., Franquelo, L.G., Bialasiewicz, J.T., Galvan, E., Portillo Guisado, R.C., Prats, M.A.M., Leon, J.I., Moreno-Alfonso, N.: Power-Electronic Systems for the Grid Integration of Renewable Energy Source: A Survey. IEEE Transaction on Industrial Electronics 53(4), 1002–1016 (2006)
10. Chinchilla, M., Arnaltes, S., Burgos, J.C.: Control of Permanent-Magnet Generator Applied to Variable-Speed Wind-Energy Systems Connected to the Grid. IEEE Transaction on Energy Conversion 21(1), 130–135 (2006)
11. Grauers, A.: Efficiency of Three Wind Energy Generator Systems. IEEE Transaction on Energy Conversion 11(3), 650–657 (1996)
12. Ahmed, T., Nakaoka, M., Nishida, K.: Advanced Control of a Boost AC-DC PWM Rectifier for Variable-Speed Induction Generator. In: Applied Power Electronics Conference and Exposition (March 2006)

13. Jang, J.-I., Kim, Y.-S., Lee, D.-C.: Active and Reactive Power Control of DFIG for Wind Energy Conversion under Unbalanced Grid Voltage. In: Power Electronics and Motion Control Conference (2006)
14. Sürgevil, T., Akpınar, E.: Modeling of a 5-kW Wind Energy Conversion System with Induction Generator and Comparison with Experimental Results. International Journal of Renewable Energy 30(6), 913–929 (2004)
15. Schiemenz, I., Stiebler, M.: Control of a Permanent Magnet Synchronous Generator Used in a Variable Speed Wind Energy System. In: IEEE International Electric Machines and Drive Conference (June 2001)
16. Johnson, G.: Wind Energy Systems. Prentice-Hall, Inc., Englewood Cliffs (1985)
17. Morimoto, S., Nakamura, T., Takeda, Y.: Power Maximization Control of Variable-speed wind generation system using permanent magnet synchronous generator. Electrical Engineering in Japan 150(2), 1573–1579 (2005)
18. Moran, L., Ziogas, P.D., Joos, G.: Design aspects of synchronous PWM rectifier-inverter systems under unbalanced input voltage conditions. IEEE Transactions on Industry Applications 28(6), 1286–1293 (1992)
19. Akhmatov, V.: Analyses of dynamic behavior of electric power systems with large amount of wind power. PhD Thesis, Orsted DTU, pp. 1–260 (2003)
20. Brando, G., Coccia, A., Rizzo, R.: Control method of a braking chopper to reduce voltage unbalance in a 3-level chopper, vol. 2, pp. 975-978
21. Hansen, A.D., Michalke, G.: Multi-pole permanent magnet synchronous generator wind turbines' grid support capability in uninterrupted operation during grid faults. IET Renewable Power Generation 3(3), 333–348 (2009)
22. Abedini, A., Nasiri, A.: PMSG wind turbine performance analysis during short circuit faults. In: 2007 IEEE Canada Electrical Power Conference, Canada, pp. 160–165 (2007)
23. Zhang, Y., Gong, J., Xie, D.J.: Inverter control strategy for direct-drive permanent magnet wind generator under unbalance of three-phase source voltage. In: 2008 11th International Conference on Electrical Machines and Systems, Wuhan, China, pp. 2497–2501 (2008)
24. Lazarov, V., Apostolov, D.: PWM inverter power transfer under unbal-anced voltage condition. In: First International Symposium on Environment Identities and Mediterranean Area, ISEIMA 2006, pp. 254–259 (2006)
25. Song, H.S., Nam, V.: Dual current control scheme for PWM converter under unbalanced input voltage conditions. IEEE Transactions on Industrial Electronics 46(5), 953–959 (1999)
26. Ng, C.H., Ran, L.: Unbalanced- Grid –Fault Ride – Through Control for a wind Turbine Inverter. IEEE Transactions on Industry Applications 44(3) (May/June 2008)
27. Hu, J., He, Y.: Modeling and control of grid-connected voltage-source converters under generalized unbalanced operation conditions. IEEE Transaction on Energy Conversion 23, 903–913 (2008)
28. Yazdani, A., Iravani, R.: A unified dynamic model and control for the voltage-source converter under unbalanced grid conditions. IEEE Transactions on Power Delivery 21(3), 1620–1629 (2006)
29. Suh, Y., Tijeras, V., Lipo, T.A.: A control method in dq synchronous frame for PWM boost rectifier under generalized unbalance condition. In: 2002 Power Electronics Specialists Conference, vol. 3, pp. 1425–1430 (2002)

30. Suh, Y., Tijeras, V., Lipo, T.A.: A nonlinear control of the instantane-ous power in dq synchronous frame for PWM ac dc converter under generalized unbalanced operating conditions. In: Proceedings of 2002 IEEE Industry Applications Conference, vol. 2, pp. 1189–1196 (2002)
31. Rodriguez, P., Timbus, A., Teodorescu, R., Liserre, M., Blaabjerg, F.: Reactive power control for improving wind turbine system behavior under grid faults. IEEE Transactions on Power Electronics 24(7), 1798–1801 (2009)
32. Rodriguez, P., Timbus, A.V., Teodorescu, R., Liserre, M., Blaabjerg, F.: Flexible active power control of distributed power generation systems during grid faults. IEEE Transactions on Industrial Electronics 54, 2583–2592 (2007)
33. Wang, Q., Chang, L.: PWM control strategies for wind turbine inverters. Wind Engineering 25(1), 33–40 (2001)
34. Dixon, J.W., Ooi, B.-T.: Indirect Current control of a unity power factor sinusoidal current boost type three-phase rectifier. IEEE Transactions on Industrial Electronics 35, 508–511 (1988)
35. Zhang, Y., Gong, J., Xie, D.J.: Inverter control strategy for direct-drive per-manent magnet wind generator under unbalance of three-phase source voltage. In: 2008 11th International Conference on Electrical Machines and Systems, Wuhan, China, vol. S, pp. 2497–2501 (2008)
36. Moran, L., Ziogas, P.D., Joos, G.: Design aspects of synchronous PWM rec-tifier-inverter systems under unbalanced input voltage conditions. IEEE Transactions on Industry Applications 28, 1286–1293 (1992)
37. Ng, C.H., Li, R., Bumby, J.: Unbalanced-grid-fault ride-through control for a wind turbine inverter. IEEE Transactions on Industry Applications 44, 845–856 (2008)
38. Stankovic, A.V., Lipo, T.A.: Novel Control Method for Input Output Har-monic Elimination of the PWM Boost Type Rectifiers Under Unbalanced Operating Conditions. IEEE Transactions on Power Electronics 16, 603–611 (2001)
39. Stankovic, A.V., Chen, K.: A New Control Method for Input-Output Harmoinic Elimination of the PWM Boost Type Rectifier Under Extreme Unbalanced Operating Conditions. IEEE Transactions on Industrial Electronics 56, 2420–2430 (2009)

Advances in Power Distribution System Management Technology

Fang Yang, Vaibhav Donde, Zhao Li, and Zhenyuan Wang

Abstract. Two important elements of the efficacy of a smart distribution grid are distribution automation and online distribution system analysis. This chapter presents recent research efforts from authors examining these two elements. In the study of distribution automation, an enhanced outage management system based on intelligent electronic devices and meter data is discussed. In the research of online system analysis, state-of-the-art high performance computation architectures (multi-core central processing unit and many-core graphics processing unit) that efficiently manage the complexity of the distribution system analysis problem on a large scale are explored.

1 Introduction

Driven by environmental concerns, the technology evolution, and politic incentives, the electric power distribution system has significantly improved. These improvements have occurred in various areas, including the measurement, monitoring, analysis, and automation of the distribution system, and are gradually transforming the traditional passive distribution system into a more proactive smart distribution grid.

The literature has thoroughly addressed the following two enhancements: the application of distribution automation control technology in fault detection, isolation, and power restoration (FDIR) [1]~[10], and the integration of an advanced metering infrastructure (AMI) with a distribution management system (DMS) [26]~[30], [34]. The development of the FDIR has effectively reduced the duration of power outages from hours to minutes, considerably raising the level of system reliability. As the AMI offers two-way communication between utility control centers and residential meters, the integration of AMI with DMS systems greatly enhances system monitoring and control capabilities.

Because of the large number of real-time or near real-time meter measurements provided by the AMI, advanced online system analysis functions such as

Fang Yang · Vaibhav Donde · Zhao Li · Zhenyuan Wang
ABB Inc. US Corp. Research Center
940 Main Campus Dr., Suite 200
Raleigh NC, 27606

L. Wang (Ed.): *Modeling and Control*, Green Energy and Technology, pp. 155–191.
springerlink.com

distribution state estimation [31] can be realized. The temptation to implement the above advanced online system analysis functions further motivates the exploration of high-performance computation technologies such as the multi-core central processing unit (CPU) and the many-core graphics processing unit (GPU) in various grid system analysis.

This chapter summarizes research on the distribution outage management and the high-performance system analysis on which authors have recently focused. Section 2 addresses the meter and intelligent electronic device (IED) data-based distribution outage management system, including field and control center-based FDIR, short-term load forecasting in the FDIR, and a meter data-based outage analysis. Section 3 first discusses the formulation of a distribution system state estimation problem based on real-time or near real-time meter and sensor measurements and then explores the possibility of online solutions to such complex and large scale problems using general high performance computation architectures (e.g., the multi-core CPU and the many-core GPU). Section 4 concludes this chapter.

2 Outage Management of the Distribution System

Over the years, utilities have been pursuing more intelligent distribution networks, i.e., network that are less passive and more active or dynamically adapting. This trend has evolved into the *Smart Grid,* which, for some electric utilities, refers to electric power systems that enhance grid reliability and efficiency by automatically anticipating and responding to system disturbances. To establish the Smart Grid on a power distribution system level, utility control centers have attempted to implement various automation technologies in system metering, protection, and control. Within these technologies, automated fault detection, isolation, and power restoration have become important parts of the Smart Grid puzzle.

Traditionally, electric distribution utilities rely on a trouble call system to detect the occurrence of faults and resulting power outages (i.e., when a fault occurs, customers who experience power outages may call the utility to report the loss of power). The distribution utility control center then dispatches a maintenance crew to the field, where they investigate the fault location and implement switching scheme(s) to conduct fault isolation and power restoration. This procedure may take several hours to complete, depending on how quickly customers report the power outage and the maintenance crew locates the fault point.

In moving closer to the Smart Grid concept, utilities have, in recent years, deployed feeder switching devices such as reclosers and circuit breakers with IEDs. IEDs are usually equipped with measuring, monitoring, protection, control, and communication functions, which enable the implementation of automated fault identification, isolation, and power restoration. In particular, they can transmit the switch status and other measurements to a substation computer or a control center.

Information from IEDs can be used to identify fault location and isolation. For instance, if a recloser is found in a lockout status after a sequence of reclosing operations, it can be concluded that the fault lies in the downstream of the recloser; then if the switch located directly downstream of the fault is opened, the fault can be isolated. While the identification and isolation of a fault may be relatively easy, automatically restoring power to the healthy feeder section during the repair of the section with the fault remains a challenge.

Studies have proposed a number of automated power restoration algorithms [1]~[10] such as heuristic search-based techniques, artificial intelligence-based algorithms, algorithms using analytical methods, and algorithms combining two or more of these techniques. Although the aim of several of these algorithms is to provide a real-time solution, most are suitable for only planning analysis, or they were intended for use in the distribution utility control center to aid system operators with appropriate decisions. Automated power restoration at the distribution system level demands an efficient, scalable method suitable for online applications.

The following section examines two complementary power restoration control schemes: field- and control center-based schemes. Both conduct a restoration switching analysis (RSA) to achieve back-feed power restoration, that is, healthy load zones that have lost power will be restored through their boundary tie switching devices from neighboring sources. The field-based scheme uses a substation computer or a programmable logic controller (PLC) to run the RSA while the control center-based scheme uses the DMS outage management system.

2.1 The Field-Based Scheme

In the field-based scheme, either a substation computer or a PLC can host the RSA engine and communicate with feeder IEDs [11],[12]. Figure 1 provides an illustration of an electric distribution network with sensors and actuators (IEDs), a power restoration controller (a substation computer or a PLC), and an optional communication link to the DMS. The network contains three sources (S1 to S3), seven loads (L1 to L7), and various switching devices (Brk1 to Brk4 and SW1 to SW6). Devices SW2, SW4, and SW6 serve as normally open tie switches. The tie switches make the network electrically radial, where each load is supplied by only one source. It is imperative that the total load supplied by each source and the current flowing through each switching device fall within their respective maximum capacities. Figure 1 shows that source S1 serves L1, L2, L6, and L7 (900A in total), which is less than its maximum capacity of 1000A. Similarly, sources S2 and S3 supply 400A and 50A of load, respectively, which are less than their respective capacities. Furthermore, the current flow through each switching device is less than its maximum load current-carrying capacity (e.g., 1000A for Brk1, 900A for SW1).

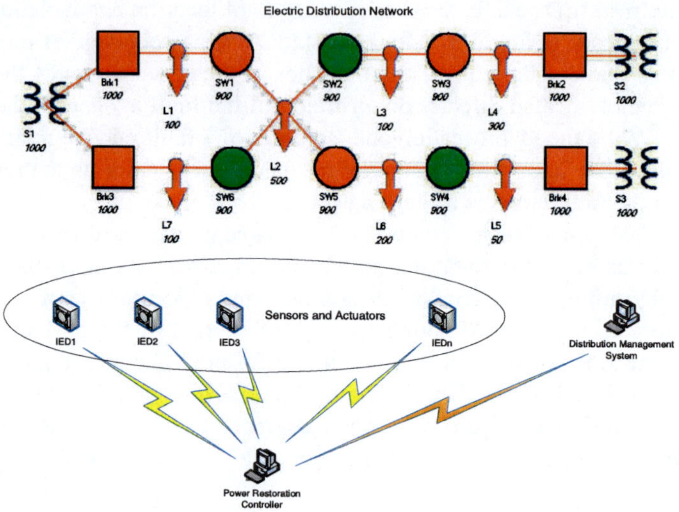

Fig. 1 Electric distribution system example [13] (©2009 IEEE)

Each switching device is associated with an IED that acts as both a sensor and an actuator driver for controlling the switch. IEDs sense the current through the switch and its terminal voltage, which enable them to locally detect a fault and send appropriate open/close commands to the switch to clear the fault. The figure shows that IEDs communicate in real-time with a power restoration controller (or a group of coordinated controllers). The controller periodically polls the voltage and current data from the IEDs. When a switch locks out after its reclosing sequence to clear a permanent fault, the fault occurrence is communicated to the controller by report-by-exception. The fault reporting triggers the fault isolation and power restoration algorithms that reside in the controller. According to the resulting power restoration plan, the controller orders various IEDs to open or close their respective switches in a timely manner. The controller may communicate with the DMS to obtain more detailed real-time information about the system, discussed in the following sections.

Consider, for instance, a permanent fault occurring at load node L1. Switching device Brk1 goes through a reclosing sequence and finally opens (locks out), isolating the fault from the upstream side. Then device SW1 opens, as ordered by the power restoration controller, isolating the fault from the downstream. Following the fault isolation, loads L2 and L6 remain unserved. The power restoration controller now has the option of closing any of the three tie switches SW2, SW4, or SW6 so that the unserved loads are restored. The controller then performs a capacity check to determine a feasible option. If SW2 is closed, source S2 would have to supply the additional load of L2 and L6, which is 700A. It is already supplying L3 and L4, which are 400A in total. Given that its maximum capacity is 1000A, it has only 600A net capacity margin, so the capacity check would not be successful in this case.

However, if either SW6 is closed, providing a backfeed from source S1, or if SW4 is closed, providing backfeed from source S3, the capacity check would be successful. In the former case, source S1 would have to supply a total of 800A (80% of its maximum capacity) and in the latter case, source S3 would have to supply 750A (75% of its maximum capacity). The best option would be the one that results in the lowest loading of the source that supplies the backfeed, which, in this example, is source S3.

The following subsections discuss two field-based power restoration schemes, hosted on substation computer and PLC, respectively.

2.1.1 Substation Computer as a Power Restoration Controller[1]

When a substation computer is used as a power restoration controller that hosts the RSA engine, it requires a distribution network model consisting of at least the following major feeder components: sources, such as distribution substation transformers; switching devices, or "switches" that act to sectionalize parts of the network, load switches, circuit breakers, and reclosers; and loads. The RSA engine is triggered following fault detection and uses a network tracing-based algorithm to identify a valid radial post-restoration network with no current or voltage violation on any network component or network node. The outcome of the RSA is a restoration switching sequence generated online according to the pre-fault network condition instead of pre-programmed rules that are usually generated offline.

It is important that back-feed power restoration not overload any part of the back-feeding network, which the RSA algorithm deals with by a recursive network tracing-based loading aggregation method involving following steps: (1) It begins at a back-feeding source (usually a transformer); (2) it traces down all the network components it supplies until it reaches the end of the node-component graph tree structure; (3) when returning to the source, it sums up the loading current at each network component, and if applicable, compares the loading current with its corresponding limit; and (4) after the tracing returns to the source, it calculates the available capacity of a source.

The RSA algorithm begins with a back-feeding isolation switch search carried out on the graph tree structure of the pre-fault network with the tripped breaker/recloser as the root. The search is conducted down the tree to find the most downstream switch that passed the fault current. This switch is called the forward-feed isolation switch. The search then moves downward to the first layer of downstream switches, referred to as back-feed isolation switches. The algorithm then applies numerous recursive steps, including the following:

- It identifies any multi-connected load nodes (also known as "T-nodes," defined in the following paragraph) via network tracing
- It determines if single-path restoration can be achieved via a single source
- If single-path restoration cannot be achieved, it then continues to search for other switches in the network in order to achieve multi-path restoration.

[1] [11] (©2009 IEEE).

A T-node is defined as the connection point of a lateral in a feeder. If an isolated network has T-nodes, its pre-restoration tree structure will define the isolation switch as the root and the potential back-feeding tie switches as the termination end. If both of the two downstream branches of a T-node are able to provide a backfeed, the algorithm has to choose one out of the two; otherwise, a circuit loop will be generated in the post-restoration network. If the two downstream branches are to be back-fed from the same source, the branch with higher loading capability all the way to the source is chosen for backfeeding.

A back-feeding path is defined as a set of connected circuit components from the back-feed source to the to-be-closed tie switch. If a source can provide the restoration power over a single path to an out-of-service load zone, the restoration is called a single-path restoration. Otherwise, the out-of-service load zone may have to be split into two or more load zones to be back fed, and the scenario is called multi-path restoration.

In the case of multi-path restoration, the algorithm attempts to determine the best reconfiguration. In some cases, the network must be divided into two sub-networks to restore all the possible unserved loads, moving one or more normally-open tie switches to other switching device locations. In other cases, all the unserved loads cannot be completely restored even if the tie switch locations are moved. Both single- and multi-path restorations may have to shed load if the back-feed source capacity or the feeder component loading capability is not sufficient.

The restoration algorithm produces radial post-restoration networks. Using the loading aggregation method, the algorithm performs a current violation check that ensures that the post-restoration loading currents of all the network components are less than their loading current limits. After a load flow analysis of the post-restoration network takes place, voltage violations can be checked as an integral part of the algorithm. The restoration validation check confirms the validity of the post-restoration network configuration to ensure that the network is radial and all the currents and voltages fall within the limits.

The RSA algorithm may also consider other requirements such as minimizing losses, minimizing the number of switching operations, and balancing the loading of the back-feed transformer.

2.1.2 PLC Platform as a Power Restoration Controller[2]

The RSA algorithms can also be hosted on a PLC, which would act as a power restoration controller. Electrical field engineers generally prefer IEC61131-3 standard language-based PLC platform for the design and implementation of RSA. Such PLC platform, however, do not provide many of the advanced features offered by substation computers with advanced programming environments and languages. Therefore, the development of efficient algorithms based on PLC languages and capable of realizing complex power restoration logic has become a critical need. One solution is graph theory-based design, which is capable of (1) dynamically updating a system configuration and a load profile, and (2) providing generic logic for fault location detection, isolation, and power restoration after the

[2] [12] (©2009 IEEE).

occurrence of a permanent fault. Such a logic engine implemented using PLC languages can easily be understood and further customized by field engineers.

The overall PLC platform-based FDIR logic engine flow diagram, shown in Figure 2, is similar to the substation computer-based RSA. The diagram includes functional modules such as system information collection, system configuration and load profile update, fault location detection, fault isolation, power restoration, and so on. The system information collection includes polling IED information to obtain switch status (open, close, lockout), counter values for reclosers, and electric parameters such as currents, voltages, and real/reactive power. The modules of system configuration and load profile update and power restoration can be accomplished by the use of depth-first and breadth-first search strategies in graph theory and analytical techniques. Communication between the master controller and slave IEDs can use various communication protocols such as Modbus, DNP, and so on.

The automatic generation of power restoration logic includes two stages: in stage one, a depth-first search technique is used to search all possible restoration sources/paths; in stage two, a reverse search procedure is developed using an analytical method that locates one or multiple restoration path(s) to restore as much load as possible while balancing the load level in each restoration path. This power restoration logic is illustrated in the following stages:

Stage 1: All possible restoration sources/paths are searched, and the search starts from the isolation switch. All the downstream nodes and connected switches are searched and stored. This search procedure will stop at normally open tie switch(es), which can potentially provide a backfeed for power restoration.

Stage 2: One or multiple restoration sources/paths are searched based on system operating constraints. In this work, note that only the capacity limits of sources (substations) and switches in terms of current magnitudes are included in the operating constraint set. For the purpose of simplifying the problem, all other operating constraints such as voltages and real/reactive power flow are assumed to be within normal limits after the power restoration. These constraints can be included in the constraint set in a similar way to determine the restoration sources/paths. Among all the possible restoration sources/paths, the equivalent capacity margin (ECM) of each restoration/path down to the normally open tie switch is first calculated. Based on such ECM information, the source/path that has the largest ECM is selected to restore load in the healthy out-of-service area. After picking up a certain amount of load, the selected source/path may have less ECM than others; if so, then another restoration source/path with the largest ECM is selected so that the restoration work can continue. This procedure is repeated until the entire load in the healthy out-of-service area is restored or all the restoration sources/paths have run out of capacity. When a joint node that connects to more than one restoration source/path is encountered, the restoration source/path with the highest ECM is selected to continually pick up the load while other restoration sources/paths stop power restoration just before the joint node, thus guaranteeing that the restoration is performed to the maximum extent for the load before the joint node and beyond.

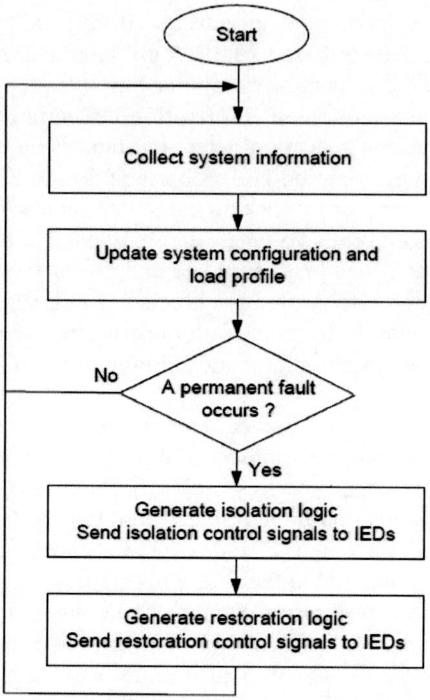

Fig. 2 The flow diagram of the PLC platform-based FDIR [12] (©2009 IEEE)

2.1.3 Single- and Multi-path Power Restoration Examples

Results from a typical RSA engine hosted on either a substation computer or a PLC are illustrated in Figure 3. On the left-hand side of the figure, single-path full restoration is used where a fault at T-node, L3, must be isolated by opening the forward-feed isolation switch, R3, and two back-feed isolation switches, R6 and R10. The back-feed sources, S3 and S4, both have sufficient capacity to pick up the out-of service load on their corresponding restoration path, and each tie switch, R9 and R12, can be closed to achieve the restoration. The post-restoration circuit topology is shown in Figure 3 (right).

Figure 4 (left) shows a multi-path full restoration example, where a fault at load node, L1, must be isolated by a forward-feed isolation switch, R1, and a back-feed isolation switch, R22). In this example, none of the back-feed sources, S2-S5, can completely pick up all the unserved loads after fault isolation. Hence, the RSA algorithm splits the network into two parts by opening R13, and all the out-of-service loads are restored by closing both R8 and R11 (from both S3 and S4). The post-restoration circuit topology is shown in Figure 4 (right).

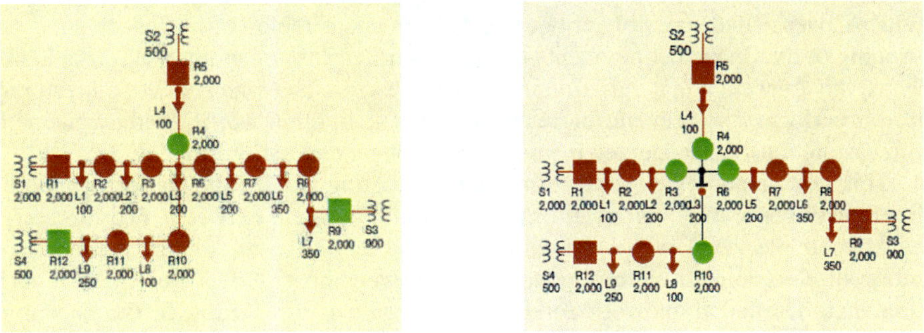

Fig. 3 A single-path restoration example (left: normal topology, right: post restoration topology) [11] (©2009 IEEE)

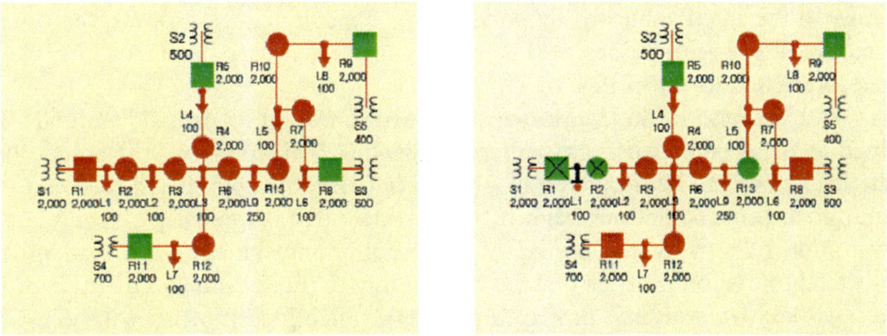

Fig. 4 A multi-path restoration example (left: normal topology, right: post restoration topology) [11] (©2009 IEEE)

2.2 Incorporation of Short-Term Load Forecasting[3]

As discussed earlier, the process of evaluating whether an alternate source is capable of supplying the unserved load requires knowledge about the current or power of the load. A straightforward approach to obtaining this information is to use the rated value of the load in the evaluation process, as done in the example in Figure 1. The advantage of this procedure is that it can be programmed offline as the load rating is known from the network configuration data. In many instances, however, this procedure may not be sufficient because the actual load typically varies with time, and it may deviate considerably from its rated value. The load rating itself may also vary from season to season and even from hour to hour. For instance, if loads L5 and L6 in Figure 1 are assumed to be industrial loads and the restoration as outlined above is carried out during the night hours when the rated load demands are typically low, it is highly likely that the demand of these industrial loads will be significantly higher during the following day, and the new larger loads may possibly overload the source that supplies it. Then if the rated demand of L5 increases to 200A (from 50A) and that of L6 to 350A (from 200A) during the day, source S3

[3] [13] (©2010 IEEE).

will be overloaded, possibly triggering further undesirable consequences such as tripping of overloaded components of the network or loss of another sensitive load. Thus, it is important that variation in the load be examined when back-feed restoration is performed to prevent future consequences. In other words, load profile and forecasting must be accounted for while designing a restoration strategy.

This section addresses short-term load forecasting and its application in an online power restoration algorithm for a distribution system network. The literature contains ample work on the general topic of the service restoration and reconfiguration of electric distribution networks. However, most studies lack documentation of a detailed framework for incorporating load forecasting in the capacity check algorithm and its implementation, which is the focus of this section. In particular, this section discusses the process of online estimation and dynamic adjusting of loads for use in the capacity check algorithm that are used to determine appropriate back-feed strategy for load restoration. Rather than depending on the less reliable offline rated values of the loads in the capacity check process, the process estimates the pre-fault values of the network loads and uses them to appropriately adjust forecasted load profiles.

A "load" as used in this work is the aggregation of all loads inside a load zone, defined as the area between two adjacent switches. Since any of the actual loads in the load zone can be turned on or off individually, the aggregated load can vary from time to time, sometimes significantly. Each load is typically associated with its variation pattern over a period of time, commonly known as a load profile. Loads tend to exhibit profiles that vary according to the season and the type of day (i.e., a workday, a weekend day, or a holiday). Such load profiles will be used to forecast the future value of the load and use the value in the application of feeder restoration.

Load profiles can be predefined once a restoration controller is configured, using information derived from the DMS. If the information is not available at the restoration controller configuration time, the load profiles can be populated with rated maximum load values and then populated online over time by the restoration controller.

Real-time current flowing through any two adjacent switches is measured by the corresponding IEDs. Based on the magnitudes of the switch current and the statuses of the open/closed switches (as known from corresponding IEDs), the pre-fault load can be estimated using variants of Kirchhoff's current laws for the radial topology of the network.

Given that such a load profile represents only a forecasted load, but the actual load will likely differ, a user predefined safety margin is incorporated in the capacity check algorithm to identify the suitable backfeed. However, if no feasible backfeed exists with the maximum value of the load over the 24-hour period, the maximum value of the load over a shorter period is used. Consider the worst case in which even after trying all possibilities (e.g., down to one hour of fault repairing time), no feasible backfeed is found to power the disconnected loads resulting from insufficient available capacity of alternate sources and devices in those paths, a load management scheme and/or a demand response scheme that limits the load to existing available capacity can be instated through the DMS. While load management can limit the load under a certain amount by shedding loads, demand response encourages customers to reduce their load willingly via certain incentives in response to system reliability concerns.

Network stability and performance constraints play an important role in determining the real-time source capacity limits that can safely be used for load restoration. For instance, a source may have 1000A maximum current handling capacity; however, the network voltage and steady state stability constraints may reduce the actual available capacity to as low as 700A. If a source attempts to supply more load than that specified by this capacity limit, the system may suffer from sagging voltages that may trigger a voltage collapse process. The power restoration controller must use the real-time capacity limit rather than the rated maximum capacity in the algorithms that check capacity.

The DMS has a detailed model for the entire distribution network (within the territory of the utility) and is equipped with various network analysis functionalities that enable the DMS to estimate the actual capacity limits of the alternate sources. As illustrated in Figure 1, the power restoration controller communicates with the DMS, which analyzes the network in real-time using tools such as power flow and contingency analysis. Using these tools, the DMS predicts the short-term (hours ahead) actual capacity limit of each source and communicates it back to the restoration controller. A virtual IED runs at the DMS server that responds to calls from the restoration controller and communicates the actual source capacity information to the controller for use in power restoration logic.

Figure 5 shows a field-based power restoration application integrated with the short-term load forecasting-based capacity check algorithm. A three-bus system network is used for illustration. Switches R4 and R5 are normally open. Each load is characterized by its 24-hour profile depicted in Figure 6. Loads are assumed to peak at 4:00PM and reach the lowest point in their variation at 5:00AM, with a sinusoidal variation.

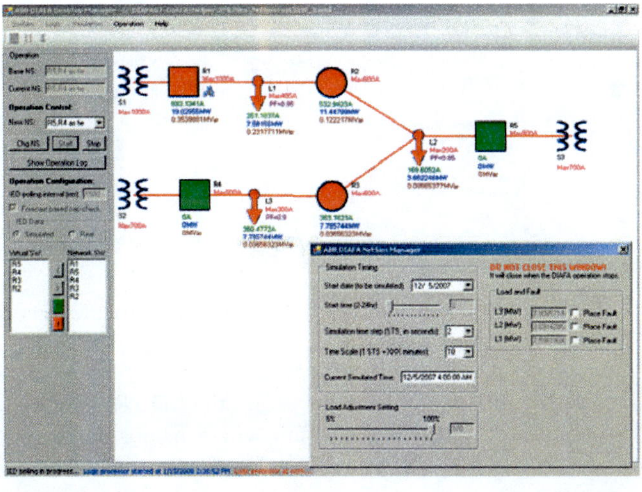

Fig. 5 Load forecasting-based capacity check integrated in a power restoration application [13] (©2010 IEEE)

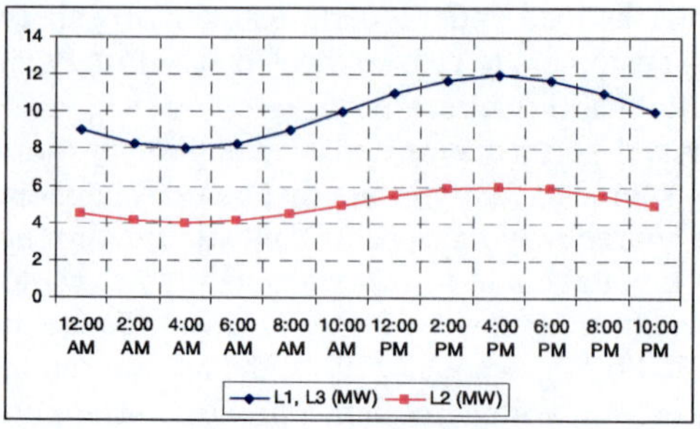

Fig. 6 Load profiles of loads: L1, L2, and L3 [13] (©2009 IEEE)

Test case 1 is carried out with the mean time to repair (MTTR) for the fault set to 5 hours, both upstream and downstream. A permanent fault is placed on the downstream terminal of R1, where load L1 is located. The fault triggers the reclosing operation of R1, which locks out to isolate the fault from the upstream, triggering automation algorithms in the power restoration control to create isolation and restoration logic. The fault isolation switch is identified as R2. Direct restoration of the lost loads (L2 and L3) can be accomplished in two ways: by closing R5 or by closing R4. The load forecasting-based capacity check requires knowledge of the maximum load at each of the load buses during the mean time to repair the fault, which is 5 hours for the present test case. Figure 6 shows that the maximum values of the loads during this time are about 9MW, 4.5MW, and 9MW. The capacity check is run using these maximum load values, which in this case is successful and R5 is issued a close command. Figure 7 (left) shows a screenshot of the restored network.

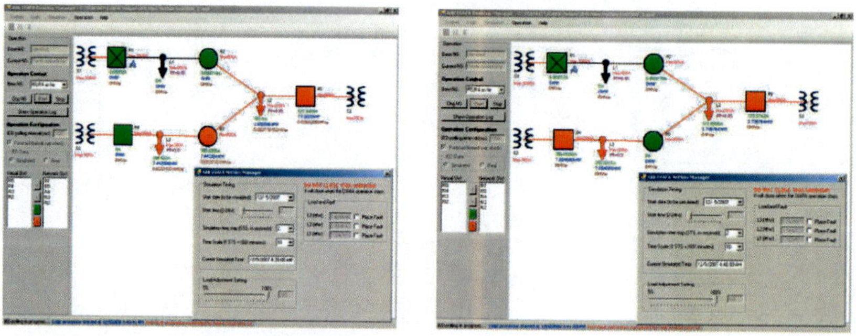

Fig. 7 Screenshot of the restored network for test cases (left: case 1, right: case 2) [13] (©2010 IEEE)

Test case 2 is carried out using the same setup; the MTTR, however, is now set to 12 hours, so the capacity check algorithm will consider the maximum of the predicted load in the 12-hour time span. In Figure 6, these values are 12MW,

6MW, and 12MW for loads L1- L3, respectively (larger than the ones from the previous test case). Consequently, no single source-based backfeed passes the capacity check since no single source is capable of fully serving the peak load. The restoration algorithm finds a feasible solution that uses a multi-source based backfeed that involves the opening of R3 and the closing of R4 and R5. Thus R3, R4, and R5 are issued appropriate commands to restore the previously unserved loads L2 and L3. Figure 7 (right) shows a screenshot of the restored network.

2.3 Control Center-Based Scheme [14]

In a control center-based scheme, the substation computer is used as a gateway to transmitting field IED data back to the outage management system at the control center and control commands from the control center to the field IEDs. The first uses industry-accepted protocols such as IEC61850, DNP3, and Modbus, to obtain the necessary data from each of the feeder IEDs, and then analyzes these data to detect if a fault has occurred in the system. In the event of a fault, the gateway sends this information upstream to the DMS. After data analysis, DMS determines the location of the fault and subsequently runs the RSA, generating the proper isolation and restoration switching actions that should be taken. The DMS then sends the switching control commands to the gateway either automatically or after an operator authorization action, whichever is preset in the control center DMS application.

With the integrated control center-based restoration scheme, the gateway detects the outage based on the IED-sensed network events and informs the DMS automatically. When the DMS receives this notice, it will run the RSA with respect to the outage area and generate power restoration schemes. The RSA is based on the detailed network model and unbalanced load flow analysis of this model to insure that the post-restoration network does not have current and voltage violations. The RSA combines a network topology tree-tracing and genetic algorithm, thereby enabling it to support both lightly and heavily loaded network conditions. If the loading of the network is light, then single-path restoration is sufficient. If it is heavily loaded, either multi-path restoration or a multi-layer RSA must be used. The concept of the multi-layer RSA is illustrated in Figure 8, where green squares represent tie switches or fault clearance/isolation switches. The tie switches that bound the unserved load area are called the first layer of tie switches for restoration.

Subsequent layers are named sequentially (e.g., second layer, third layer). Under heavy load conditions, closing only the first layer restoration switches may not be sufficient for meeting the power requirements of the unserved loads. Thus, load transfers from the zone between the first and second layers to that between the second and the third layers may be necessary.

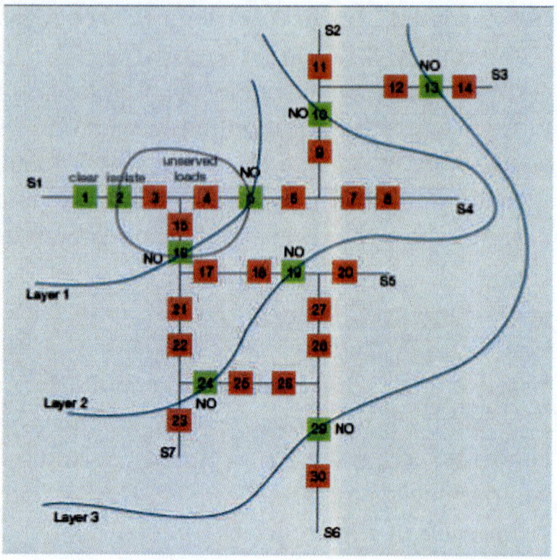

Fig. 8 The multi-layer RSA concept [14]

2.4 *Outage Analysis Using Smart Meter Data*

Two types of faults can occur in the distribution system: the short circuit fault and the open conductor fault. They both result in power outages for customers located downstream of the fault. A short circuit fault can be isolated from upstream by the opening of protective devices such as circuit breakers, reclosers, or fuses, depending on the utility protection coordination mechanism. Circuit breakers and reclosers are usually equipped with IEDs that can communicate status changes in these protective devices to the substation computer or the control center. If these devices operate to open, the field- or control center-based FDIR technique can quickly detect the fault and conduct isolation as well as power restoration.

However, if a short circuit fault occurs in a lateral with a fuse as the protective device, the fuse may burn out to isolate the fault. Fuses in the distribution system are generally not equipped with communication capabilities. Therefore, the trouble call system is the only means by which a utility can detect a fault occurrence. The same scenario takes place when the distribution system has an open conductor fault in which no protective devices operate to open, and the utility control center has to depend on trouble calls from customers to report an outage.

After receiving trouble calls from customers, the utility control center first needs to conduct an outage analysis based on the locations of the trouble calls. The outage analysis aims to identify the outage area and therefore infer the outage source (such as a burned out fuse or an open conductor). Fast and accurate outage analysis is critical so that a utility can efficiently dispatch a crew to repair the circuit and to complete the service restoration task.

Most DMS outage management systems adopt the upstream tracing method [15][16] to determine the outage source and area. Specifically, they use the locations of trouble calls and the network topology to find the common upstream protective device, which is assumed to be the outage source, and the entire downstream area of this device is concluded to be the outage area.

The upstream tracing method is most effective in a single outage situation caused by one protective device operation. However, when multiple outages occur in the same neighborhood, or an open conductor fault occurs, this method inevitably fails to locate outage source, so it enlarges the outage area. Figure 9 illustrates both a distribution circuit with multiple outages and an open circuit fault.

In the left side circuit, two short-circuit faults occur on two laterals with fuses F2 and F3, which sense the short-circuit current and burn out. Based on trouble calls received from the downstream of F2 and F3, the upstream tracing method infers that the common protective device CB of all the received trouble calls is the outage source, and the outage scope includes the entire downstream area of CB instead of just the downstream areas of F2 and F3. A similar conclusion is reached in the right side circuit, when an open conductor fault occurs on the feeder section between F1 and F2, which causes a power outage inference in the entire downstream area of CB based on both the trouble call locations and the upstream tracing method.

In both situations, the normal circuit section downstream of CB, such as the lateral line with F1, the feeder line downstream of CB in the left side circuit, part of the feeder line between CB and the open conductor in the right side circuit are assumed to experience a power outage. Based on such an inaccurate assumption, the utility cannot pinpoint the exact fault location until it sends a crew into the field.

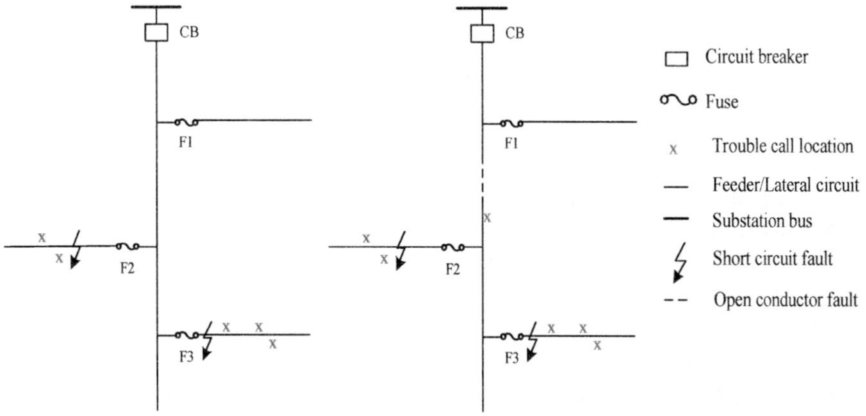

Fig. 9 Outage analysis example

Recently, with the wide deployment of the AMI in the distribution system, the smart meter is capable of providing an alternative means of the outage detection besides trouble calls. The last gasp sent out by smart meters in an outage area can notify the control center of the power outage occurrence in real time or near real time, which is much faster than the arrival of trouble calls at the control center. Besides the quick outage notification, smart meters have an on-demand polling capability that can be used to check the energized status of the meter. This feature is particularly useful in resolving any uncertainty about the outage area during multiple outages or open conductor fault situations, as described in Figure 9. For instance, after receiving trouble calls from the downstream of F2 and F3, the utility control center can further conduct on-demand polling to check if the smart meters located downstream of F1 and/or on the feeder line close to CB is energized or not. If these meters are still energized, then the CB cannot possibly be concluded to be the outage source.

Some researchers have begun to explore the application of the smart meter in outage management [17]~[20]. Effectively utilizing meter last gasp information and the on-demand polling function will expedite the initiation of the outage management process, provide accurate outage analysis results, result in fast power restoration, and improve the end customer reliability level and service satisfaction.

2.5 Discussion

Two of the most important Smart Grid applications for power distribution systems include FDIR and outage analysis. FDIR can be hosted at either a field level or a control center level. At the heart of FDIR lies the RSA algorithm, which identifies the optimal back-feed solution and may involve single- and multi-path restoration options. If the network components are too stressed and the multi-path restoration cannot restore all the out-of-service loads, the algorithm will try to shed a minimal load while restoring as many other loads as possible. Multi-layer RSA addresses more complex situations in which single-layer RSA cannot provide a back-feed restoration solution. An outage analysis provides the operator with valuable information about power outages that cannot be detected by IEDs. Information from AMI systems can be integrated into outage analysis to provide the operator with a clearer picture of the network status.

The FDIR algorithm can be improved further by incorporating short-term load forecasting information. While using the real-time source capacities obtained from DMS, this online procedure uses an estimation of feeder loads to adjust load profiles appropriately or generate load profiles online in case they are not available. These characteristics of the power restoration control process ensure that unserved loads due to fault isolation are restored in a reliable manner and no network elements are overloaded.

3 High Performance Computing for Online Smart Grid Analysis[4]

In modern distribution systems, many emerging features, such as the intermittency and uncertainty of renewable energy resources and bi-directional power flows, inevitably exacerbate system unpredictability and make distribution system analysis, protection, and control are more complex and dynamic. As a result, effective

[4] [31] (©2011 IEEE).

system management becomes difficult without online monitoring and analysis applications such as power flow and state estimation. Online monitoring and analysis of the utility grid, influenced by various unpredictable factors, require real-time system measurements and high performance computation (HPC) architecture. As of today, large amounts of real-time or near real-time measurements are available to a utility control center through the SCADA and AMI. These measurements, particularly meter measurements from the low-voltage network, tend to significantly increase the dimension and complexity of the underlying mathematic problems of distribution system monitoring and analysis applications. Therefore, the HPC architecture is expected to play a critical role in efficiently resolving these challenging problems and making them applicable in real time.

This section first addresses the distribution system state estimation based on meter and sensor data and then presents the authors' research work in the area of high performance computation architecture that improves the efficiency of the state estimation and power flow solution. This section is organized as follows: section 3.1 introduces the meter and sensor data-based distribution system state estimation solution; section 3.2 examines the computation architecture of the multi-core CPU and many-core GPU; section 3.3 analyzes the features of power system calculations by evaluating a serial implementation of the conjugate gradient normal residual (CGNR) algorithm used as a linear equation solver; section 3.4 provides a detailed design and prototype of the CGNR algorithm based on both the multi-core and many-core computation architectures; section 3.5 conducts a performance evaluation of the above parallel prototypes against large-scale datasets of state estimation and power flow problems; and section 3.6 concludes the discussion.

3.1 Distribution System State Estimation

Historically, the distribution system consists of very limited measuring devices, most of which are located at the feeder head and close to controllable equipment such as voltage regulators and shunt capacitor banks. Under these circumstances, many distribution management systems obtain overall system operating states using an approximate method in which first, load estimation is executed to allocate the total demand measured at the feeder head to each service transformer proportionally according to the transformer nominal capacity rating and second, a distribution system load flow is solved to provide system state values. The assumption that each service transformer loading level is proportional to its capacity rating is nearly always inaccurate in practice.

Nowadays, as a result of the wide deployment of AMI and other sensors such as IEDs in distribution systems, not only are a large number of measurements offered by the meters and sensors available to the utility control center, but the measurements also retain a certain level of system information redundancy. The redundant measurements enable the application of advanced distribution system state estimation algorithms such as the weighted least squares (WLS) method and others that have been broadly applied for transmission system state estimation.

However, compared to the transmission system model, the distribution system model required in the state estimation is much more complicated. For one, it should take into account many detailed and realistic system and component characteristics, including multi-phase unbalanced construction and operating practice, various transformer connection and configuration types (Y/Y, Y/Δ, Δ/Y, Δ/Δ, grounded or ungrounded, primary, or secondary with leading or lagging, primary, or secondary regulation capability, an open-delta connection, and so on), various load and capacitor connection types (Y or Δ), and various combinations of load models (constant power load, constant current load, and constant impedance load). Figure 10 to 13 provide representative model examples of various distribution network components, including distribution lines, loads, capacitor banks, and transformers.

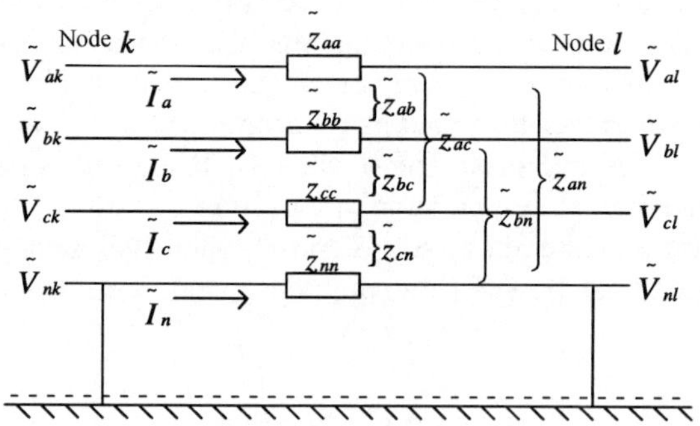

Fig. 10 A four-wire grounded three-phase line

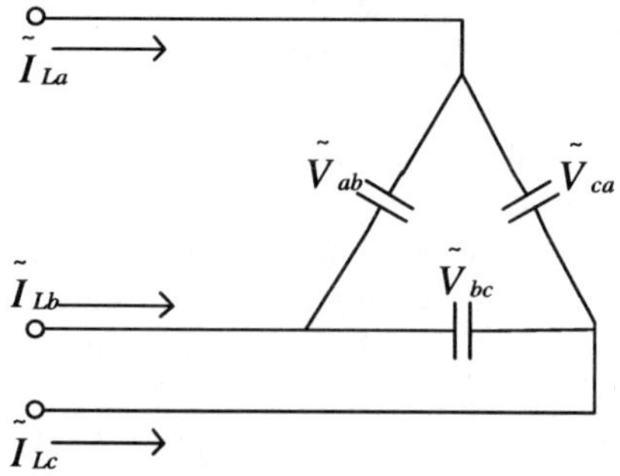

Fig. 11 A delta-connected capacitor bank

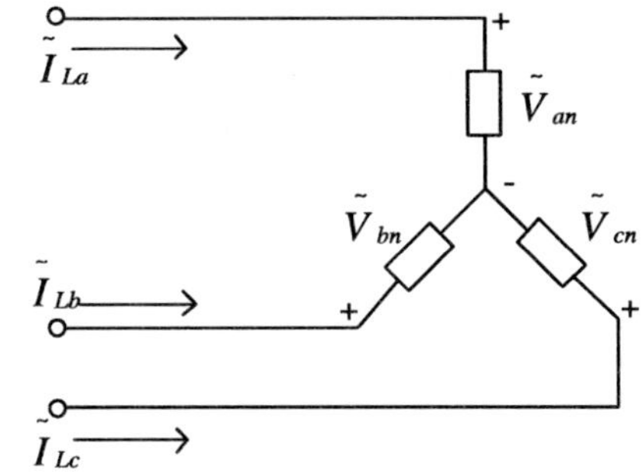

Fig. 12 A wye-connected load

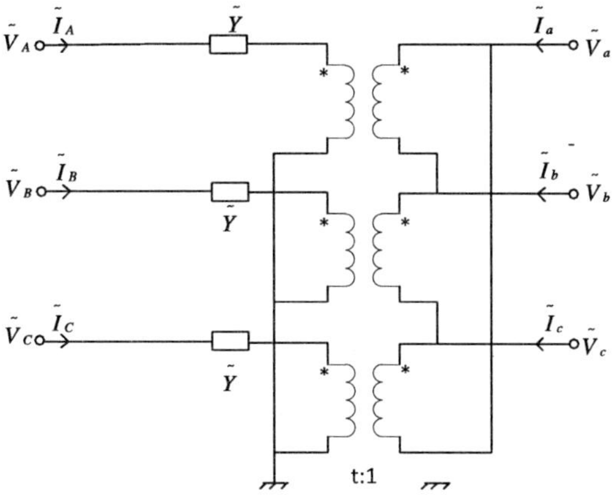

Fig. 13 A wye-delta connected transformer

Distribution system state estimation is a computational procedure that uses a redundant set of measurements and a detailed multi-phase network model to compute a statistical estimate of system operating states. Generally, system states, which consist of node voltage magnitudes and phase angles, are used to uniquely define system operating conditions. In the set of available measurements, pseudo load power measurements provided by meters are major constituents. Also available are measurements from other sensors, including bus voltage magnitudes, branch power/current flows, and so on.

The literature [21]~[25] has introduced various methods for distribution system state estimation Most of them are based on the WLS (weighted least squares) approach and its variations such as the equality constraints formulation and augmented matrix method, which are all solved by the LaGrange multiplier method. The augmented matrix WLS formulation is selected here to illustrate the distribution state estimation problem. Eq. (1) represents an augmented WLS formulation in which residual vector r is introduced into the objective function, and all the measurement functions, including the zero injection measurement function, are treated as equality constraints.

$$\begin{aligned} & Min.J = \frac{1}{2} r^T W r \\ & s.t.\ c(x) = 0 \\ & \qquad r - z + h(x) = 0 \end{aligned} \tag{1}$$

where

W: a diagonal matrix with the diagonal element w_{ii} being the weight for measurement z_i

r: residual vector $r = z - h(x)$
x: system state vector
z: measurement vector including all actual measurements
h: measurement function vector corresponding to all the actual measurements
c: virtual zero injection measurement function vector

Typically, the LaGrange multiplier method is used to solve the optimization problem. In this case, the LaGrange function will have two sets of LaGrange multipliers, λ and μ :

$$L(x, \lambda) = J(x) - \lambda^T c(x) - \mu^T (r - z + h(x)) \tag{2}$$

Adhering to the first-order optimality conditions, we can derive the following equations:

$$\begin{aligned} & \frac{\partial L}{\partial x} = C^T \lambda - H^T \mu = 0 \\ & \frac{\partial L}{\partial \lambda} = -c(x) = 0 \\ & \frac{\partial L}{\partial r} = Wr - \mu = 0 \\ & \frac{\partial L}{\partial \mu} = r - z + h(x) = 0 \end{aligned} \tag{3}$$

By eliminating r ($r = R\mu$, $R = W^{-1}$) and linearizing the remaining three equations in (4) around a given system state x^k, we can obtain the following linear equation set:

$$\begin{bmatrix} H^T & 0 & C^T \\ 0 & C & 0 \\ R & H & 0 \end{bmatrix} \begin{bmatrix} \mu^{k+1} \\ \Delta x^{k+1} \\ \lambda^{k+1} \end{bmatrix} = \begin{bmatrix} 0 \\ c(x^k) \\ z - h(x^k) \end{bmatrix} \tag{4}$$

The formulated linear equation set is solved iteratively to update the system state variables, and according to the author, it usually takes an average of 10 or more iterations before the convergence criterion is satisfied. For a realistic distribution circuit with thousands of buses and the distribution system state estimation method using the detailed multi-phase network model, the dimensions of the above augmented linear equations can range between 10k~100k. In light of the fact that a general distribution system control center could cover as many as several hundred distribution substations and each substation could have multiple such distribution circuits, from a utility perspective, the distribution management system will require considerable computational effort to solve the distribution system state estimation problem.

Due to the heavy computational burden in solving large-scale power system problems, many research efforts have tested the potential of high performance computation architectures. Traditionally, solving a large-scale power system problem relies on a mainframe computer or specially-designed computation architecture. For instance, [35] attempted to solve such problem using a computer with 128 1.5GHz titanium-2 processors and 256G of memory, but the high cost and special deployment conditions prevented its use for practical industrial applications. In such a situation, research directions should focus on finding an affordable computation strategy from mainstream computation architectures.

Mainstream computation architectures in the current market technically fall into two categories: the multi-core CPU and the many-core GPU. Both are shared memory computation architectures in which several computational cores sharing the main memory bandwidth and banks are glued into one die. The current market is dominated by a quad-core CPU, but six-, eight-, and even ten-core CPUs are finding their way into product lines [36], [37]. Graphic processor units (GPUs) are originally used to boost the performance of graphic applications. Since 2007, in a movement initiated by NVidia, GPUs are now accessible to high performance computation applications through the CUDA or OpenCL interfaces [38].

Under various power system applications, state estimation as well as power flow are two fundamental applications, at the heart of which lies a solution of linear equations $Ax=b$, such as Eq. (4), with A typically being a highly sparse matrix. Although direct methods based on LU and QR have been traditionally applied to power system problems [39], iterative methods may prove to be strong contenders when it comes to solving large scale-power system problems because of the structure of iterative methods, which involves simple mathematical calculations that are more amenable to parallelization using a parallel platform [40], [41]. Hence, the following sections aim to investigate a high-performance iterative solver by utilizing the existing mainstream computation architecture in the context of power system state estimation and power flow applications.

3.2 A Closer Look at the CPU and GPU Architecture

From the perspective of architecture, both the multi-core CPU and many-core GPU package several cores into one die, sharing the same memory channel. To maintain high performance, this type of computation architecture needs to have a high-speed memory bandwidth, feeding data to and getting results from computational cores.

Generally, the performance of the computation architecture is determined by one of the following two perspectives: the computational power and speed of the data channel, which are separately measured by GFLOPS and memory bandwidth. GFLOPS refers to the billions of floating point operations per second handled by a processor, and the memory bandwidth is defined as the data transfer rates between computation cores and the main memory.

Figure 14 illustrates the primary differences between the multi-core CPU and the GPU: a multi-core CPU has a larger cache system and stronger control capabilities; by contrast, a GPU has a significant number of computational cores (ALUs). Table 1. Multi-core CPU vs. many-core GPU may help to clarify the above difference between the multi-core CPU and the many-core GPU. Generally, the total cache of a multi-core CPU is 10M (L1 cache + L2 cache + L3 cache), 6 to 7 times as large as that of the GPU system; while the GPU has more computational cores (i.e., GTX480, which has 480 cores) than the multi-core CPU (generally less than 10 cores).

As more transistors are devoted to data processing rather than caching and flow control, a GPU offers significantly more powerful computation capabilities than a multi-core CPU does. As of 2010, the latest version GPU from NVIDIA (GTX480) offers 576 GFLOPS, 20 times as much as the latest Intel CPU (i.e., Intel Westmere having six cores).

However, increasing the computing power by assembling many computational cores into one die escalates memory I/O requests and overuses the existing memory channel. In particular, the increasing number of memory I/O requests blocks the memory channel, depriving the computational cores of data and eventually eliminating the benefits of increased computation power brought by the added computation cores. Thus, the performance bottleneck for computation architectures that share memory, including both multi-core CPU and many-core GPU, is memory bandwidth. The literature refers to this issue as the "memory wall".

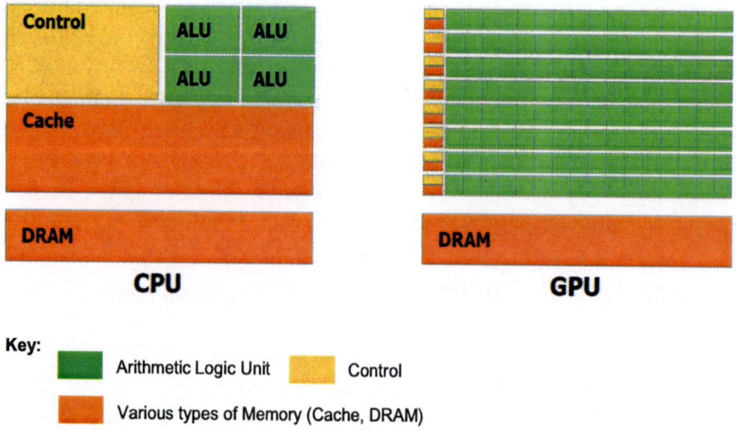

Fig. 14 The models of multi-core CPU and GPU [43]

For a multi-core CPU, mitigating the overuse of the memory channel demands that various types of cache (L1, L2 and L3 caches) be placed along the path between the individual cores and the main memory. As the most frequently used data can be entered into the caching system, reading/writing data can be done through the cache rather than through the main memory. This strategy effectively reduces the consumption of costly memory bandwidth. However, efficiently utilizing the above caching infrastructure necessitates an application with a regular memory usage pattern and relatively small memory footprints. Table 1 illustrates the caching infrastructure of the main-stream CPUs produced by both Intel and AMD.

Table 1 Multi-core CPU vs. many-core GPU [31] (©2011 IEEE)

	# of Physical Cores	L1 Cache	L2 Cache	L3 Cache	Memory Bus
Intel Bloomfield Core i7 950 (June, 2009)	4	32KB (instruction) 32KB (data)	256KB/ Core	8M (shared by all cores)	25 GB/s
Intel Yorkfield Core 2 Quad Q9400 (Aug. 2008)	4	4 x 32K(instruction) 4 x 32 (data)	2 x 3M shared by two cores	0 M	10 GB/s
AMD Phenom II 1090T (Apr 2010)	6	N/A	3M	6 M shared	37 GB/s
AMD Phenom X4 9950 (Mar 2008)	4	64 KB + 64 KB data + instructions	2M	2M	33 GB /s
Fermi NVDIA GTX480 (April 2010)	480	64 KB (16 + 48)	768KB	768KB	180 GB /s

Key: I/O bus includes front side bus (FSB), Quick Path structure and Hyper Transport Bus

Unlike the CPU, the GPU adopts a unique strategy to hide the memory latency: driving many threads to concurrently access memory in different locations. In practice, the above strategy of reducing memory latency particularly beneficial for data parallel computation problems, in which the same mathematical operation can be concurrently applied to many data stored in different memory locations. Fortunately, most matrix calculations (i.e., matrix multiplication) that dominate power system calculations fall into this category.

In the past decade, implementing the GPU strategy of hiding memory latency through concurrent access is not only technically easier but also leads to higher memory bandwidth growth rates when compared to the multi-core CPU. Since 2003, the GPU memory bandwidth has been grown from 20 GB/s to 180 GB/s; during the same period, the bandwidth of multi-core CPU's has only expanded from 10 GB/s to 25 GB/s.

In summary, the efficient use of the limited memory bandwidth is the key to unleashing the computation power of both the multi-core CPU and many-core GPU computation architectures. Previous experience has demonstrated that an application with a regular memory access pattern is suitable for the multi-core computation architecture because most memory footprints of the application can be cached; however, an application with simple logic controls but bulk concurrent data calculations, suits the many-core computation architecture, in which concurrent transportation of data effectively hides the memory latency.

3.3 *Features of the Numerical Methods Applied to Power System Calculations*

To identify the features of performance issue in power system calculations, this work evaluates the serial version of the conjugate gradient normal residual (CGNR) [44] in the context of the power system state estimation application. The CGNR method is selected because it is a general technique for solving a WLS based system of linear equations and it is built upon on the concepts of the conjugate gradient (CG) method. From a computational perspective, the CGNR (CG) serials algorithms involve some basic matrix operations such as matrix-vector multiplication, vector dot products, and vector updates.

3.3.1 CGNR and CGNR + Jacobi Pre-conditioner

We consider a problem, in which we seek to solve the system of linear equations $Ax = b$, where A is an n by m real matrix, b a known real vector of dimension n by 1 and x an unknown real vector of dimension m by 1. Here, $n > m$ because the state estimation problem is over-determined (more equations than unknowns). For a power system state estimation problem, matrix A is the measurement Jacobian.

In the WLS normal equations formulation of the power system state estimation problem, A^TA describes the gain matrix that is symmetrical positive definite (SPD) by structure, so the conjugate gradient (CG) method is applicable. Note that the gain matrix requires matrix-matrix multiplication of the measurement Jacobian A with its transpose, which is hidden by CGNR.

The CGNR with Jacobi pre-conditioning method is called when the diagonal of the gain matrix is used as pre-conditioner matrix M (the left part of Figure 15). The inverse of M consists of only the reciprocal of the diagonal of M; thus Jacobi pre-conditioning is an attractive iterative methods because it involves very few additional computations compared with other pre-conditioning methods, but greatly reduces the number of iterations [35].

3.3.2 Evaluation of Serial Implementations of CGNR and CGNR + Jacobi Pre-conditioner

The serial version of CGNR and CGNR with the Jacobi pre-conditioner, presented in Figure 15, has been implemented using C in the Windows platform. The behavior of iterative methods in the context of the power system state estimation application can be observed, if they are executed using a group of power system state estimation data matrices listed in Table 2 and the following measurements are recorded: the number of iterations to convergence and the time spent during iterations. Iterations are declared as converged when the 2-norm of a residual vector is less than a predefined threshold value – 1e-3.

Given an initial guess $x^{(0)}$
Compute $r^{(0)} = b - Ax^{(0)}$, $z_0 = A^T r_0$,
$p_0 = z_0$
For i = 1, 2, ... until convergence Do:
$w_i = Ap_i$
$temp = z^T z$
$\propto_i = \frac{temp}{w^T w}$
$x_{i+1} = x_i + \propto_i p_i$
$r_{i+1} = r_i - a_i w_i$
$z_{i+1} = A^T r_{i+1}$
$\beta_i = z_{i+1}{}^T z_{i+1} / temp$
$p_{i+1} = z_{i+1} + \beta_i p_i$
EndDo

CGNR

Given an initial guess $x^{(0)}$
Compute $r^{(0)} = b - Ax^{(0)}$, $\bar{r}_0 = A^T r_0$,
$z_0 = M^{-1}\bar{r}_0$, $p_0 = z_0$
For j = 1, 2, ... until convergence Do:
$w_j = Ap_j$
$temp = (z_j, \bar{r}_j)$
$\propto_j = \frac{temp}{w^T w}$
$x_{j+1} = x_j + \propto_j p_j$
$r_{j+1} = r_j - a_j w_j$
$\bar{r}_{j+1} = A^T r_{i+1}$
$z_{j+1} = M^{-1}\bar{r}_{j+1}$
$\beta_i = z_{j+1}{}^T \bar{r}_{j+1} / temp$
$p_{j+1} = z_{j+1} + \beta_j p_j$
EndDo

CGNR with Jacobi preconditioner

Fig. 15 CGNR and CGNR with a Jacobi pre-conditioner [31] (©2011 IEEE)

A. The Test Environment

Hardware

- Intel Core 2 Quad Q9400 (2.66GHz)
- L1 Cache: 4 x 32 KB instruction caches, 4 x 32 KB data caches
- 2 x 3M 12-way set associative caches (each L2 caches is shared between 2 cores)
- Seagate 1T hard drive, 7200 rpm
- 4G DDR2 memory
- PCI-E x 16 (GPU interface)
- NVIDIA GTX480 GPU
- GPU GF100
- Memory 1.5G (DDR5)
- Bus width (384 bit)
- Bandwidth 182.4 GB/s

Software

- Windows 7 Enterprise
- Visual Studio 2008 Professional

B. The Test Data Set

Table 2 describes the test matrices in terms of dimension, number of nonzero elements, condition number, and percentage of nonzero elements. The number of conditions is specified for index matrix *A*, which represents the measurement Jacobian. It should be emphasized that for CGNR, we are dealing with the gain matrix that has a squared condition number even though it accepts matrix *A* as an input, its convergence behavior depends on the condition number of the gain matrix ($A^T A$).

Table 2 The power system state estimation data for evaluating the serial implementation CGNR [31] (©2011 IEEE)

Data set (SE measurement Jacobian H)	Size	Non Zero Elements	Condition number	Ratio of Non-Zero Elements
3 bus system	22 x 6	88	26.2	66.7%
14 bus system	73 x 27	329	118.78	16.7%
14 bus system diagonal extension	36500 x 13500	164,500	137.87	0.033%
14 bus system non-diagonal extension	36500 x 13500	205,296	1571.9	0.041%
PSSE2	28634 x 11028	115,262	1.2422 x 10 ^6	0.036%

The 3-bus system and the 14-bus system test matrices are generated from a simple 3 bus system and the standard IEEE 14 bus system. To test the scalability of the algorithm, we diagonally extend the 14-bus system measurement Jacobian to form a large sparse matrix, in which the original 14-bus matrix is duplicated 500 times along the diagonal. In practice terms, such a diagonally extended matrix stands for 500 independent 14-bus systems and its condition number is the same as of the 14-bus system measurement Jacobian. A few of-diagonal elements are introduced in a test case with a worse condition number. The obtained matrix is referred to as the 14-bus system diagonal extension with inter-area dependencies, the condition number of which is 1571, 9 to 10 times larger than that of the original 14-bus system. In addition, we also consider another large power system state estimation matrix PSSE2 [48].

C. Sparse Matrix Storage Scheme

The chosen matrix storage scheme is a combination of the coordinate (COO) format [49] and the compressed sparse row (CSR) format [49]. The combination of COO and CSR efficiently uses both the search based on individual locations (COO) and the search based on rows (CSR), resulting in the following data structure illustrated in Figure 16.

$$A = \begin{bmatrix} 1 & 7 & 0 & 0 \\ 0 & 2 & 8 & 0 \\ 5 & 0 & 3 & 9 \\ 0 & 6 & 0 & 4 \end{bmatrix}$$

$$row = [0\ \ 0\ \ 1\ \ 1\ \ 2\ \ 2\ \ 2\ \ 3\ \ 3]$$

$$col = [0\ \ 1\ \ 1\ \ 2\ \ 0\ \ 2\ \ 3\ \ 1\ \ 3]$$

$$data = [1\ \ 7\ \ 2\ \ 8\ \ 5\ \ 3\ \ 9\ \ 6\ \ 4]$$

$$ptr = [0\ \ 2\ \ 4\ \ 7\ \ 9]$$

Fig. 16 Sparse matrix storage scheme (COO + CSR)[31] (©2011 IEEE)

D. Test Results and Analysis

Table 3 lists the test results, which include the number of iterations and time taken until convergence by the CGNR and the CGNR with Jacobi pre-conditioner when executed using the test matrices in Table 2. We use 1e-3 residual thresholds as a termination criterion. The data type used is the double precision float type. Our observations are summarized as follows:

- For a set of linear equations $Ax=b$, the degree of the ill-conditioning of the matrix A influences the convergence rate of an iterative method. A ill-conditioned matrix A results takes more iterations to converge.
- The computational time spent by iterative algorithms is not as affected by the size of the problem as by the matrix conditioning. In our evaluation, even though the 14-bus system with a diagonal extension, the size of which is 36500 x 13500 (larger than PSSE2), takes a reasonable time (<0.1s) to converge. Note that this matrix is well-conditioned.
- The Jacobi pre-conditioner greatly reduces the number of iterations to converge and therefore effectively reduces the computation load. In our test, CGNR with the Jacobi pre-conditioner using PSSE2 data and a residual threshold of 1e-3 as the stopping criterion takes 27,605 iterations to converge, which is 14.4% of the number of iterations taken by the CGNR without any pre-conditioning.

Table 3 Number of iterations and time spent by the CGNR, the CGNR with the Jacobi pre-conditioner for different test matrices [31] (©2011 IEEE)

Methods (Algorithm) Residual = 1e-3		3 bus system	14 bus system	14 bus system diagonal extension	14 bus system diagonal extension with inter-area dependencies	PSSE2 Data
CGNR	Iteration	5	28	33	1369	188,271
	Time (sec)	0.000159	0.000318	0.077579	3.801	271.45
CGNR + Jacobi	Iteration	3	24	25	845	27,605
	Time (sec)	0.000007	0.000117	0.064539	2.428825	40.844019

E. Performance Evaluation of the Serial CGNR Implementations

To roughly estimate performance in terms of GFLOPS and memory bandwidth, we manually calculate the number of double precision operations, the number of accessing memory, and the time consumed by each iterative algorithm (Table 3). The estimation assumes that the data are directly picked up from memory with no caching capability. For instance, the statement $c = a + b * c$ can be counted as two double precision operations (double precision add and multiplication), and four memory operations (three reading operations and one writing operation). In the estimation process, only the double precision operations and memory operations that occur during the iterative loop are counted.

Table 4 illustrates the overall performance of the serial implementations of the CGNR and the CGNR + Jacobi pre-conditioner in terms of GFLOPS and memory bandwidth. The CGNR with the Jacobi pre-conditioner takes 0.36 GFLOPS, which is far below the peak GFLOPS of the Intel Quad Core 9400 (42.56 GFLOPS (Double Precision)) [47]. The average memory bandwidth is around 6.2 GB/s, which is more than half of the peak memory bandwidth of the Intel Core 2 Quad Core system (8.5GB/s [47]).

Table 4 The effective performance of selected serial implementations in terms of GFLOPS and memory bandwidth [31] (©2011 IEEE)

	Total GFLOP	Total Mem Op. (DP)	Spent Time (s)	GFLOPS	Memory Bandwidth (GB/s)
CGNR	1.5G	3.11G x 8	3.83	0.39	6.4
CGNR + Jacobi	0.932G	1.98G x 8	2.53	0.36	6.2

Key:
DP – Number of Double Precision Float Point Calculations
Test Data: 14 bus system diagonal extension with inter-area dependencies
Residue: 1e-3

The above test results show that the serial implementations of iterative methods consume a small part of CPU computation power (less than 1% of the peak CPU computation power), but a large portion of memory bandwidth (around 60% of the peak memory bandwidth) to transport double precision data back and forth between the CPU and its memory. Consumption of a large portion of the memory bandwidth demonstrates that an iterative method in the context of the state estimation application is highly memory bounded.

3.4 Parallelization of Iterative Methods

In this section, we discuss the parallelization of iterative methods using multi-core and many-core computation architectures. CGNR + Jacobi pre-conditioning and CG + Jacobi pre-conditioning are selected for the parallelization evaluation based on the test results in the former section.

3.4.1 The Parallelization Strategy

As previously discussed, both the CG + Jacobi and the CGNR + Jacobi pre-conditioner (Figure 17) have a common framework that includes an initialization part and an iterative part constructed by several common operations (e.g., matrix vector multiplication, vector inner products, and vector updates). Here, the vector update operation is defined as $u_i=v_i+\alpha w_i$, where u, v, w are n x 1 vectors, and α is a scalar.

Table 5 illustrates the time spent by the common operations of serial implementation of the above two algorithms in the following two test scenarios: CG + Jacobi against a power flow test data set and CGNR + Jacobi against a power system state estimation data set. The fact that matrix-vector multiplication takes 66.1%, which constitutes 82.3% of the total time separately in the above two test cases, demonstrates the importance of matrix vector multiplication. Thus, improving its performance should be the first priority of parallelism.

Table 5 Time spent by three basic operations [31] (©2011 IEEE)

Stopping criterion residual = 1e-3

	Total Times (sec)	Matrix-Vector Multiplication (sec)	Vector and Vector Multiplication (sec)	Vector Updates (sec)	Iteration times
CG + Jacobi	2.557	1.69 (66.1%)	0.46 (17.9%)	0.40 (15.6%)	715
CGNR + Jacobi	2.544	2.0945 (82.3%)	0.2272 (8.9%)	0.2 (7.8%)	845

Test Data Set:
CG + Jacobi: 1138 bus system with 56 times extension for power flow.
CGNR + Jacobi: 14 bus system diagonal extension with inter-area dependencies for state estimation

3.4.2 The Multi-core CPU Implementation of Matrix Vector Multiplication

We aim to boost the performance of matrix-vector multiplication operations using multi-core technologies. To simplify analysis, we parallelize only the matrix-vector multiplication of $w_i = Ap_i$ (in the left part of Figure 17) using a series of operations that can be concurrently executed as follows: $A_0 p_i, A_1 p_i, \ldots, A_n p_i$ ($A_0, A_i, \ldots, A_n$ are the individual rows of matrix A). The remaining parts of the algorithm use the serial implementations.

The multi-core CG + Jacobi has been implemented using C++ on the windows platform. To reduce the overhead of creating and deleting a thread, we use a thread pool provided by the Microsoft platform. Managed by the thread pool, the overhead of creating and destroying a thread is greatly reduced since the thread pool initializes a pool with a certain number of threads and reuses them. Thread synchronization is done by a shared variable and an event handler.

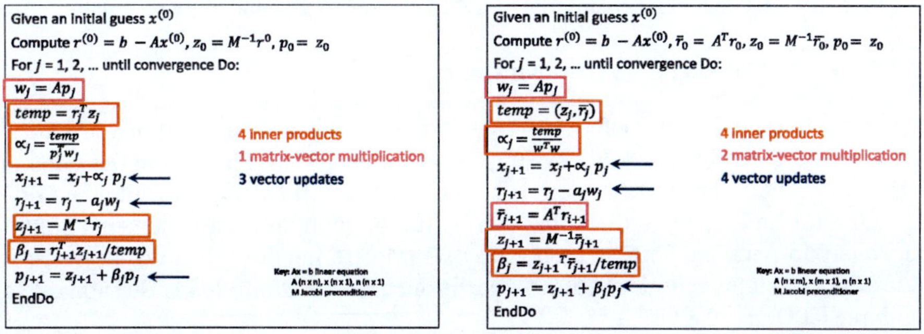

Fig. 17 CG with the Jacobi pre-conditioner and the CGNR with Jacobi pre-conditioner [31] (©2011 IEEE)

The workflow of the multi-core CG + Jacobi is as follows: The main thread enqueues individual operations A_0P_i, A_1P_i, …, A_nP_i into the buffer of the thread pool, where the above operations can be currently executed by the certain number of threads offered by the thread pool; then the main thread idles for an event indicating that all the operations A_0P_i, A_1P_i, … , A_nP_i have finished. Once the main thread receives such an event, it ends its idling status and continually conducts calculations in the remaining part of the algorithm. For each thread, a shared variable is used to synchronize its execution: Once a thread finishes the calculation of A_jP_i, it adds one to the shared variable through the atomic add operation, and then checks if the shared variable has reached n (the size of the problem). When the shared variable reaches n, it will trigger an event announces that the main function that calculations of all the threads are complete. The above multi-core CG + Jacobi preconditioner is tested against the data set 1138_56, a power flow Jacobian test data extended from the 1,138 system provided by NIST.

It is apparent from Table 6 that the multi-core version CG + Jacobi has a higher CPU utilization rate (94%) than the single-core version. However, the actual performance of the multi-core CG is only one-tenth as fast as that of the CG + Jacobi.

Table 6 The performance and CPU utilization of the multi-core and single-core CGs [31] (©2011 IEEE)

	CPU Utilization	Consumed Time (sec)
CG Multi-Core version	94%	32.168
CG Single-Core version	25%	3.39

Key: The test is against the 1135_56 data set

Because of the irregular scarcity pattern of the input matrix, it is cannot be easily cached by the infrastructure of the multi-core CPU in the test system. Thus, most of the data have to be picked up from the main memory. Observations in the above serial version algorithm evaluation demonstrate that the serial version iterative method consumes >60% memory bandwidth. The experimental results confirm the above observation that the four cores concurrently consuming the same memory bandwidth block the memory data channel, causing the individual cores to idle and starve for data.

3.4.3 Many-Core GPU Implementation

The many-core GPU computation architecture consists of a host (the original CPU system) and a device (the GPU system), in which a C/C++ function in the GPU is called a "kernel". The CPU system is suitable for workflow control while the GPU is suited for bulk computations that can be run on numerous data elements simultaneously in parallel. So that each type of hardware is allowed to do the work it does best, the application should be partitioned and suitable loads distributed to both the CPU and the GPU.

Because the execution of an iterative algorithm consists of a series of iterations, the latter of which relies on the results of the former, we cannot concurrently execute individual iterations. Hence, a parallel strategy utilizes parallelism of the GPU threads to parallelize individual iterations as much as possible; at the same time, the strategy should synchronize the execution of the GPU threads using a global barrier performed by the CPU to maintain the execution sequence between the individual iterations.

As common operations of the iterative method (Figure 18) are GPU friendly, in actual implementation, these operations are offloaded to the GPU for high performance computation. At the same time, the CPU coordinates the execution of the operations that are offloaded to the GPU.

Figure 18 illustrates the implementation of CGNR + Jacobi pre-conditioning under the CPU-GPU computation architecture, in which the CPU and the GPU collaborate to maximize the hardware utilization of both the CPU and the GPU. To minimize the overhead of data transportation, we transport only initial data from the CPU to the GPU at the beginning of the calculation and then transport the results of calculation at the end. During the calculation process, only the synchronization control signal is transported, but the data stays in the GPU.

Given an initial guess $x^{(0)}$
Compute $r^{(0)} = b - Ax^{(0)}$, $\bar{r}_0 = A^T r_0$, $z_0 = M^{-1}\bar{r}_0$, $p_0 = z_0$
For j = 1, 2, ... until convergence Do:
$w_j = Ap_j$
$temp = (z_j, \bar{r}_j)$
$\propto_j = \frac{temp}{w^T w}$
$x_{j+1} = x_j + \propto_j p_j$
$r_{j+1} = r_j - a_j w_j$
$\bar{r}_{j+1} = A^T r_{i+1}$
$z_{j+1} = M^{-1}\bar{r}_{j+1}$
$\beta_i = {z_{j+1}}^T \bar{r}_{j+1}/temp$
$p_{j+1} = z_{j+1} + \beta_j p_j$
EndDo

Create and initialize the memory in GPU, including $x, r, z, p, w, \alpha, \beta, b, A^T, A$
Initialize $r^{(0)} = b - Ax^{(0)}$, $\bar{r}_0 = A^T r_0$, $z_0 = M^{-1}\bar{r}_0$, $p_0 = z_0$
For j = 1, 2, ... until convergence Do:
GPU: $w_j = Ap_j$
CPU: Sync
GPU: $temp = (z_j, \bar{r}_j)$
CPU: Sync
GPU: $temp1 = w^T w$
CPU: Sync
GPU: $x_{j+1} = x_j + \alpha_j p_j$
CPU: Sync
GPU: $r_{j+1} = r_j - \alpha_j w_j$
CPU: Sync
GPU: $\overline{r_{j+1}} = A^T \overline{r_{j+1}}$
CPU: Sync
GPU: $z_{j+1} = M^{-1}\overline{r_{j+1}}$
CPU: Sync
GPU: $\beta_i = {z_{j+1}}^T \overline{r_{j+1}}/temp$
CPU: Sync
GPU: $p_{j+1} = z_{j+1} + \beta_j p_j$
EndDo
CPU: get the result from CPU

Fig. 18 The CGNR with Jacobi and its CPU + GPU execution [31] (©2011 IEEE)

3.4.4 GPU Kernel Implementation

As the parallelization of the vector-update operation is straightforward, this section focuses primarily on matrix-vector multiplication and the vector-vector dot product.

GPU version matrix vector multiplication

The GPU implementation of the matrix vector multiplication is based on [49], in which matrix vector multiplication $Ax=b$ is partitioned into a series of independent operations $A_i x = b_i$ (A_i refers to the row vector of A, b_i is the ith element of vector b), concurrently executed by individual GPU threads. In this partition, the number of concurrently executed threads is equal to the row number of matrix A. For example, in the test scenario of PSSE2, the calculation of $Ax = b$ is done by 28634 threads, each calculating an $A_i x = b_i$ but concurrently with all the other threads.

The vector inner product

GPU implementation of the vector inner product is based on [50], which defines the vector inner product as the sum of the elements of $c_i = a_i \times b_i$ (a_i and b_i are the ith element of vectors a and b). Generally, the GPU version of the vector-vector dot product consists of two steps:

Concurrently conducting $c_i = a_i \times b_i$ (a_i and b_i are the ith element of the vector)
Summing c_i ($i = 1 \dots n$)

Theoretically, the parallel vector-vector dot product algorithm requires that all the calculations ($c_i = a_i \times b_i$) be done before summing c_i, which in turn requires the CPU to globally synchronize the executions of $c_i = a_i \times b_i$.

3.5 Evaluations of Many-core GPU Implementation

This section evaluates the GPU implementation of CG series algorithms with Jacobi pre-conditioning, including CGNR + Jacobi preconditioning and CG + Jacobi preconditioning, in the context of power system state estimation and power flow applications. To ensure comparable results, we use the same test system as that in the previous sections.

Power System State Estimation

Table 7 shows the test results of the CPU and the GPU versions of the CGNR with Jacobi pre-conditioner using the test data sets listed in Table 2. Regardless of the conditioning of the matrix, the GPU version shows an improvement in performance by a factor of between 7 and 10 over the CPU version.

Table 7 A comparison of the serial and GPU implementations of CGNR + Jacobi in the context of power system state estimation application [31] (©2011 IEEE)

	14 bus system diagonal extension		14 bus system diagonal extension with inter-area dependencies		PSSE2	
	# of iterations	Spent Time (s)	# of iterations	Spent Time (s)	# of iterations	Spent Time (s)
CPU	25	0.063972	845	2.43	27605	42.873
GPU	25	0.006327	845	0.3619	27605	7.13

Power Flow

Table 8 lists the various test matrices used for comparing the performance of CG with Jacobi pre-conditioning on the CPU to that on the GPU. Each of these matrices represents a power flow admittance matrix [51] equivalent to the power flow Jacobian. The first row represents a 494-bus system that is duplicated 28, 56, and 336 times in rows two, three, and four, respectively. Similarly, rows 5 through 7 represent a 685-, 662-, and 1138-bus system duplicated 336 times to obtain corresponding power flow admittance matrices. The results in Table 8 demonstrate that for a large system, the GPU solution is much faster than the corresponding CPU solution; however, for a small system such as the 494-bus network (Table 8), the GPU solution is slower than the CPU solution due to the overhead of the GPU computation platform, such as the expense of launching thousands of threads.

Matrices that are typically used in power flow calculations, namely, the active power flow Jacobian or equivalently, the system admittance matrices, are better conditioned than those used in other power system applications such as state estimation. Well-conditioned matrices are more favorable for pre-conditioned iterative methods such as CG using the Jacobi pre-conditioner. Furthermore, as the GPU architecture is suitable for parallelizing large-scale matrix-vector multiplication in iterative methods, it is also a promising approach to efficiently solving power flow problems on a large-scale power system (Table 8). The GPU architecture may also be effective for power flow-based contingency analysis because it involves multiple power flows for various contingency situations.

Table 8 The performance comparison between the serial and GPU implementations of CG+Jacobi preconditioner in the context of power flow application [31] (©2011 IEEE)

	Size	Time spent by CPU (ms)	Time spent by GPU (ms)	SpeedUp
494 bus	494 x 494	7.53	38.8	-5.15
494_28	13,832 x 13,832	226.6	40.0	5.6
494_56	27,664 x 27,664	451.4	40.8	11.0
494_336	165,984 x 165,984	5,400	141.7	38.1
685_336	230,160 x 230,160	4,030	124	32.5
662_336	222,432 x 222,432	3,700	74.9	49.3
1138_336	382,638 x 382,638	25,000	687.6	36.4

3.6 Discussion

The deployment of Smart Grid applications in a large scale power grid results in more complex operation and control. Providing operational assistance to the control center operator requires an online system analysis based on high frequency and the mass scale measurements provided by the AMI and other data collection systems, increasing the computational load.

Studies have addressed HPC in the context of power system applications handled by mainframe computers, typically placed in a special environment such as a university or a research lab. However, to evaluate the feasibility of handling a large-scale power system calculations using state-of-the-art personal computers in the market, we have conducted numerical experiments using main stream computation devices, the multi-core CPU and many-core GPU (as of 2010) and large-scale power system data for power system state estimation and power flow applications and reported our findings in this section.

Results of this evaluation show that the solution to a system of linear equations, the key part of most power system applications is highly memory bonded when performed using iterative methods. In addition, they show that the multi-core CPU and GPU computational architectures impact performance differently. Unlike the multicore CPU, the GPU can significantly improve the performance of CG-based iterative solver when matrices are well conditioned and large, as typically encountered in a power flow formulation on a large-scale power system network.

As this work has demonstrated the potential of the GPU architecture on certain class of power system applications, future work should further investigate other applications to determine whether they would also benefit from this architecture and test them using real-world data.

4 Conclusions

This chapter presents recent improvements in distribution system management technology. It covered the topics that included outage management of the distribution system and the application of high performance computation technology to Smart Grid analysis.

Distribution system operators consider effective outage management one of the most important applications in the distribution management system to reduce the duration of power outages and improve the level of system reliability. The automatic fault detection, isolation, and power restoration technology based on IED information can quickly identify fault location and automatically implement fault isolation and power restoration to a healthy feeder section while crews are repairing the faulty section. In addition, the last gasp message of a meter can quickly notify the utility control center of the occurrence of faults that cannot be detected by IEDs. Enhanced by the FDIR and outage analysis based on IED and meter data, the outage management system can significantly reduce the outage duration time experienced by customers from hours to minutes.

With the development of the Smart Grid, the power grid is becoming more and more complex and dynamic. However, a more thorough understanding of such a system in real-time has necessitated online system analyses consisting of high-frequency and large-scale measurements that require a heavy computation load. Therefore, to provide utility control centers with less cost prohibitive solution, the work presented here has carried out high performance computations in the context of power system applications using the multi-core CPU and the many-core GPU. The results of the computations demonstrated that the GPU can significantly improve the performance of the CG-based iterative solver particularly when matrices are well conditioned and large.

In this chapter, a snapshot of authors' recent research on the distribution system management is presented, which primarily focuses on the areas of outage management and application of HPC on Smart Grid analysis. As part of the entire Smart Grid paradigm, the technology improvements in these two areas together with others in various aspects of the smart gird are driving the traditional passive distribution system forward to the modern proactive Smart Grid with advanced features including the self-healing attribute and the real-time system monitoring, analysis, and management capabilities.

References

[1] Taylor, T., Lubkeman, D.: Implementation of Heuristic Search Strategies for Distribution Feeder Reconfiguration. IEEE Transactions on Power Delivery (January 1990)

[2] Moon, Y., Cho, B., Park, H., Ryu, H., Ha, B., Lim, S.: Fault Restoration Algorithm Using Fast Tracing Technique based on the Tree-Structured Database for the Distribution Automation System. In: IEEE PES Summer meeting (2000)

[3] Huang, C.: Multi-objective Service Restoration of Distribution Systems Using Fuzzy Cause-Effect Networks. IEEE Transactions on Power Systems 18(2) (May 2003)

[4] Miu, K.N., Chiang, H.D., Yuan, B., Darling, G.: Fast Service Restoration for Large-Scale Distribution Systems with Priority Customers and Constraints. IEEE Transactions on Power Systems 13(3) (August 1998)

[5] ABB, Patent Application E20050360. Advanced Feeder Architecture with Automated Power Restoration

[6] S&C Electric, Patent Application US2007/0005193A1. Method and Apparatus for Control of an Electric Power Distribution System in Response to Circuit Abnormalities

[7] NovaTech. Distribution Automation Orion Application Note, `http://www.novatechweb.com/orion.cfm`

[8] Peponis, G., Papadopolous, M.: Reconfiguration of Radial Distribution Networks: Application of Heuristic Methods on Large-Scale Networks. In: IEE Proceedings Generation Transmission Distribution, vol. 142(6) (November 1995)

[9] Hsu, Y., Huang, H., Kuo, H.: Distribution System Service Restoration Using a Heuristic Search Approach. IEEE Transactions on Power Delivery 7(2) (April 1992)

[10] Toune, S., Fudo, H., Genji, T.: A Reactive Tabu Search for Service Restoration in Electric Power Distribution Systems. In: Proceedings of IEEE International Conference on Evolutionary Computation (May 1998)

[11] Wang, Z., Donde, V., Yang, F., Stoupis, J.: A Deterministic Analysis Method for Back-feed Power Restoration of Distribution Networks. In: IEEE PES General Meeting, Calgary (July 2009)

[12] Yang, F., Li, Z., Donde, V., Wang, Z., Stoupis, J.: Graph Theory-Based Feeder Automation Logic for Low-End Controller Application. In: IEEE PES General Meeting, Calgary (July 2009)

[13] Donde, V., Wang, Z., Yang, F., Stoupis, J.: Short-Term Load Forecasting Based Capacity Check for Automated Power Restoration of Electric Distribution Networks. In: IEEE PES Transmission and Distribution Conference and Exposition (2010)

[14] Wang, Z.: I-FA-NET Design Document. ABB Inc. US Corp. Research Center (2008)

[15] Kearney, S.: How Outage Management Systems can Improve Customer Service. In: Transmission & Distribution Construction, Operation & Live-Line Maintenance Proceedings, pp. 172–178 (1998)
[16] Yang, L.: Knowledge Based System for Distribution System Outage Locating Using Comprehensive Information. IEEE Transactions on Power Systems 17(2), 451–456 (2002)
[17] Tram, H.: Technical and Operation Considerations in Using Smart Metering for Outage Management. In: IEEE PES Transmission and Distribution Conference and Exposition (2008)
[18] Kuang, H., Wu, Z.: Application of AMR Based on Power Line Communication in Outage Management System. In: International Conference on Sustainable Power Generation and Supply (2009)
[19] Munoz, A.M., Oterino, D., Carmona, A., de la Rosa, J.J.G.: Automated Meter Reading Systems in Outage Management. Compatibility in Power Electronics (2007)
[20] Yan, L., Fischer, R.A., Schulz, N.N.: Distribution System Outage and Restoration Analysis Using a Wireless AMR System. In: IEEE PES Winter Meeting (2002)
[21] Meliopoulos, A.P.S., Zhang, F.: Multiphase Power Flow and State Estimation for Power Distribution Systems. IEEE Transactions on Power Systems 11(2), 939–946 (1996)
[22] Baran, M.E., Kelley, A.W.: State Estimation for Real-Time Monitoring of Distribution Systems. IEEE Transactions on Power Systems 9(3), 1601–1609 (1994)
[23] Lin, W.M., Teng, J.H.: State Estimation for Distribution Systems with Zero-Injection Constraints. IEEE Transactions on Power Systems 11(1), 518–524 (1996)
[24] Lu, C.N., Teng, J.H., Liu, W.H.E.: Distribution System State Estimation. IEEE Transactions on Power Systems 10(1), 229–240 (1995)
[25] Wang, H., Schulz, N.N.: A Revised Branch Current-Based Distribution System State Estimation Algorithm and Meter Placement Impact. IEEE Transactions on Power Systems 19(1), 207–213 (2004)
[26] Yang, F.: Advanced Metering Infrastructure Technology. Pre-study Report, No. PT-07045, ABB USCRC (2007)
[27] Uluski, R.W.: Interactions between AMI and DMS for Efficiency /Reliability Improvement at a Typical Utility. In: IEEE PES General meeting (2008)
[28] Ipakchi, A.: Implementing the Smart Grid: Enterprise Information Integration. In: Grid-Interop Forum (2007)
[29] Advanced Distribution Infrastructure. GE's AMI and DMS integration Solution, http://www.gepower.com/prod_serv/products/metering/en/going_ami_new.htm
[30] EnergyIP, Siemens's AMI and DMS Integration Solution, http://www.energy.siemens.com/us/pool/us/services/powertransmission-distribution/mdms/downloads/MDMS-overview.pdf
[31] Li, Z., Donde, V., Tournier, J.C., Yang, F.: On Limitations of Traditional Multi-core and Potential of Many-core Processing Architectures for Sparse Linear Solvers used in Large-Scale Power System Applications. In: IEEE PES General Meeting (2011)
[32] Roytelman, I., Landenberger, V.: Real-time Distribution System Analysis – Integral Part of DMS. In: IEEE PES Power Systems Conference and Exposition (2009)
[33] Huang, Z., Nieplocha, J.: Transforming Power Grid Operations via High Performance Computing. In: IEEE PES General Meeting, Pittsburgh, PA (2008)

[34] Li, Z., Wang, Z., Tournier, J.C.: A Unified Solution for Advanced Metering Infrastructure Integration with a Distribution Management System. In: IEEE International Smart Grid Communication Conference (October 2010)

[35] Nieplocha, J., Chavarría-Miranda, D., Tipparaju, V., Huang, H., Marquez, A.: A Parallel WLS State Estimator on Shared Memory Computers. In: Proceedings of Power Engineering Conference (2007)

[36] Intel Corp., http://www.intel.com

[37] AMD, http://www.amd.com

[38] NVIDIA, http://www.nvidia.com/object/cuda_opencl_new.html

[39] Vempati, N., Slusker, I.W., Tinney, W.F.: Enhancements to Givens Rota-tions for Power System State Estimation. IEEE Transactions on Power System 6(2) (May 1991)

[40] Pai, M.A., Sauer, P.W., Kulkarni, Y.: Conjugate Gradient Approach to Parallel Processing in Dynamic Simulation of Power Systems. In: Proceedings of the Automatic Control Conference, pp. 1644–1647 (1992)

[41] Pai, M.A., Dag, H.: Iterative Solver Techniques in Large Scale Power System Computation. In: Proceedings of the 36th Conference on Decision & Control, San Diego, California USA (December 1997)

[42] Shewchuk, J.R.: General Introduction to Conjugate Gradient Method, http://www.cs.cmu.edu/~quake-papers/painless-conjugate-gradient.pdf

[43] NVIDA, NVIDIA CUDA C Programming Guide 3.1 (March 2010)

[44] Conjugate Gradient Method. Wikipedia, http://en.wikipedia.org/wiki/Conjugate_gradient_method

[45] Saad, Y.: Iterative Methods for Sparse Linear Systems, 2nd edn. SIAM, Philadelphia (2008)

[46] The Computation Power of Intel Core 2 Quad Core Q9400, http://www.intel.com/support/processors/sb/CS-023143.htm#3

[47] The Memory Bandwidth of a General Computation Architecture, http://en.wikipedia.org/wiki/Front-Side_bus

[48] Cheng, Y.: PSSE2, http://www.cise.ufl.edu/research/sparse/matrices/YCheng/psse2

[49] Bell, N., Garland, M.: Implementing Sparse Matrix-Vector Multiplication on Throughput-Oriented Processors. In: Proceedings of Supercomputing, November 9 (2009)

[50] Harris, M.: Optimizing Parallel Reduction in CUDA. NIVIDIA Technical Report (2008)

[51] NIST, http://math.nist.gov/MatrixMarket/

[52] Power Systems Test Case Archive, http://www.ee.washington.edu/research/pstca/pf14/pg_tca14bus.htm

Power System Emission Control Using Electrical Energy Storage Systems

Hamid Falaghi and Maryam Khosravi Mahmooee

Abstract. Increasing air pollution in recent decades has raised many environmental concerns and affected the operation of electric power plants. On the other hand, energy storage systems with their fast-developing technologies are interesting options to be used with various goals in planning and operation of power systems. In this chapter, the power system operation with the Battery Energy Storage Systems (BESS) is analyzed and mathematically modeled in order to manage and control emission of power plants. Also, the life-time parameters of the BESS are considered in the problem modeling. The proposed model is optimized using a hybrid optimization approach which is based on combination of the λ-iteration based economic load dispatch and genetic algorithm. The proposed approach is applied to a 69-bus system and the optimal parameters of the BESS are obtained. These parameters include the charge and discharge power and the stored energy of the BESS as well as its corresponding idle hours. The obtained results show that BESS can be used as an effective option for emission control of power plants.

1 Introduction

Population growth and tendency toward industrializing have increased the required amount of electricity. The existing energy production processes in the fossil fueled power stations produce large amounts of contaminant such as SO_2, CO_2, and NO_X. These pollutants affect the environment so drastically that special constraints are required in power system operation. Moreover, Economic Dispatch (ED) is one of the major studies of the power system operation in which the total production cost of electricity is minimized by determining the output power of each committed generating unit. But operating at absolute minimum cost can no longer be the only criterion for ED due to the public awareness and demand toward environmental problems. These concerns have enforced the utility operators to consider environmental constraints as well as economic factors in their decision

Hamid Falaghi · Maryam Khosravi Mahmooee
Department of Electric Power Engineering,
The University of Birjand,
Birjand, Iran

L. Wang (Ed.): *Modeling and Control*, Green Energy and Technology, pp. 193–207.
springerlink.com

makings. Also, several studies have been performed so far to reduce the emission level of the power plants [1-3].

On the other hand, several advanced energy storage systems are being developed, demonstrated and recently commercialized. They can provide new opportunities for storage systems to be used in the power industries. These technologies can help the power system with its main goal of supplying the load at all times by storing low priced energy when the demand is low and for a later use when energy cost is high hence supplying a share of the peak demand. This process can increase the total profitability of the system. Also, energy storage systems can be used instead of additional peaker power generators whose efficiencies are relatively low and air emissions are high. Fundamentally, the electrical energy storage systems can cover a wide spectrum of goals such as reliability improvement, power quality application, load leveling, peak shaving, and etc [4]. Energy storage systems are already a common part of many utilities and extensive researches such as [5-6] have been done to discuss about various applications of storage technologies in the power systems. However, they are not yet integral to generation, distribution, and transmission operations [7] and limited studies have been done in applying the energy storage systems in the operational studies such as economic dispatch and unit commitment [8, 9]. Regarding the latest progresses in the constructing technologies of the energy storage systems and their promising future, it is necessary to revise the application of these systems in the mentioned studies.

In general, there are limited experiences and familiarities with electricity storages and their benefits. Even if benefits are understood, in most cases it is quite challenging to estimate the magnitude of the benefits. Most utility planning and electrical evaluation tools and financial and accounting evaluation criteria do not accommodate storage evaluation [7]. Therefore, development of an improved utility operation models which include energy storage system as a component is necessary. To fulfill this purpose, in this study the economic dispatch problem is investigated in the presence of battery energy storage systems (BESSs). In the present study, the benefits associated with BESSs are the arbitrage benefits and pollution mitigation. To control the power plants emission level, special amount of BESS are introduced to the system which determine the discharging periods. Supplying the BESS with its required amount of energy is performed through an optimization process. This optimization process is done by considering the environmental restrictions. According to [10] it is assumed that environmental protection agency (EPA) can determine the permissible produced amount of emission in each hour. The EPA values are determined according to weather conditions (like wind speed or temperature), health issues, weather quality parameters and etc. Finally, these values are transmitted to the power dispatching center as environmental constraints.

It should be noted that in some cases energy storage systems are physically unable to serve some special applications. For example, the storage cannot tolerate numerous deep discharge and/or significant cycling [7]. Thus, the operation of the BESS can affect the problem and it is necessary to be modeled. In the present study, the ED is performed for a defined period of time and the emission levels and BESS restrictions are considered as operation constraints. Each of these constraints should be checked hourly. Regarding these constraints, the optimal operation of the battery determines the logic needed for charging and discharging periods of the battery. To solve the mentioned model, a hybrid approach based on the λ-iteration ED method and genetic algorithm (GA) is proposed as optimization tool. This optimization approach eliminates most of integrating difficulties of the BESS in the ED problem. In the following sections, the proposed model and its corresponding formulations are described in details. In this study, any pollutant can be considered as the environmental restriction, but here NO_X is the main goal of the emission control system due to its very harmful effects.

2 Battery Energy Storage Systems

Energy storage systems are not common elements of today's power grids and there are many challenges in introducing them to the power systems. However, due to many benefits and various applications which these technologies can provide, their promising future in electric power systems are obvious. An overview of different energy storage technologies are introduced in [4, 7, 11]. These technologies have various structural characteristics that make them appropriate or inappropriate for a special application. Technologies which are suitable for large-scale application in the power industry are introduced in [4] and decision criteria for selecting the proper energy storage types are also discussed.

A major factor in choosing a beneficial storage type in power system is the capacity. The capacity of the storage system is measured in both nominal power (in kW) and nominal energy (in kWh). The values of these parameters determine the high-power and high-energy specifications of the storage system which affects the storage application in the system. With high-energy characteristic, sodium-sulfur battery, super capacitor, and Vanadium Redox Battery (VRB) are recommended as the proper technologies for energy management applications [4]. The VRB is a flowing-electrolyte type battery that stores chemical energy in external electrolyte tanks sized in accordance with application requirements [7]. Electrolyte is pumped from storage tank to stack cell where chemical energy is converted to electrical energy (during discharge period) or electrical energy is converted to chemical energy (during the charge period). In the case of VRB, power and energy ratings are independent of each other [7]. This type of battery has some specifications that make it suitable for large-scale application in the power system. Some of these characteristics are high discharge power, and long discharge period. It has fast response in state-changing from charge to discharge mode (because of the very short reaction time). Due to separate electrolyte reservoirs, electrolyte replacement is convenient, electrolytes cannot react, and the system does not have any self-discharge. More important than all for emission control system it is environmentally friendly because of recyclable materials [4, 7, 11]. Therefore in this study, VRB is selected based on its magnificent characteristics.

3 Problem Formulation

The operating fuel cost is the main cost incurred in power generation. The primary objective of the ED problem is to determine the optimal combination of power generations that minimize the total generation cost while operating constraints are satisfied.

In the proposed model, regarding the presence of the BESS in the ED, the charge/discharge state of the battery has a key role. While the storage system is being charged, it is modeled as an increase in the load and during the discharge hours it is modeled as a decrease in the load.

3.1 Objective Function

The fuel cost function of each generating unit is represented as a quadratic function. Therefore for N generators and over a period of time T, the objective function can be formulated as follow:

$$\min \quad F_{tot} = \sum_{t=1}^{T}\sum_{i=1}^{N}(a_i + b_i P_{it} + c_i P_{it}^2) \quad (\$) \tag{1}$$

where, F_{tot} is total operating cost of the thermal power plants in \$, P_{it} is output power of the i-th generating unit at the t-th hour in MW, a_i, b_i, c_i are fuel cost coefficients of the i-th generating unit, N is number of online generating units, and T is the operational period in hour.

3.2 Constraints

In this problem, constraints are divided into three groups of operational constraints, environmental constraints, and battery constraints. These constraints are respectively defined in the following sections.

3.2.1 Operational Constraints

In minimizing the fuel cost of the ED, the total generating power of the units should be equal to the system load demand plus the transmission losses. The storage effect on the load of the system was described before. Regarding this effect, if P_{Dt} is the total demand of the system at the t-th hour, we will have:

$$\sum_{i=1}^{N} P_{it} = P_{Dt} + P_{Lt} \qquad \text{for } t = 1,\ldots,T \tag{2}$$

where, P_{Lt} shows the transmission network loss at the t-th hour. For a given system load demand, the transmission network loss is a function of power generation at each generating unit. The exact valve of P_{Lt} can be determined using a power flow solution. However in the most popular approach, the transmission network loss is approximated by function of the unit power outputs and the transmission loss matrix coefficients called B matrix. By using this popular method, in the present modeling P_{Lt} is formulated as follow:

$$P_{Lt} = \sum_{i=1}^{N} \sum_{j=1}^{N} P_{it} \cdot B_{ij} \cdot P_{jt} + \sum_{i=1}^{N} B_{0i} \cdot P_{it} + B_{00} \qquad (3)$$

In this equation, B_{ij} , B_{00} , and B_{0i} are the coefficients of the transmission power loss.

Also, the power generated by each generating unit i at each hour t should be within its lower limit $P_i^{\min}$ and upper limit $P_i^{\max}$. So, the following constraints for each generating unit should be satisfied. These constraints can be expressed mathematically as:

$$P_i^{\min} \leq P_{it} \leq P_i^{\max} \qquad \text{for } i = 1,\ldots,N \quad \text{and} \quad t = 1,\ldots,T \qquad (4)$$

3.2.2 Environmental Constraints

The amount of the produced emission at the t-th hour is correspondent to the output power of the generating units at that hour and can be described as follow:

$$E_t = \sum_{i=1}^{N} d_i + e_i P_{it} + f_i P_{it}^2 \quad (\text{kg/hr}) \qquad (5)$$

where, E_t is the total produced amount of NO_X at the t-th hour in kg/hr, and d_i, e_i, f_i are the emission coefficients of the i-th generating unit.

The total amount of NO_X produced in each hour should be equal or less than its permissible value announced by the environmental protection agency. Thus,

$$E_t \leq EPA_t \qquad \text{for } t = 1,\ldots,T \qquad (6)$$

In this constraint, EPA_t is the permissible amount of the produced NO_X at the t-th hour.

3.2.3 Battery Constraints

The BESS or generally any storage system can be defined with three important parameters named the nominal energy capacity, the nominal power capacity, and the round-trip efficiency of the battery. These parameters are respectively shown by E_s , P_s , and η in this section. The round-trip efficiency of the BESS is the product of charging and discharging efficiencies as follow:

$$\eta = \eta_c \times \eta_d \qquad (7)$$

where, η_c and η_d are respectively the charging and the discharging efficiencies. The introduced parameters of the BESS and their values set different constraints on the battery operation.

Regarding the energy capacity, this value limits the energy that can be stored in the battery. The stored amount of energy in the battery depends on the previous charging and discharging powers of the battery and also its initial stored energy. The charging and discharging powers are the amounts by which the storage system is being charged and discharged. These amounts are respectively assumed to be positive and negative values. At each operational hour, the stored energy is constrained to a value equal or lower than the energy capacity of the battery. This constraint can be formulated as:

$$SE_t \leq E_s \qquad \text{for } t = 1,\dots,T \tag{8}$$

where, SE_t is the stored energy at the t-th hour and can be defined as:

$$SE_t = SCE_t - SDE_t \tag{9}$$

In this equation, SCE_t is the sum of charged energy stored in the battery and SDE_t is the sum of discharged energy extracted out of the battery at the t-th hour. These amounts are calculated from the beginning of the operational period and can be obtained based on the followings:

$$SCE_t = \sum_{j=1}^{t} PCB_j \times \eta_c + INEB \tag{10}$$

$$SDE_t = -\sum_{j=1}^{t} PDB_j \times \eta_d \tag{11}$$

In each of these equations, the charging and discharging efficiencies are applied. PCB_j and PDB_j are respectively the charging and discharging powers of the battery at the j-th hour. These values are usually limited by the storage converter system and range between zero and the nominal power of the battery P_s. $INEB$ is also the initial stored energy in the battery at the beginning of the operational period which can vary between zero and the nominal battery energy capacity.

In the first time operation of a battery, it is usually charged for a time larger than the nominal charging period and $INEB$ is equal to E_s. However, after the first operation, the $INEB$ amount can have any other positive value equal or less than E_s depending on the operational strategy of the battery. For example, if the starting point of the operational cycle of the battery is midnight, the battery is not usually fully charged because of the off-peak hours ahead.

For a better maintenance of the battery during its entire life-time, special operational constraints should be considered. Keeping the battery in low energy levels for a long period of time will reduce its life time. Therefore, the maximum discharged energy from the battery should be less than its depth of discharge (DoD) defined as the maximum permissible amount of extracting energy from the battery. Thus, the stored energy of the battery should satisfy the DoD limit as follow:

$$(1-DoD)\times E_s \leq SE_t \qquad \text{for } t=1,...,T \tag{12}$$

It should be noted that the amount of energy remained in the battery at the end of the operational period is determined by the battery operational strategy other than its life-time concerns. For instance, It is obvious that fully charging the battery at the end of the operational period increases the current operational cost while reduces the next one. Storing energy for the next operational period greatly depends on the self-discharge of the battery and the round trip efficiency. Other than daily load-leveling and pollution mitigation, VRB, with negligible self-discharge or other cheaper batteries like metal-air [11], can help the system to be prepared for the seasonal or annual peak of the load. The energy remained in the battery at the end of the operational period T or REB_T is determined using the following equation:

$$REB_T = \sum_{t=1}^{T} SE_t \tag{13}$$

It is important to mention that relating different operational periods are done using the $INEB$ and REB_T. The initial energy of the battery for the next operation will be equal to the remained energy of the battery in the present operation.

4 A Hybrid Optimization Approach Based on the λ-Iteration Method and Genetic Algorithm

The presence of the storage system in the ED problem relates the hours of the operation. In this study, the ED is performed for a whole day and a combination of the λ-iteration method [15] and GA is used to solve the defined problem. The proposed optimization approach has two stages as follows.

4.1 First Stage

In the first stage, by using the real load profile, the generating power of each unit and the related emission production are calculated on an hourly basis. By comparing the produced emission at each hour with its related permissible level, the hours in which the emission constraint is violated, can be determined. At these hours, the BESS can play the role of an emissionless generating unit. To satisfy the environmental constraint at the mentioned hours, the system load is decreased step by step in an iterative process. The step value depends on the BESS specification and in this study was selected to be 0.1 MW. Then, the ED is performed for the reduced load to determine the output power of each generating unit. Next, the total produced amount of NO_x is calculated and compared to its relevant permissible level at that hour. If the environmental constraint is not satisfied, the load reduction process continues. This iterative process continues until all environmental constraints are satisfied in the entire operational period. Consequently, the minimum discharge powers and discharge hours of the battery are determined according to the amount of decreased load at the related hours.

4.2 Second Stage

In the second stage, by using the GA, the optimal solution of the problem is obtained in an evolutionary process.

Basically, the GA is a metaheuristic search mechanism based on the Darwinian principle of natural evolution. A GA starts with a number of solutions to a problem, encoded as strings. The string that encodes each solution is a 'chromosome' and the set of solutions is called the 'population'. Each chromosome consists of 'genes'. The initial population can be generated randomly, or may consist of a number of known solutions, or a combination of both. Each chromosome represents a feasible solution. The objective function is then evaluated for these chromosomes. If the best chromosome satisfies the optimization criteria, the process terminates, assuming that this best chromosome is the solution of the problem. If the optimization criteria are not met, the creation of new generation starts. Pairs, or chromosomes are selected randomly and subjected to crossover and mutation operations. The resulting chromosomes are selected according to their fitness for the production of the new chromosomes.

In the present study, the initial population comprises a limited number of chromosomes. Each chromosome has T genes as shown in Fig. 1.

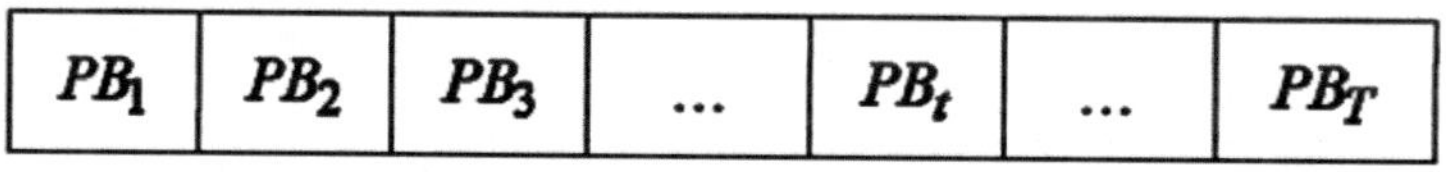

Fig. 1 Chromosome structure

The t-th gene of the chromosome represents the power of the BESS at the t-th hour and must satisfy the following constraint.

$$-P_s \leq PB_t \leq UB_t \tag{14}$$

In this inequality constraint PB_t represents the power of the BESS at the t-th hour and P_s shows the nominal power of the battery. Negative and positive values in each gene show the discharging and charging powers of the battery, respectively. Zero genes determine the idle states of the BESS. UB_t represents the upper limit of the battery power at the t-th hour. If load at the t-th hour is decreased due to the emission constraint in the first stage, the battery should be at discharge mode and UB_t is set to the amount of load reduced at that hour with a negative sign. Otherwise, UB_t is set to P_s .

The genes are randomly chosen by a random number generator according to the acceptable range based on constraint (14). Then, the mentioned battery constraints should be checked. Those chromosomes that satisfy the constraints are kept. This stage is continued till the desired number of initial population is obtained.

For each of these chromosomes and by having the load profile, a modified load profile can be constructed. In these load profiles, the power considered for charging the battery (positive value) is served as an increase in the load and the power considered to be discharged from the battery (negative value) is served as a decrease in the load, and zero amounts have no effect on the load profile. Then the modified load profile is used in the λ -iteration based ED process and the total operating cost is calculated. Violation of the environmental constraint in each hour is considered as an extra cost in the objective function; but if the chromosome does not satisfy the operational constraint of the battery, it will be omitted. After the off-spring production by mathematical crossover and mutation operators, those chromosomes that don't satisfy the mentioned constraints are omitted from the population and the rest are kept to calculate their objective function values. In the reproduction stage, a dynamic method is used to determine the probability of the crossover and mutation operators to improve the performance and convergence of the GA [16]. The roulette wheel and elitist selection methods are used to select the new population. This process is continued until finally, the best solution with the minimum objective function value is obtained.

5 Numerical Analysis and Results

In this study, the 69-bus test system in [10] is used to show the performance and applicability of the proposed method. This system has 11 fossil fueled generating units with total capacity of 3570 MW. Fuel cost and emission functions of the power plants are of the quadratic form. Coefficients related to fuel cost functions and generating limits of each power plant, and also their emission coefficients are given in Table 1 and Table 2, respectively.

The optimization process is performed for the load data given in Table 3. The BESS is consisted of five strings of VRB batteries with the following specifications:

$$P_s = 10MW,\ E_s = 40MWh,\ \eta = 0.85.$$

Table 1 Fuel cost coefficients and generating limits of the generating units

Unit No.	Fuel Cost Coefficients			Generating Limits	
	a_i	b_i	c_i	P_{min}	P_{max}
1	0.00762	1.92699	387.85	20	250
2	0.00838	2.11969	441.62	20	210
3	0.00523	2.19196	422.57	20	250
4	0.00140	2.01983	552.50	60	300
5	0.00154	2.21181	557.75	20	210
6	0.00177	1.91528	562.18	60	300
7	0.00195	2.10681	568.39	20	215
8	0.00106	1.99138	682.93	100	455
9	0.00117	1.99802	741.22	100	455
10	0.00089	2.12352	617.83	110	460
11	0.00098	2.10487	674.61	110	465

Table 2 Emission Coefficients of the generating units

Unit No.	Emission Coefficients		
	d_i	e_i	f_i
1	0.00419	-0.67767	387.85
2	0.00461	-0.69044	441.62
3	0.00419	-0.67767	422.57
4	0.00683	-0.54551	552.50
5	0.00751	-0.40006	557.75
6	0.00683	-0.54551	562.18
7	0.00751	-0.40006	568.39
8	0.00355	-0.51116	682.93
9	0.00417	-0.56228	741.22
10	0.00355	-0.41116	617.83
11	0.00417	-0.56228	674.61

Table 3 The 24-hour load data of the system

Time	Demand (in MW)	Time	Demand (in MW)	Time	Demand (in MW)
1	1590	9	2150	17	2150
2	1450	10	2270	18	2280
3	1345	11	2320	19	2345
4	1200	12	2320	20	2270
5	1200	13	2150	21	2150
6	1345	14	2225	22	1925
7	1450	15	2250	23	1740
8	1750	16	2170	24	1615

The following cases are investigated in this numerical analysis and the results are shown in each case:

Case 1: Economic Dispatch without BESS
Case 2: Economic Dispatch in the presence of BESS with DoD = 100%
Case 3: Economic Dispatch in the presence of BESS with DoD = 75%

In case 1, ED is performed without the presence of BESS. In this case, the power plants' output powers are determined and the relevant operational cost and produced amount of NO_X are calculated and given in Table 4. Although minimum fuel cost is obtained in the absence of the BESS but as it is depicted in Fig. 2, the produced amounts of NO_X exceed the permissible amounts of EPA in 9 hours.

In case 2, the mentioned approach is used in the presence of the BESS and then the ED is performed for the system under study. In this case DoD is equal to 100 percent. The proposed method provides the opportunity of controlling the produced amount of emission while minimizing the fuel cost as well. In this method, the scheduling start-point of the BESS can be any hour of the day. By using the time shifting technique in [8] and considering the *INEB* , the beginning of the 4-th hour is selected as the start point of scheduling the BESS of the day. In the presence of BESS, Fig. 3 shows the produced amount of NO_X during the operational period.

By comparing Fig. 2 and Fig. 3, it is obvious that using the proposed method limits the produced amount of NO_X below the permissible amount of EPA in each hour.

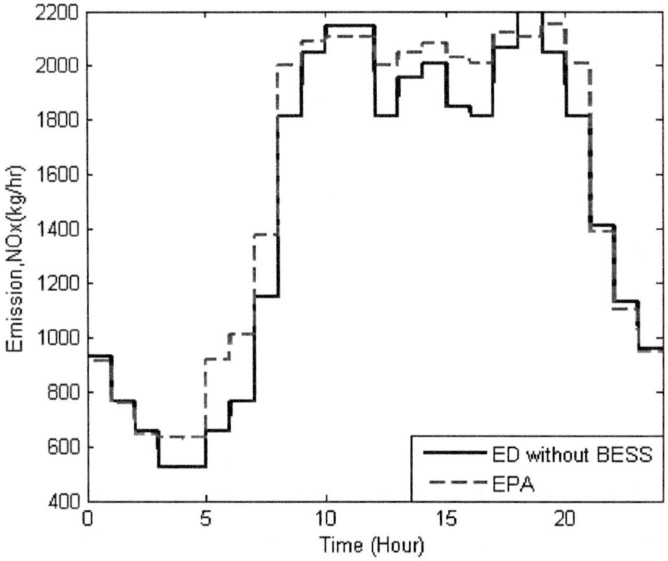

Fig. 2 Produced amount of NO_X by pure ED

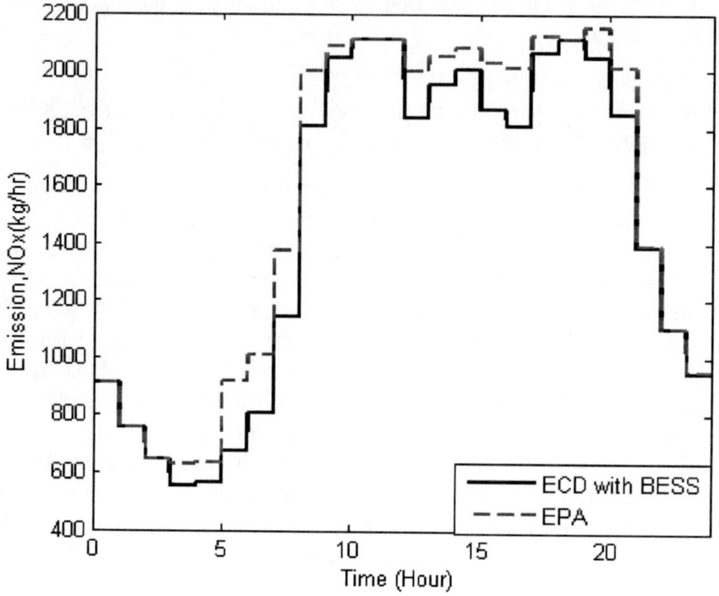

Fig. 3 Produced amount of NO_X by ED and in the presence of the BESS

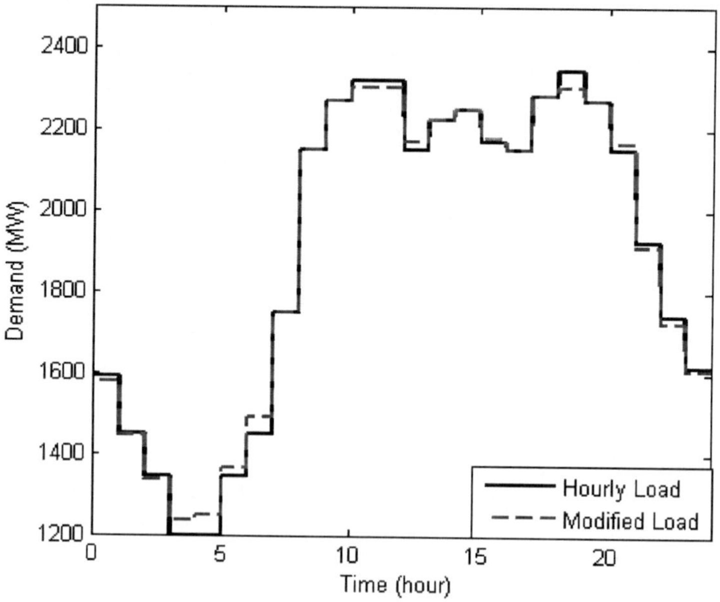

Fig. 4 Hourly load and modified load of the system after the presence of the BESS

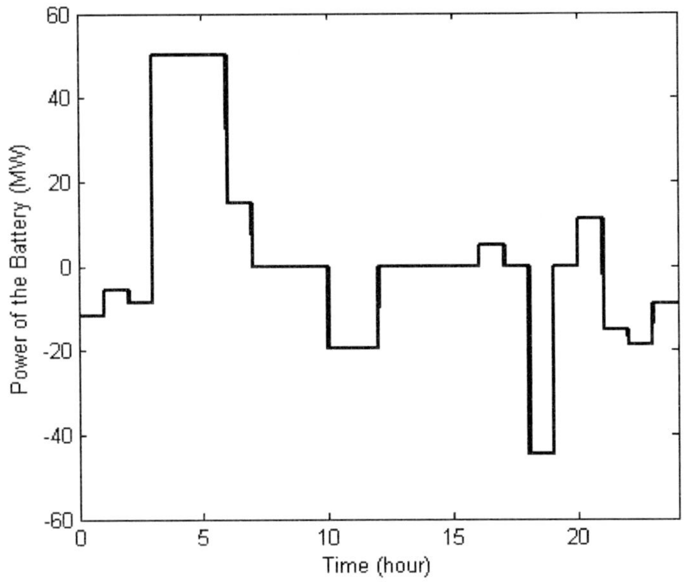

Fig. 5 Charge, discharge and idle hours of the battery in optimal solution obtained for 100% of depth of discharge

In the presence of BESS the daily load profile supplied by generating units is modified according to Fig. 4. The real load profile is shown for comparison in this figure.

By considering Fig. 2 and Fig. 5 and comparing them to Fig. 3, it is realized that in charging hours of the battery, the required power is supplied from the fossil fueled power plants and the relevant amount of the produced NO_X is increased. In fact in this method, by optimal charging and discharging of the battery, the amount of the load that causes violation of the EPA constraint is transferred to other hours of the day and finally the EPA constraint is satisfied in the whole hours.

As it is mentioned before, satisfying the DoD constraint increases the battery life time. So, in case 3 the proposed method is performed with DoD equals to 75 percent. There are two important points about DoD which are:

1. Batteries with DoDs equal to 100 percent, like nickel-cadmium or sodium-sulfur batteries [12], do not need to keep extra stored energy and these batteries can be discharged completely. Therefore, they don't have extra operational or investment charges on the system.
2. $(1-DoD)\times E_s$ is the required amount of energy which needs to be stored in the battery. This amount satisfies the DoD constraint and it will be unavailable during the battery operation and imposes extra charges to the system. The more the DoD of the battery is, the less this amount is.

If the DoD is equal to 100 percent, the *INEB* is considered to be zero and REB_T can be zero too. However if the DoD is equal to 75 percent, the minimum required energy is considered for the *INEB* and REB_T to satisfy the DoD constraint in the beginning and at the end of the operational period.

Optimal solutions obtained from the proposed method in cases 2 and 3, are shown in Fig. 5 and Fig. 6. The produced amount of NO_X and their relevant costs in the mentioned cases are given in Table 4. As it can be seen in Table 4, considering the depth of discharge slightly affects the total operational cost which can be neglected compare to the total operational cost of the day. In other words, by using the proposed method, without considerable changes in the total fuel cost, the produced amounts of NO_X are kept within the desirable limits.

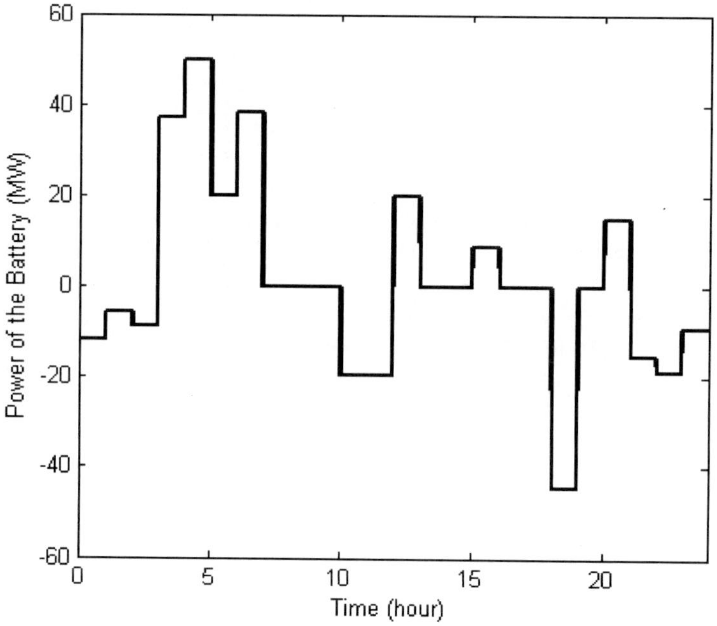

Fig. 6 Charge, discharge and idle hours of the battery in optimal solution obtained for 75% of depth of discharge

Table 4 Comparison of the total cost and total emission of different cases

Case	Total Produced Amount of NO_x for the operational Period (in kg)	Total Operational Cost (in $)
Case 1- Pure ED	35173	256275
Case 2- ED with BESS, DoD = 100	35091	256314
Case 3- ED with BESS, DoD = 75	35129	256343

6 Conclusion

Nowadays, concerns about the produced amount of environmental pollutants have affected the operation of the power system. Moreover, technology improvements in electrical energy storage systems and tendency toward using these storages in power system raised the necessity of reviewing the operational studies of the power system. In this study, to investigate the effect of battery energy storage system on power system economic dispatch and its optimal operation, a heuristic optimization based approach is proposed. This approach was applied in a numerical example and its performance in emission controlling of the power system was shown. In case of further studies, the result can be used for optimal sizing of the battery energy storage systems.

References

1. Venkatesh, P., Lee, K.Y.: Multi-objective evolutionary programming for economic emission dispatch problem. In: Proc. of the IEEE Power and Energy Society General Meeting (2008)
2. Sirinivas, K., Patevard, C., Dos, D.B.: A new elitist multi-objective stochastic search technique and its application to economic-emission dispatch problem in power systems. In: Proc. of the IEEE Congress on Evolutionary Computation, pp. 2988–2995 (2007)
3. Das, D.B., Patvaradhan, C.: New multi-objective stochastic search technique for economic load dispatch. In: IEE Proc.-Gener. Transm. Distrib., pp. 1350–2360 (1998)
4. Nourai, A.: Large-scale electricity storage technologies for energy management. In: IEEE Power Engineering Society Summer Meeting, pp. 310–315 (2002)
5. Joseph, A., Shahidehpour, M.: Battery storages systems in electric power systems. In: Proc. of the IEEE Power Engineering Society General Meeting (2006)
6. Nourai, A., Kogan, V.I., Schafer, C.M.: Load leveling reduces T&D line losses. IEEE Trans. Power Delivery 23, 2168–2173 (2008)
7. Mears, L., Gotschall, H.: Handbook of Energy Storage for Transmission and Distribution Applications. EPRI, DOE, Palo Alto, CA, Washington (2003)
8. Lo, C.H., Anderson, M.D.: Economic dispatch and optimal sizing of battery energy storage systems in utility load-leveling. IEEE Trans. Energy Convers 14, 824–829 (1999)
9. Lu, B., Shahidehpour, M.: Short-term scheduling of battery in a grid-connected PV/battery system. IEEE Trans. Power Syst. 20, 1053–1061 (2005)
10. Palanichamy, C., Babu, N.S.: Day-night weather-based economic power dispatch. IEEE Trans. Power Syst. 17, 469–475 (2002)
11. Ter-Gazarian, A.: Energy storage for power systems. Institution of Engineering and Technology, Peter Peregrinus Ltd., London (2007)
12. Divya, K.C., Østergaard, J.: Battery energy storage technology for power systems-an overview. Elec. Power Syst. Res. 79, 511–520 (2009)
13. Huang, K.L., Li, X., Liu, S., Tan, N., Chen, L.: Research progress of vanadium redox flow battery for energy storage in China. Renewable Energy 33, 186–192 (2008)
14. Kaiser, R.: Optimized battery-management system to improve storage life-time in renewable energy systems. J. Power Sources 168, 58–65 (2007)
15. Saadat, H.: Power system analysis. McGraw-Hill, WCB (2004)
16. Zhang, J., Huang, T., Zhang, H.: The reactive power optimization of distribution network based on an improved genetic algorithm. In: Proc. of the IEEE Transmission and Distribution Conference and Exhibition (2005)

Modeling and Control of a Hybrid Smart Micro Grid Using Photo-Voltaic Arrays and Proton Exchange Membrane Fuel Cells

Sukumar Kamalasadan*, James T. Haney, and Chad M. Tanton

Abstract. Conventional power systems are in need for renovation. The main source of energy for the present power system is fossil fuel based, which are depleting in supply day-by-day and at the same time releasing dangerous emissions into the atmosphere. The renewable energy resources such as photovoltaic arrays have attracted great attention in recent years. However, the intermittent nature of this renewable energy and others similar in nature impedes their bulk use and penetration level to the power grid, especially when implemented as a grid connected system. It can be beneficial if a hybrid smart micro grid such as a Photo Voltaic (PV) system and fuel cell can be designed and developed.

This chapter discusses the modeling and control of a PV array and PEM fuel cell based hybrid smart micro grid system and the interconnection to power grid as a smart grid option. The main goal is to develop a nonlinear model for PV and fuel cell, integrate that with power conditioning devices such as DC/DC converters and three-phase Voltage Source Inverter (VSI) and analyze the performance of this hybrid micro grid connected to power grid acting as an infinite bus based on novel controllers. Overall this chapter discusses the development of a simulation environment that has unique nonlinear models with controllers and shows the ability of the proposed architecture to work as a hybrid micro-grid test bed that is useful for testing real-life implementation.

1 Introduction

Today's most commonly used fossil fuels are becoming more expensive and the resources are quickly running dry. Also, the power plants that burn these fossil fuels are responsible for over half of the green house gas emissions in the U.S. [1].

Sukumar Kamalasadan · James T. Haney · Chad M. Tanton
Department of Electrical and Computer Engineering
The University of North Carolina at Charlotte,
Charlotte, NC-28223
e-mail: skamalas@uncc.edu

* Corresponding author.

L. Wang (Ed.): *Modeling and Control*, Green Energy and Technology, pp. 209–247.
springerlink.com

Besides economic and environmental incentives to change the way electricity is produced, technological advances is also leading to change. A vast growing technology to solve some of the above problems is Distributed Generation (DG). DG differs from centralized generation/distribution by "distributing" generation sources along the existing power transmission system. The number of DG systems across the U.S. has not yet reached significant levels, but a study conducted by *Electric Power Research Institute* indicated that by the end of 2010, 20% of new generation would be distributed [2]. DG power systems not only provide environmental benefits, but also offer a more efficient way of generating and distributing electricity, as well as enabling the integration of renewable energy sources [3]. Unfortunately, DG's may cause more problems than it may solve due to design and interconnection issues [4]. One method to control DG is to treat it as a micro grid (MG). A micro grid allows the grouping of certain micro-sources, such as, solar arrays, fuel cells, wind turbines, and micro-turbines. MG's allow different generation sources to be added or removed without any modification of the existing system and for local control and operation of DG, thus, reducing or eliminating the need for central dispatch [5]. MG's can be directly connected to the existing electric power grid, or disconnected and used in island mode. During disturbances in the MG or loads connected to it, MG can be disconnected without harming the integrity of the existing grid. It can also switch intentionally to islanding when fault occurs in the power grid, keeping local reliability higher than what grid provides [5].

This chapter provides a PV and Fuel Cell (FC) based smart micro grid model that can be integrated to the power grid. A PV system consists of a PV generator (with PV modules and arrays), battery bank, power control and conditioning system. Power produced varies with temperature, cloud coverage, season change, and day-night cycle. Thus, a stand-alone PV system is intermittent in nature and seen as a negative load because of its uncontrollable power characteristics [6]. To compensate for PV system down times, a FC can be connected together with the PV system. FC is an efficient, cost-effective, renewable energy source that is considered one of the most promising sources of electric power [7] [8]. FCs are not only more efficient than conventional power plants, but they are also more environmentally clean, have extremely low emission of nitrogen oxide and sulfur, and are quiet [9]. The FC produces power to the loads during PV system downtime, or when the PV system produces sufficient power for the loads, it can charge the battery bank or produce power to be sent back to the electrical grid. FCs consists of a reformer, fuel cell stack, and power-conditioning unit. Reformer generally produces hydrogen to supply the FC by processing a hydrocarbon fuel such as propane, methane, or methanol.

In this chapter, first, a nonlinear model of PEM fuel cell, including the reformer and fuel cell stack is discussed and developed. Then nonlinear modeling of the solar PV module and array is designed and discussed. Further, a Maximum Power Point Tracking (MPPT) controller for PV array and a load controller for PEM fuel cell system will be designed and modeled. The nonlinear detailed model of these two renewable energy based micro-systems will then be integrated using a three phase Voltage Source Inverter (VSI) model and power conditioning DC-DC converter models for each system. Finally this hybrid system will be integrated with the power grid and the performance of the hybrid micro grid as a part of the smart-grid will be analyzed with the power system treated as an infinite bus. Three cases for interconnection of the micro-grid a) Island mode b) Connected to the power grid with P and Q negative c) Connected to the power grid with P and Q positive will be evaluated and the load following nature of the micro grid with effective control will be discussed. Then the operation of unique controllers for load following with priority to PV system and optimal power tracking of the micro grid will be analyzed. The chapter concludes by discussing the impact of the micro grid on the power system as a load following customer driven micro-system and the improvement of micro-system parameters using smart controllers.

2 Modeling and Control of PEM Fuel Cell Micro Grid

A fuel cell generator in the form of a micro grid can be represented as shown in fig. 1 block diagram. Main components consist of fuel cell stack, power conditioning devices such as DC/ DC Converters and inverter and controllers for fuel cell and VSI Inverter. Each of these blocks is discussed next.

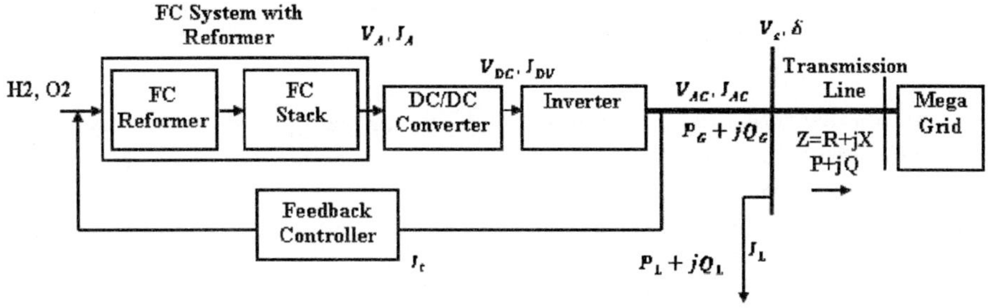

Fig. 1 Block Diagram of a Fuel Cell Generator

2.1 *Fuel Cell Generator*

A FC system is comprised of two main components, the reformer and fuel cell stack. The reformer is the component that takes hydrocarbons, in most cases, such as methanol or propane, separates the hydrogen molecules from the gases and injects that into the anode side of the fuel cell stack. In the anode, hydrogen is forced through a catalyst, usually made of platinum, which separates the hydrogen molecules into electrons and protons. The protons are allowed to permeate through a proton conducting polymer membrane called a Proton Exchange Membrane (PEM), which separates the anode and cathode. The electrons are forced around the electrically insulated PEM through an external circuit which in turn supplies electrical power to it. An elementary block diagram is given in fig. 2. Several models have been proposed for the fuel cell [1]-[12]. However, most of these models shows the basic equations for the fuel cells with some modeling assumptions or do not consider the power conditioning devices to use this system as a micro grid. The proposed model of the fuel cell is an electromechanical one with no major modeling assumptions and reflecting the nonlinear relation between chemical reactions in the fuel cell. Also the model presented here is based on simulating the relationship between output voltage and partial pressures of hydrogen, oxygen, and water. It gives insight to the transient and linear response of a PEM fuel cell with a generating capacity of one thousand watts.

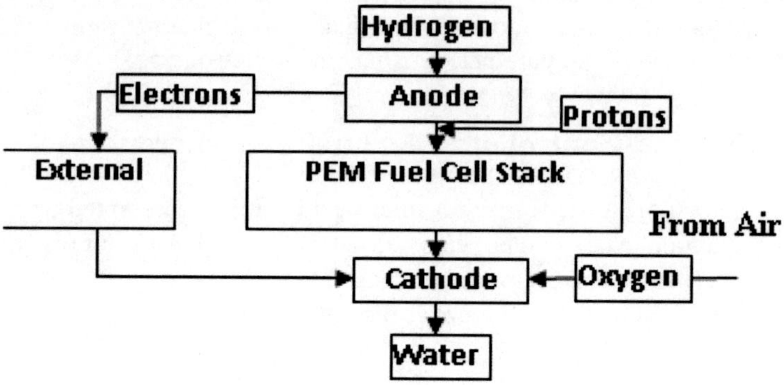

Fig. 2 Elementary Flow Block Diagram of a PEM Fuel Cell

The reformer provides relatively pure hydrogen to the fuel cell, using a hydrocarbon that is readily available such as methanol, propane, or natural gas to name a few. For houses and stationary power generation, fuels like natural gas or propane are preferred, simply because many power generator stations and houses are already hooked up to natural gas supplies by pipeline or have propane tanks on the property. However, this chapter discusses methanol reforming because methane has a higher energy density and is easier to transport. The process starts with the vaporization of liquid methanol and water. Heat produced in the reforming process is used to accomplish this. This mixture is passed through a heated chamber that contains a catalyst. As the methanol molecules hit the catalyst, they split into carbon monoxide (CO) and hydrogen gas (H_2). The water vapor splits into hydrogen gas and oxygen; this oxygen combines with the CO to form CO_2. It is important to eliminate as much of the CO from the exhaust gas, and this way, very little CO is released, as most of it is converted to CO_2.

The fuel cell system continuously consumes hydrogen and the reformer is continuously generating hydrogen according to power demand. To model a reformer a second-order transfer function as in (1) is developed. Controlling the hydrogen flow rate, according to power demand, is essential to the operation of a fuel cell system. Thus a PI controller is introduced to the system to control the hydrogen, oxygen, and water vapor flow rates. This feedback control is designed by taking the fuel cell output current and feeding it back to the input while converting the hydrogen into molar form based on (2)-(12):

$$\frac{q_{H2}}{q_{methanol}} = \frac{CV}{\tau_1\tau_2 s^2+(\tau_1+\tau_2)s+1} \tag{1}$$

The relationship between the hydrogen flow and the feedback current can be represented as

$$q_{H2} = \frac{N_o I}{2FU} \tag{2}$$

The PI controller is used to calculate the amount of hydrogen needed from the reformer which is then used to determine the methane flow rate into the reformer. This can be expressed as:

$$q_{methanol} = \left(k_3 + \frac{k_3}{\tau_3 s}\right)\left(\frac{\mathrm{N_o I}}{\mathrm{2FU}} - \mathrm{q_{H2}} + \mathrm{q_{methref}}\right) \tag{3}$$

Mathematical representation of the oxygen flow rate can be seen below:

$$q_{O2} = \frac{1}{r_{H-O}} \tag{4}$$

This model is dependent on the partial pressures of hydrogen, oxygen, and water vapor that is used to develop the fuel cell stack. Some additional parameters are used to get the model highly accurate and nonlinear. However following design assumptions are made for the development of the model:

1. The operating fuel cell temperature will remain under 373 K.
2. The reaction product entering the reformer is in the liquid phase.
3. The fuel cell stack output voltage can be obtained by lumping together the individual cell parameters to represent a fuel cell stack
4. In the fuel cell gas flow channels the pressures are kept constant.
5. Both the hydrogen and oxidants are humidified to prevent damage to the proton exchange membrane.

According to assumptions 3 and 4, the fuel cell output voltage can be obtained by the sum of the Nernst voltage, the activation over voltage, the ohmic over voltage and the concentration over voltage. The mathematical representation of this detailed model will then be (5)-(9):

$$V_{cell} = E - V_{act} - V_{ohm} - V_{conc} \tag{5}$$

where,

$$V_{conc} = \frac{RT}{zF} ln\left(1 - \frac{I}{I_{limit}}\right) \tag{6}$$

$$V_{ohm} = I(R_{ohmo} + k_{RI} I - k_{RT} T) \tag{7}$$

$$V_{act} = -B ln(CI) \quad (8)$$

$$E = E_o + \frac{RT}{2F} ln[p_{H2}(p_{O2})^{0.5}] - E_d \quad (9)$$

To calculate E properly, E_d is used which is called the reversible voltage delay. This delay is considered so that thee time delay of the fuel and oxidant can be included. Under steady state conditions E_d is equal to zero due to the fact that the fuel and oxidants are assumed to continue in steady state. Also, notable in the equation above are the partial pressures of hydrogen and oxidants. These are used to help account for the delay of the hydrogen and oxidants. The three mathematical representation of this delay can be expressed as:

$$E_d = \lambda_e I \left(\frac{\tau_e s}{\tau_e s+1}\right) \quad (10)$$

$$p_{H2} = \frac{I}{k_{H2}}\left(\frac{\tau_{H2} s}{\tau_{H2} s+1}\right) \quad (11)$$

$$p_{O2} = \frac{I}{k_{O2}}\left(\frac{\tau_{O2} s}{\tau_{O2} s+1}\right) \quad (12)$$

2.2 *Simulation Results and Model Comparison*

The proposed equations are all implemented using MATLAB® and Simulink®. There is only two input and one outputs to the fuel cell generator and they are feedback current, hydrogen flow rate and DC output voltage, respectively. The feedback current will vary depending on the load demand and the hydrogen flow rate will vary depending on the feedback current. The voltage is a function of the feedback current, partial pressures of hydrogen and oxygen, and the over voltage potential inside the fuel cell. For the fuel cell generator, the module was created as a block. The equations described above are utilized to produce the output DC voltage. The model is set up so the number of cells and the number of stacks can be changed easily to accurately model the desired size of a power plant. A detailed model of a fuel cell reformer and generator inside the FC generator block can be seen in fig. 3 and fig. 4.

The fuel cell generator model shown in fig. 4 is then tested with step changes in the feedback current as shown in fig. 5. For this test case, the generator is

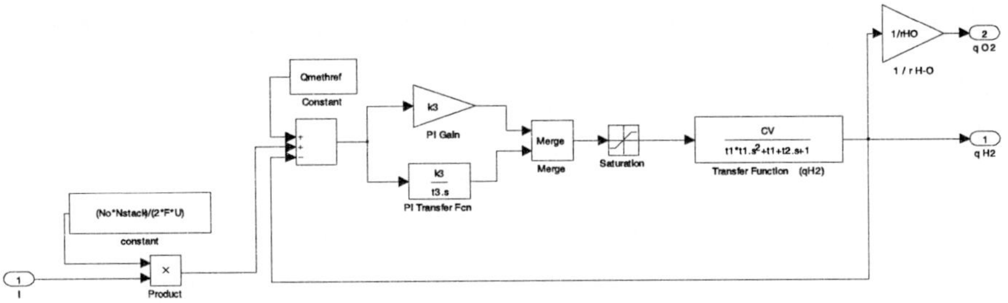

Fig. 3 Detailed Model of a Fuel Cell Reformer

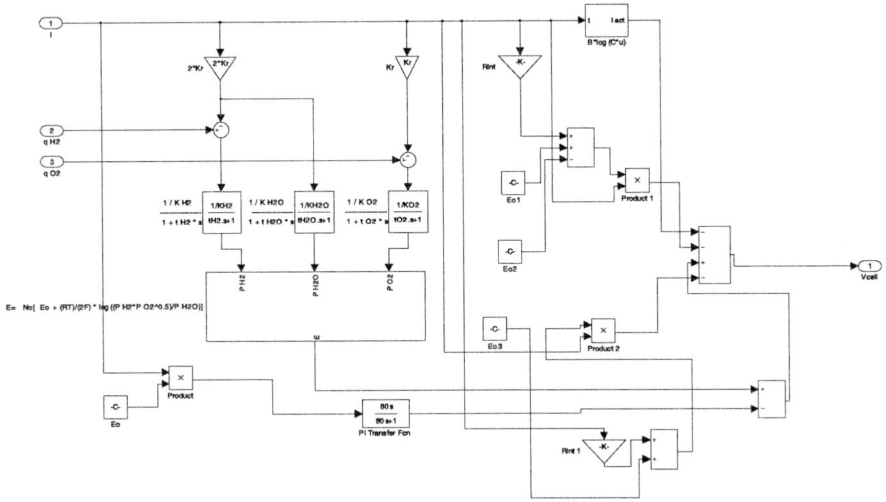

Fig. 4 Detailed Model of a Fuel Cell Stack

assumed to be in the active mode. These abrupt changes in the feedback current are for testing the dynamic response of the system, and do not necessarily represent changes in a residential load. The load model is chosen to reflect all possible variations of feedback current. The change of current, voltage, and flow rate of hydrogen are illustrated in fig. 5. The results of this simulation show that as the feedback current increases FC voltage decreases. The increase in current increases methane flow rate and hydrogen flow rate as well. All of which has a small time delay before reaching steady state. This is because the reformer and the FC power generator has time delay constants. The model output is then validated with an actual 1KW fuel cell system experimental test bed. It was concluded that, this model is able to accurately simulate the dynamic response of a PEM fuel cell generator.

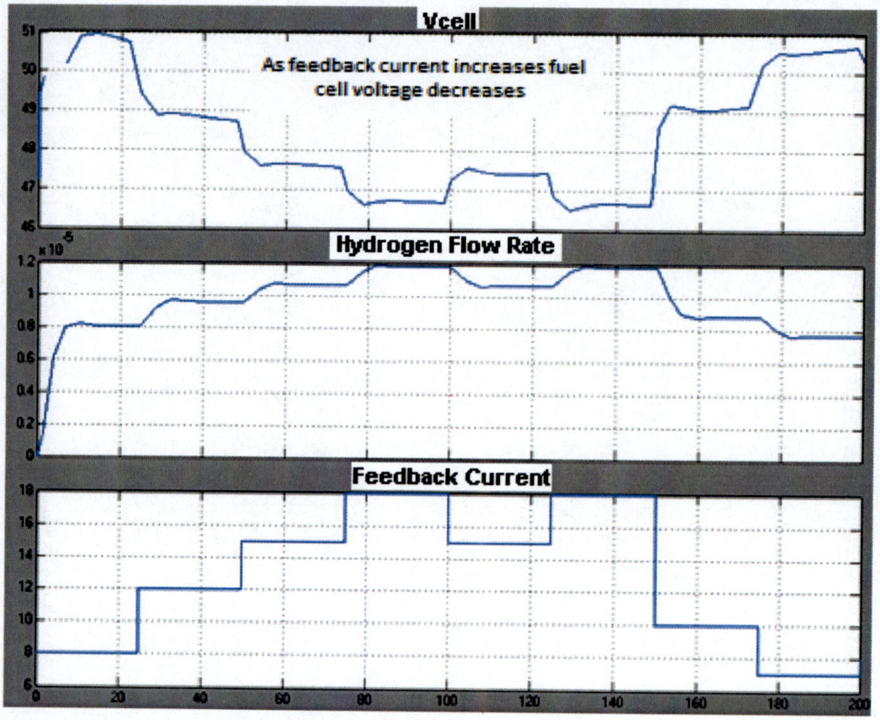

Fig. 5 Results from Fuel Cell Generator Simulation

2.3 DC/DC Convertor

Most models for fuel cells usually neglect the effect of DC/DC converters [13]. In this chapter a state space model and transfer functions for the combination of PEM FC and related DC/DC converter is discussed. The DC/DC converter was designed using the small signal state space model for the boost DC/DC converter evolved from [14]. To this end a multivariable model is introduced with one with two inputs and outputs, as in fig. 6. This convertor contains one control signal

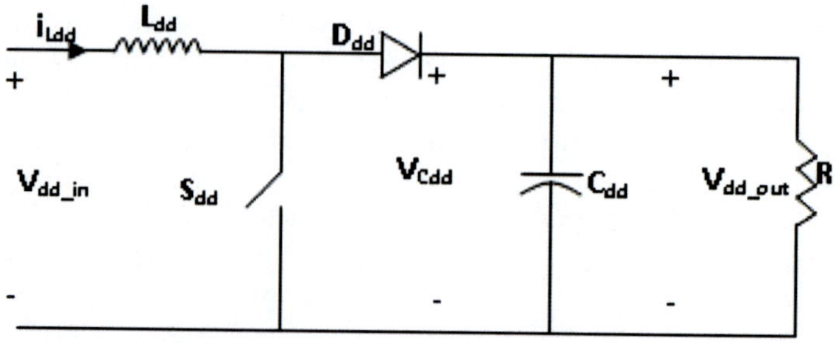

Fig. 6 Boost DC/DC Converter

which is used to shape the duty cycle. The traditional PWM to calculate the duty cycle is not needed in this model because of the enabled feedback loop.

The small signal state space model of the Boost DC/DC Converter can be obtained by using the following equations.

$$\begin{bmatrix} \frac{di_t}{dt} \\ \frac{dv_t}{dt} \end{bmatrix} = \begin{bmatrix} 0 & \frac{-(1-D)}{L} \\ \frac{1-D}{C} & \frac{-1}{RC} \end{bmatrix} * \begin{bmatrix} i_L \\ v_C \end{bmatrix} + \begin{bmatrix} \frac{v_C}{L} \\ \frac{-i_L}{C} \end{bmatrix} * U \tag{13}$$

When solved for current and voltage in the frequency domain, results are as follows:

$$i_L = \left(\frac{-v_C(1-D)}{L} + \frac{v_C}{L}U\right)\frac{1}{s} \tag{14}$$

$$v_C = \left(\frac{i_L(1-D)}{C} - \frac{v_C}{RC} - \frac{i_L}{C}U\right)\frac{1}{s} \tag{15}$$

Additionally switching losses and input port series resistance were incorporated in this model. This means the output current and voltage is calculated based on (15) in order to show the switching current losses and the input port series resistance voltage losses for the converter.

$$v_O = v_C - v_{RL} \tag{16}$$

$$i_O = \frac{V_{cell}(1-D)}{R_L} - \frac{v_O(1-D)^2}{R_L} - I_{sw} \tag{17}$$

A varying duty cycle is used in this model to ensure that the output is constant. To ensure that the model was achieving a high efficiency, an algorithm for calculating efficiency with a varying duty cycle is used.

$$Eff = \frac{1}{1+\frac{R_L \; i_O}{(1-D)^2 v_O}} \tag{18}$$

To achieve higher efficiency, the capacitor and inductor values are calculated using the following equation.

$$L = \frac{I_{ref}}{2fI_{Omax}} D_{max}(1-D)^2 \tag{19}$$

$$C = \frac{I_{ref}D_{max}}{fV_{ripple}} \tag{20}$$

Where, $$D_{max} = 1 - \frac{V_{max}}{V_{ref}} \tag{21}$$

Using these equations to find the appropriate inductor and capacitor values is critical in building a boost converter that is both functional and efficient. Fig.7. shows the test results for the DC/DC converter model. The input voltage has been varied from 25-35V and the output is observed. DC output voltage stayed around 1 p.u. with ripple within the limit and with high efficiency.

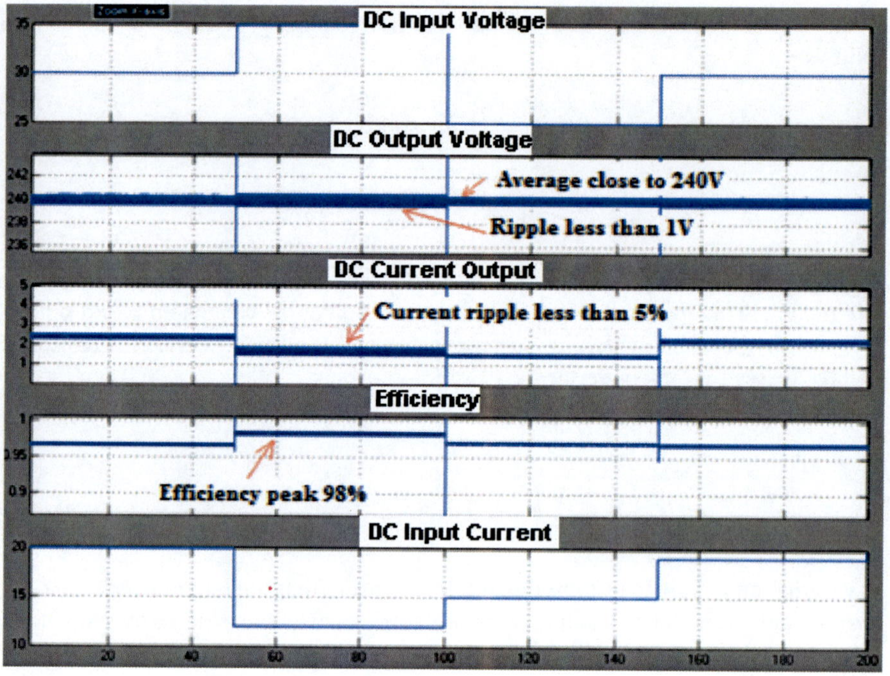

Fig. 7 Results from DC/DC Converter Simulation

2.4 *Testing Fuel Cell Generator with DC/DC Convertor*

Next a variable resistive load is used to test the dynamic response of the fuel cell and converter together. The resulting plots (fig.8) show that the designed DC/DC converter provides a conditioned output. A varying converter input voltage and current results in a constant converter output voltage. The duty cycle changes to accommodate the load demand make the efficiency as high as possible. Unlike as seen in the simulation in fig.5., hard switching in the load demand did not cause as sharp of an oscillation in the output voltage and current due to FC generator's time response.

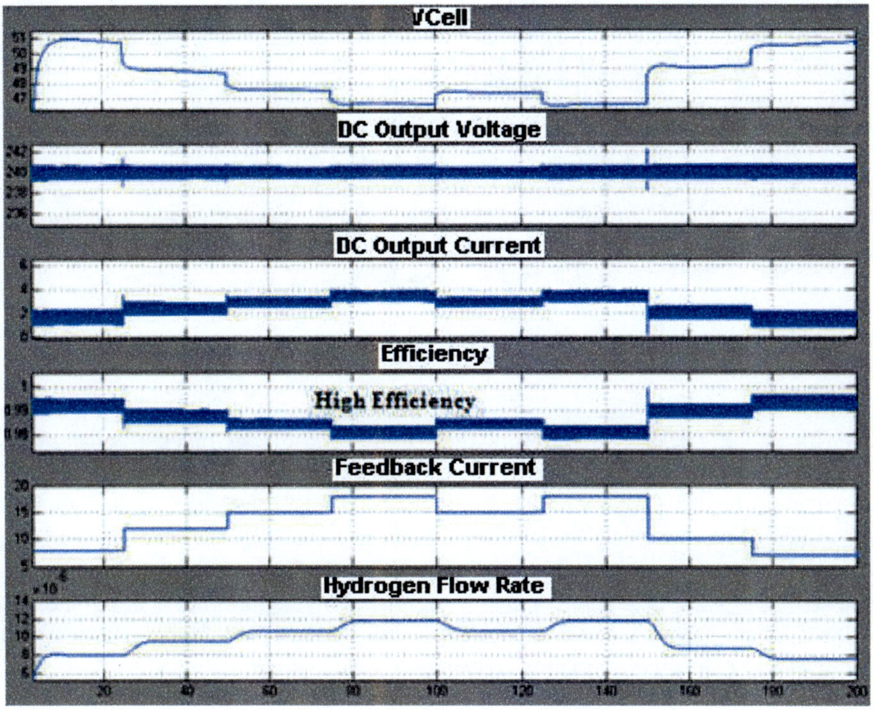

Fig. 8 Results from Fuel Cell Generator and DC/DC Converter Simulation

2.5 Inverter

For larger power applications of a FC the output of a generator would need to be connected to an AC load. This could be done with an inverter. Several studies [14]-[16] have shown that the switching function concept is a powerful tool in understanding and optimizing the performance of the inverter. Using the switching function concept, the power conversion circuits can be modeled according to their functions rather than circuit topologies. Therefore, it can achieve simplification of the overall power conversion functions. The inverter can be modeled as a black box with the input and output ports. The dc and ac variables can be input and output according to the operation mode. Then a transfer function is developed to describe the task to be performed by the circuits. In this chapter, the sinusoidal PWM (SPWM) technique is considered as a control strategy. Based on the SPWM (Fig. 9.), two switching functions are required for a Voltage Source Inverter (VSI).

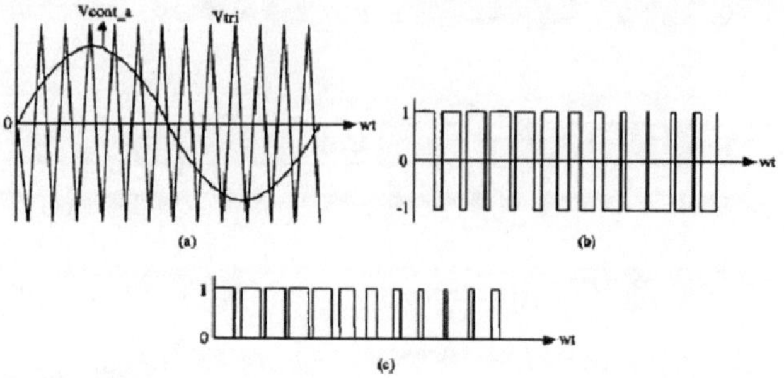

Fig. 9 (a) Carrier and Control Signals (b) Switching Function 1 {SF_1} (c) Switching Function 2 {SF_2}

A three phase DC/AC inverter with switches and diodes takes a constant DC voltage input and inverts that into three phases of AC voltage. It can be configured as in fig. 10: Two switching functions SF_1 and SF_2 are used to design S_1-S_6 as in fig.9. The switching function SF_1 expresses the V_{ao}, V_{bo}, and V_{co} and it is used to calculate the inverter line-to-line voltages (V_{ab}, V_{bc}, V_{ca}) and phase voltages (V_{an}, V_{bn}, V_{cn}). On the other hand, the switching function SF_2 designates the voltage across the switch and the load currents (I_a, I_b, I_c) are derived as ratios of voltages and respective impedances using the switching function SF_2. Mathematical representations and are given as follows:

$$SF_1 = \sum_{n=1}^{\infty} A_n \sin(n\omega t) \tag{22}$$

$$SF_2 = B_o + \sum_{n=1}^{\infty} B_n \sin(n\omega t) \tag{23}$$

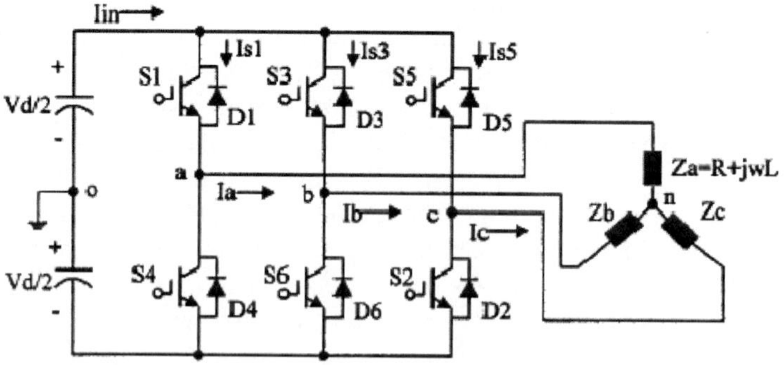

Fig. 10 Circuit Configuration of a Three Phase Voltage Source Inverter

Each switching function has three phases (a, b, and c) all 120 degrees apart. Using the switching function $SF_{1\_a,b,c}$ the V_{ao}, V_{bo}, and V_{co} can be calculated.

$$V_{ao} = \frac{V_d}{2} SF_{1_a} \tag{24}$$

$$V_{bo} = \frac{V_d}{2} SF_{1_b} \tag{25}$$

$$V_{co} = \frac{V_d}{2} SF_{1\_c} \tag{26}$$

Then the inverter line-to-line voltage can be derived as follows:

$$V_{ab} = V_{ao} - V_{bo} \tag{27}$$

$$V_{bc} = V_{bo} - V_{co} \tag{28}$$

$$V_{ca} = V_{co} - V_{ao} \tag{29}$$

In order to calculate the phase voltage V_{no} is needed as shown below:

$$V_{no} = \frac{1}{3}(V_{ao} + V_{bo} + V_{co}) \tag{30}$$

$$V_{an} = V_{ao} - V_{no} \tag{31}$$

$$V_{bn} = V_{bo} - V_{no} \tag{32}$$

$$V_{cn} = V_{co} - V_{no} \tag{33}$$

Assuming the load consists of a balanced R-L load, the load currents can be determined as a ration of the phase voltage and the respective impedance as given below:

$$I_a = \frac{V_{an}}{Z_a} = \frac{V_{an}}{R+j\omega L} \tag{34}$$

$$I_b = \frac{V_{bn}}{Z_b} = \frac{V_{bn}}{R+j\omega L} \tag{35}$$

$$I_c = \frac{V_{cn}}{Z_c} = \frac{V_{cn}}{R+j\omega L} \tag{36}$$

The switch currents are calculated by the product of the load currents with the corresponding switching function $SF_{2\_a,b,c}$.

$$I_{S1} = I_a * SF_{2_a} \tag{37}$$

$$I_{S3} = I_b * SF_{2_b} \tag{38}$$

$$I_{S5} = I_c * SF_{2\_c} \tag{39}$$

Using these equations VSI can be modeled accurately. The developed model is then tested for various loading conditions. Input in the form of P and Q are dynamically changed and the VSI output (voltage and current) and efficiency is evaluated. Fig.11 shows such a simulation. The results indicated that the model stabilizes with high efficiency, and accurate output voltage with changing inputs.

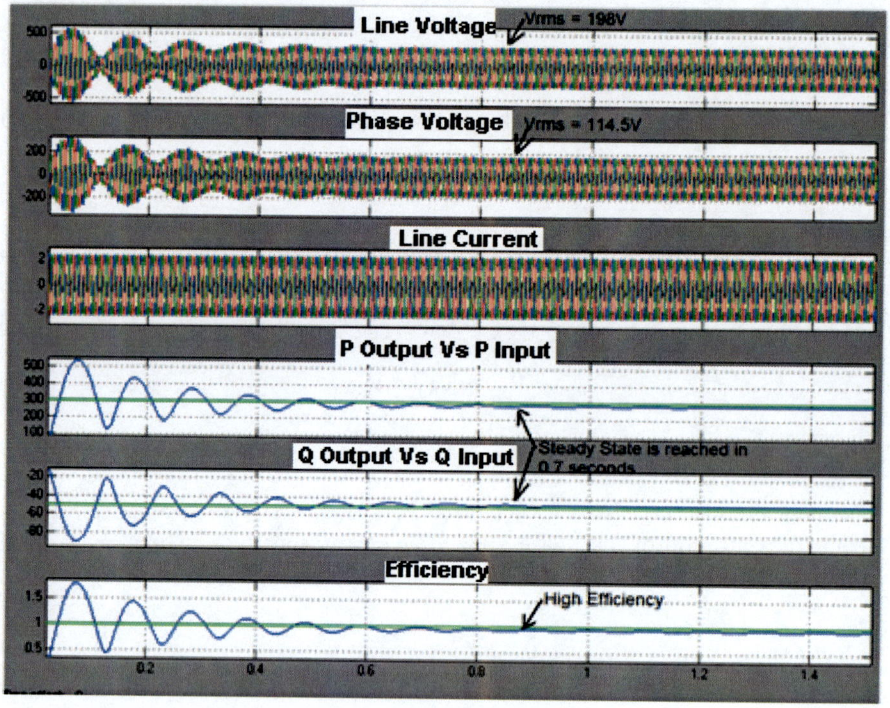

Fig. 11 Results from Light Loaded DC/AC Inverter Simulation

2.6 Fuel Cell Generator and VSI with Controllers

To meet the requirements for interconnecting a fuel cell system to a utility grid and control the real and reactive power flow between them, it is necessary to shape and control the inverter output voltage in amplitude, angle, and frequency.

In this section, a PWM controller is designed for the inverter to satisfy voltage regulation as well as to achieve real and reactive power control. To this end, a d-q transformation is used to transfer a stationary (abc) system to a rotating (dq0) system. The transformation decreases the number of control variables from 3 to 2 (component 0 will be zero) if the system is balanced. The d-q signals can also be used to achieve zero tracking error control. In order to determine the error, or difference between the reference signal and the actual signal, a reference voltage is needed. To calculate the reference signal a set of equations are used below [14- 16].

$$V_{ref} = \frac{Z^2}{E^2}(P^2 + Q^2) + E^2 + 2PZ\cos(\theta_Z) + 2QZ\sin(\theta_Z) \tag{40}$$

$$\delta = \theta_Z - \cos^{-1}\left\{\frac{ZP}{EV_S} + \frac{E}{V_S}\cos(\theta_Z)\right\} \tag{41}$$

Where V_{ref} and δ would then need to be transformed into dq coordinates (abc/dq), and this is done by using the matrix equation as in (42). As a part of the overall control architecture a voltage and current controller was designed. Fig. 12 illustrates the design and control topologies of a VSI inverter implementation.

$$\begin{bmatrix} V_d \\ V_q \\ V_0 \end{bmatrix} = \begin{bmatrix} \cos(\theta) & \cos\left(\theta - \frac{2\pi}{3}\right) & \cos\left(\theta + \frac{2\pi}{3}\right) \\ \sin(\theta) & \sin\left(\theta - \frac{2\pi}{3}\right) & \sin\left(\theta + \frac{2\pi}{3}\right) \\ \frac{1}{\sqrt{2}} & \frac{1}{\sqrt{2}} & \frac{1}{\sqrt{2}} \end{bmatrix} \begin{bmatrix} V_a \\ V_b \\ V_c \end{bmatrix} \tag{42}$$

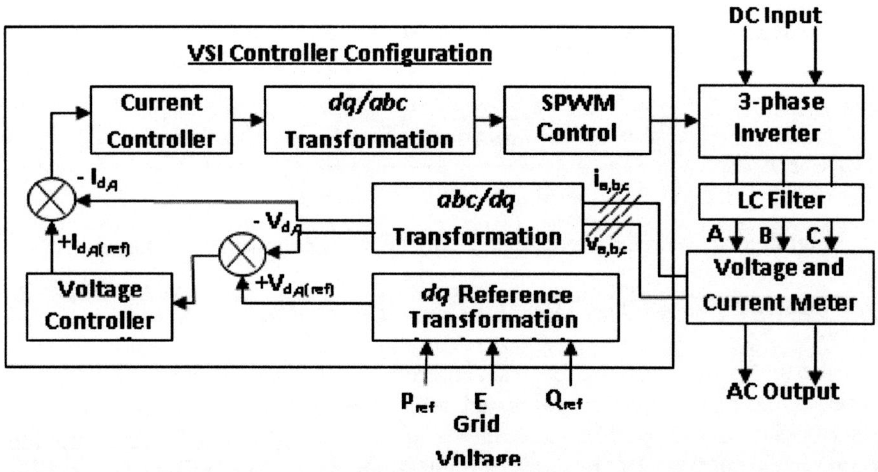

Fig. 12 Flow Diagram of a VSI Controller

Once the signal passes through the voltage and current controllers it must be passed through a dq/abc transformation so the signal can be used in the SPWM. The transformation is designed based on (43). A voltage and current controller is designed for internal control of the inverter. Fig. 13 shows the block diagram of the current and voltage controller that has been used. After the design and modeling of the inverter, the system is tested with inverter and power conditioning devices. Fig. 14 shows the detail analysis of testing the inverter DC/DC converter and the fuel cell in the closed loop. The inverter model works well with changing load current producing highest efficiency and nominal voltage. During steady state conditions the RMS voltage (line and phase) indicates steady state values and the line current changes with respect to the changes in P and Q demand. After testing the inverter, overall system is analyzed as discussed next.

$$\begin{bmatrix} V_a \\ V_b \\ V_c \end{bmatrix} = \begin{bmatrix} \cos(\theta) & \sin(\theta) & \frac{1}{\sqrt{2}} \\ \cos\left(\theta - \frac{2\pi}{3}\right) & \sin\left(\theta - \frac{2\pi}{3}\right) & \frac{1}{\sqrt{2}} \\ \cos\left(\theta + \frac{2\pi}{3}\right) & \sin\left(\theta + \frac{2\pi}{3}\right) & \frac{1}{\sqrt{2}} \end{bmatrix} \begin{bmatrix} V_d \\ V_q \\ V_0 \end{bmatrix} \tag{43}$$

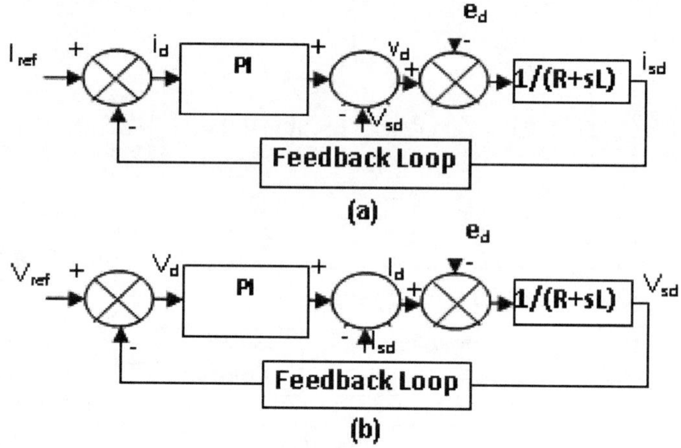

Fig. 13 (a) Current Controller Block Diagram (b) Voltage Controller Block Diagram

2.7 *Testing the System with VSI and Controllers*

As discussed, two controllers (feedback controller for FC and inverter controller) are constructed using the equations discussed before. These controller are constructed together because the PI feedback controller requires an AC feedback voltage from the inverter to calculate the DC feedback current. Both the inverter and fuel cell controller must then run together. In this test, the simulation will be run with a static and dynamic load demand and the performance of the FC system is discussed.

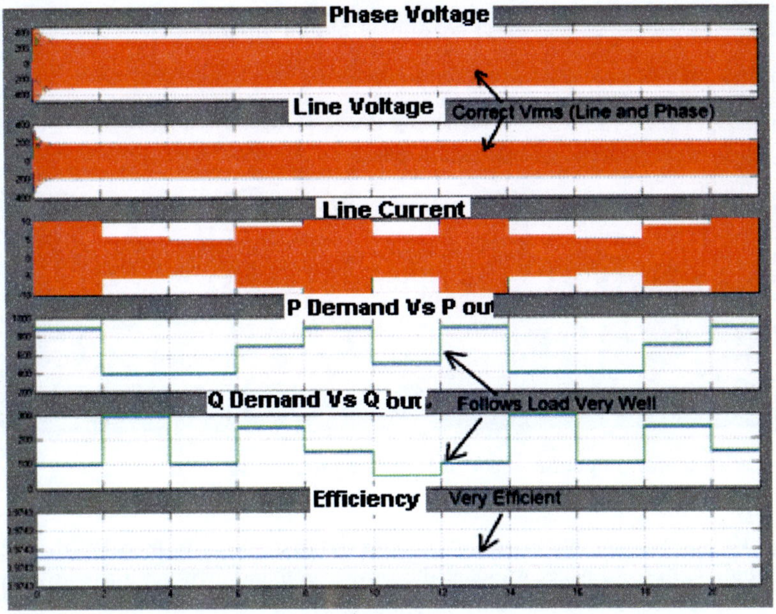

Fig. 14 Fuel Cell, DC/DC Converter, and DC/AC Inverter Simulation

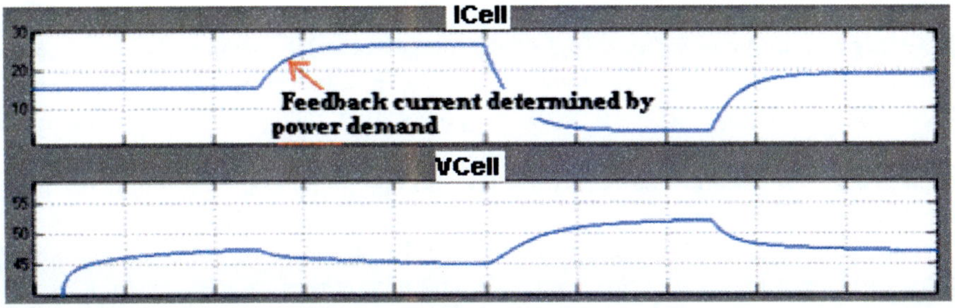

Fig. 15 Fuel Cell Controller Output Feedback Current

Output feedback current simulation results discussed in fig. 15 shows that the control signals respond to the real and reactive power demanded from the fuel cell system to the utility grid. Using the control signals the power delivered to the grid can be controlled as desired, while the dc bus voltage is maintained well within the prescribed range. The results also show that the fuel cell system is capable of following the load and can remain stable under the occurrence of severe faults. It is noted that a two-loop inverter control scheme has an advantage over a voltage-only control scheme for the inverter especially in the presence of fault protection and system stability.

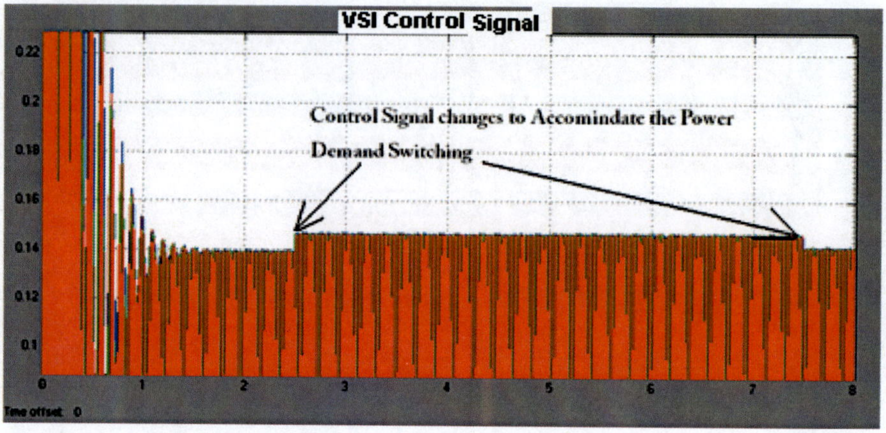

Fig. 16 VSI Control Output Feedback Signal Going to SPWM

Fig. 16. shows the control output feedback signal going to the SPWM. The control signal changes to accommodate the power demand switching as shown in the figure with respect to the load current change.

3 Modeling and Control of Photo Voltaic System

PV energy source can be considered the most essential resource because of the ubiquity, abundance, and sustainability of solar radiant energy [17]. The sun radiates approximately 1.8×10^{11} $MW$of power that is intercepted by the Earth [17]. One method of utilizing solar energy is through PV cells in the form of electrical energy. A group of cells forms a PV module and a combination of PV modules is called a solar panel, while a group of solar panels is called a PV array.

A photovoltaic system consists of the whole assembly of solar cells, connections, protective parts, supports, etc [18]. The advantages of a PV system far outweigh its disadvantages. Some of the advantages include low maintenance, unattended operation, long life, no fuel, no fumes, easy to install, and modularity. Disadvantages are high initial setup costs and low output in cloudy weather. This section discusses PV system modeling and integrating the developed model as a micro grid.

3.1 Overall System Configuration and Mathematical Formulation

The basic configuration of the PV system is shown in Fig. 17. The following subsections describe each block in the figure.

3.1.1 PV Generator

A PV generator consists of a PV array comprising of PV modules. Modeling details of PV modules and PV arrays are discussed next.

3.1.2 PV Cell

A solar module is the individual piece of equipment that encompasses numerous solar cells connected in parallel or in series. By putting solar cells together as a module, the current and voltage properties of a module increase. A solar cell can be designed as a *p-n* junction device with no voltage directly applied across the junction. It converts photon power into electrical power and delivers this power to a load [19]. The *p-n* junction is fabricated in a thin wafer of semiconductor, usually silicon. In the dark, *p-n* has the characteristics of a diode and blocks the flow of the current which results in no voltage. A solar cell equivalent circuit model is shown in Fig. 18. This is a single-diode model which includes a current source, a parallel diode and a series resistor R_s. The value of the source current depends on the light falling on the cell. A stronger light results in a higher electric current. The nonlinear I-V characteristics of the cell are determined by the diode.

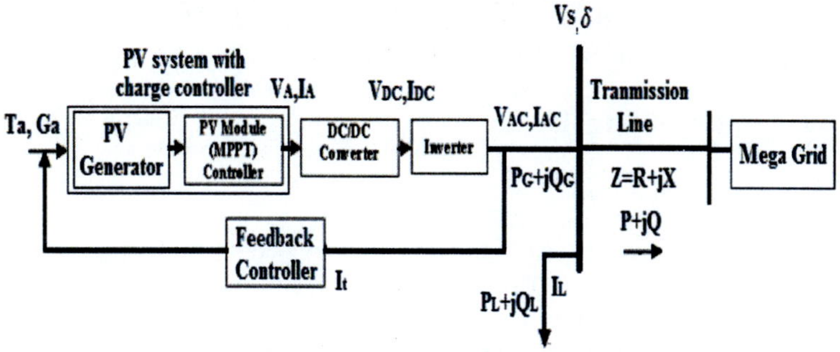

Fig. 17 Block Diagram of a PV Micro grid

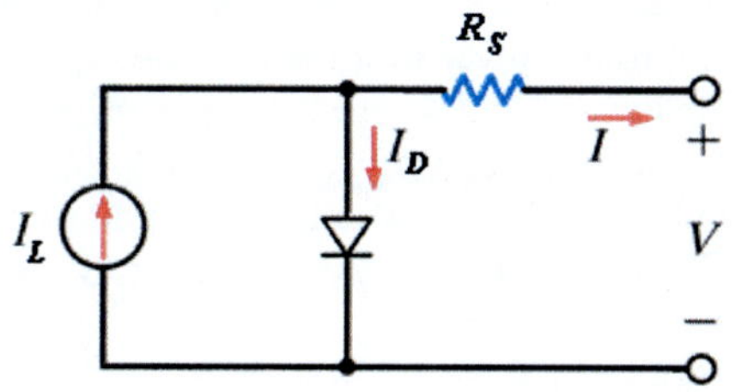

Fig. 18 Equivalent Solar Cell Model

The photocurrent, I_L, produces a voltage drop across the resistive load which forward biases the *p-n* junction. The forward-bias voltage produces a forward-bias diode current, I_D. The net current is the difference between the photocurrent and the diode current as represented by the following equation:

$$I = I_L - I_D = I_D - I_S\left(exp\frac{e(V+IR_S)}{mkT_C} - 1\right) \tag{44}$$

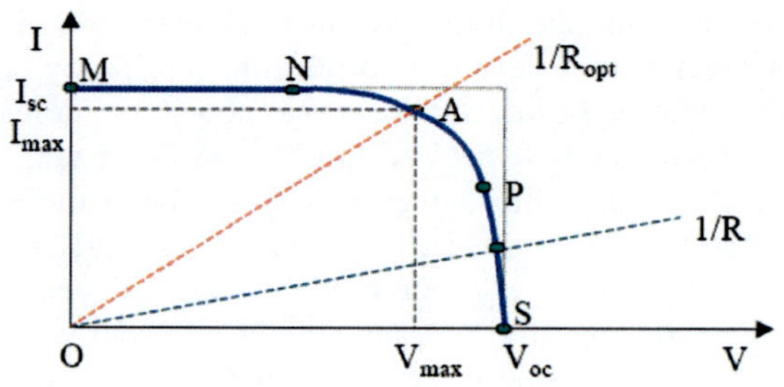

Fig. 19 Typical I-V Characteristic Curve of a Solar Cell

Where I_S is the saturation current, e is electric charge, V is the voltage across the cell, m is the diode ideal factor, k is Boltzmann's constant, and T_C is the temperature of the cell. A plot of the I-V characteristics of a PV array for a certain temperature is shown in fig. 19.

3.1.3 PV Modules and Array

The schematic shown in fig. 20 illustrates how solar cells are connected within a solar module. The output power from a single PV cell is relatively small (approximately 0.5 W) [20]. To maximize voltage and power, a solar array is a must. A PV array consists of numerous solar modules that are combined in series and parallel to form panels, and these panels are connected together to form the entire PV array. The architecture can be thought of similar to fig. 19 with modules replacing the cells. So, in a micro grid, there may be numerous solar arrays connected to provide the required power to a load. The equations used to solve the module current are presented below, where the superscripts: C represents the solar cell, M represents the solar module, and A represents the solar array. Each solar module contains a manufacturer's data sheet that contains parameters for standard operating conditions. The parameters used are related to the datasheet to test and validate the model as well.

Maximum power for cell

$$P^C_{max,0} = \frac{P^M_{max,0}}{N_{SM}N_{PM}} \tag{45}$$

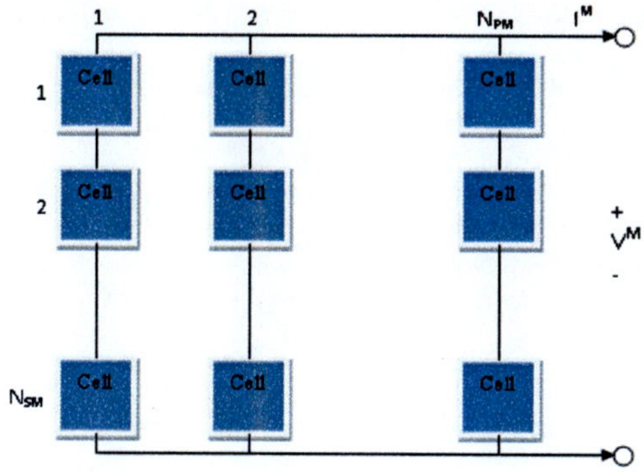

Fig. 20 PV Module with N_{SM} Series Branches and N_{PM} Parallel Branches

Open circuit voltage for cell

$$V_{OC,0}^{C} = \frac{V_{OC,0}^{M}}{N_{SM}} \tag{46}$$

Short circuit current for cell

$$I_{SC,0}^{C} = \frac{I_{SC,0}^{M}}{N_{PM}} \tag{47}$$

Thermal voltage of the cell

$$V_{t,0}^{C} = \frac{mkT_0^C}{e} \tag{48}$$

Open circuit voltage of the module

$$v_{OC,0} = \frac{V_{OC,0}^{C}}{V_{t,0}^{C}} \tag{49}$$

Fill factor

$$FF = \frac{P_{max,0}^{C}}{V_{OC,0}^{C} I_{SC,0}^{C}} \tag{50}$$

Fill factor at standard conditions

$$FF_0 = \frac{v_{OC,0} - \ln(v_{OC,0} + 0.72)}{v_{OC,0} + 1} \tag{51}$$

Equivalent series resistance

$$r_s = 1 - \frac{FF}{FF_0} \tag{52}$$

Serial resistance of a cell

$$R_S^C = \frac{r_s V_{OC,0}^C}{I_{SC,0}^C} \tag{53}$$

Cell parameters for operating conditions with inputs are: VM, Ta, Ga
Irradiation Parameters

$$C_1 = \frac{I_{SC,0}^C}{G_{a,0}} \tag{54}$$

Short circuit operating current

$$I_{SC}^C = C_1 G_a \tag{55}$$

Working temperature of cell

$$T^C = T_a + C_2 G_a \tag{56}$$

Open circuit voltage

$$V_{OC}^C = V_{OC,0}^C - C_3(T^C - T_0^C) \tag{57}$$

Thermal voltage of cell

$$V_t^C = \frac{mk(273+T^C)}{e} \tag{58}$$

Calculated solar module current using cell parameters

$$I^M = I_{SC}^C N_{PM}\left[1 - exp\left(\frac{V^M - N_{SM}V_{OC}^C + I^M R_S^C \frac{N_{SM}}{N_{PM}}}{V_t^C N_{SM}}\right)\right] \tag{59}$$

Short circuit current for the module

$$I_{SC}^M = I_{SC}^C N_{PM} \tag{60}$$

Open circuit voltage for module

$$V_{OC}^M = V_{OC}^C N_{SM} \tag{61}$$

Equivalent serial resistance for the cell

$$R_S^M = R_S^C \frac{N_{SM}}{N_{PM}} \tag{62}$$

Thermal voltage of module

$$V_t^M = V_t^C N_{SM} \tag{63}$$

Module current

$$I^M = I_{SC}^M \left[1 - exp\left(\frac{V^M - V_{OC}^M + I^M R_S^M}{V_t^M}\right)\right] \tag{64}$$

Module voltage

$$V^M = \ln\left(\frac{I_{SC}^M - I^M}{I_{SC}^M}\right) V_t^M + V_{OC}^M - I^M R_S^M \tag{65}$$

Module power

$$P^M = V^M I^M \tag{66}$$

Also, array currents and voltage are calculated from the equations below.
Total current of array

$$I^A = \sum_{i=1}^{N_{SA}} I_i \tag{67}$$

Simplified total current of array

$$I^A = N_{SA} V^M \tag{68}$$

Power generated by array

$$P^A = V^A I^A \tag{69}$$

3.1.4 PV Charge Maximum Power Point Tracking (MPPT) Controller

Renewable sources, such as solar arrays, produce varying power output. When the PV arrays are interconnected with the electrical power grid and to the loads, a constant and consistent power is required. Solar power varies with irradiation (sun power), temperature, and shadowing. A proper working power point should be maintained for both the grid and the loads. So to extract the required power from the PV array, it is important to operate the power output at the corresponding maximum power point (MPP). The MPP is unique for each solar module, it is not a fixed point, and it may change based on the required current for a particular load. Thus a Maximum power point tracking (MPPT) is required to match the PV output to load demand. In other words, the electrical tracking of the MPP is accomplished through power switching converters as an interface between the load and the source. These converters are controlled to provide a matching between the load impedance and the varying PV source impedance [20]. For the model presented in this chapter, an algorithm is used which works on a simple concept known as perturb-and-observe (P&O) method. The P&O algorithm is a commonly used MPPT algorithm. It is used for its simplicity and ease of implementation. However, there are some limitations with the P&O method, such as, oscillations around the MPP in steady state, slow response speed, and tracking in the wrong way under rapidly changing conditions like weather [19-21]. This method works best for slow changing or constant weather conditions. However, as long as the power increases, the voltage increases or decreases. Once the power decreases, the process is reversed. This method can be understood by the flow chart in fig. 21.

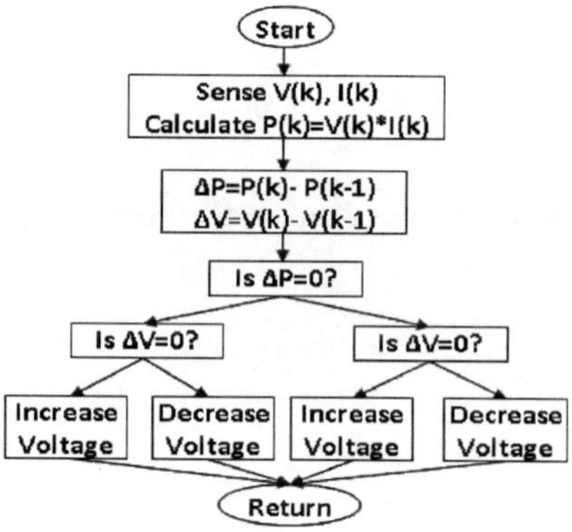

Fig. 21 Flowchart for MPPT algorithm used as the charge controller

3.1.5 DC/DC Converter and Inverter

Overall micro grid consists of a DC/DC converter and inverter similar to the ones discussed earlier.. With such a system the PV micro grid is tested for dynamic load conditions. The results are illustrated next.

3.2 Modeling and Simulation Results

The modeling of the PV components is developed using Simulink®. First, PV modules are developed using the equations dicussed earlier. Then a PV array is designed. Fig. 22 shows the developed PV array model in Simulink®. This developed PV array model is integrated with a DC/DC converter and inverter models. Fig. 23 shows an overall model with feedback controller with a static load. Figs. 24 show sample plots for the power with respect to constant temperature and irradiations. Since modules and arrays produce fairly similar results, only the plots of a module are displayed. Solar modules are affected by both the irradiation and the temperature values. Increasing irradiation causes a rise in current and voltage, thus resulting in an increased power. Increasing temperature causes a rise in the voltage, while current stays constant and power increases slightly. These plots are similar to a real system.

The plots shown in Figs. 25-26 illustrate voltage, and efficiency obtained from the DC/DC converter respectively. The efficiency is high at the beginning and at the end of the plot, but very sporadic. At these values, irradiation and temperature are low which drive the voltage of the PV array to be high. At mid-points, when the irradiation and temperature is high efficiency is the lowest.

The DC/AC inverter was built and tested using Simulink®. Irradiation was simulated from 100-1000W/m^2 and temperature was simulated from 20-25°C. Since in real time, and low irradiation values would only power small loads, small P,Q loads were simulated with real power (P) simulated from 0-400W, and reactive power, Q, simulated from 0-80var. Fig. 27 illustrates the irradiation and temperature ranges that are used as an input to test the micro grid. PV voltage is dependent on the current being fed into it. As the load increases, the voltage increases. After the inverter stabilizes, 0.1s time sampling of the inverter line and phase voltage is obtained. It was observed that the voltages are 120 degrees out of phase.

PV current I_g is the same current that is being fed from the inverter. After stabilization, a steady current is delivered to meet the load demand. Fig. 28 illustrates the PV current and the output current from the DC/DC converter. The inductor current illustrates the opening and closing of the inductor used in the converter.

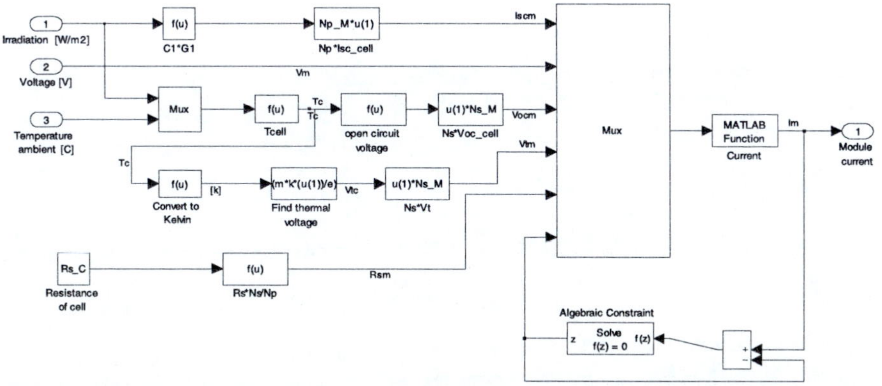

Fig. 22 PV Array model

Fig. 29 illustrates the plot of inverter line current and the feedback current (I_t) based on the load current. When connected to the load, the converter efficiency is above 99% while the inverter efficiency kept above 90%. Overall the efficiency stays very high except during system stabilization. The developed model is tested for various load conditions as a micro grid as well. One such case is shown in fig. 30. From the plots, it can be seen that the actual P,Q output obtained from the inverter is slightly less than the desired P,Q demand. This result is consistent with the practical systems developed and tested in a laboratory set-up.

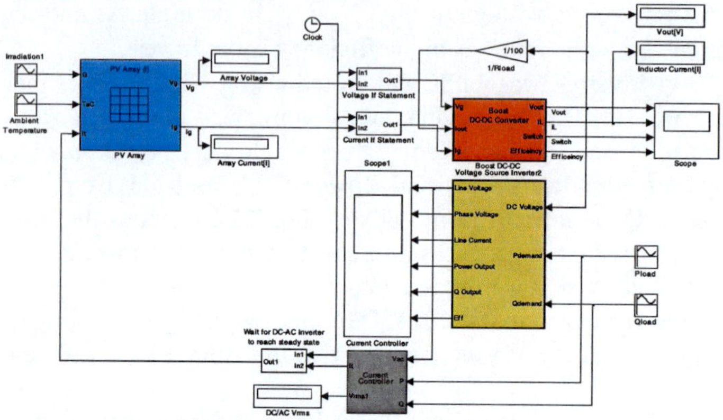

Fig. 23 Load Connected Inverter Model

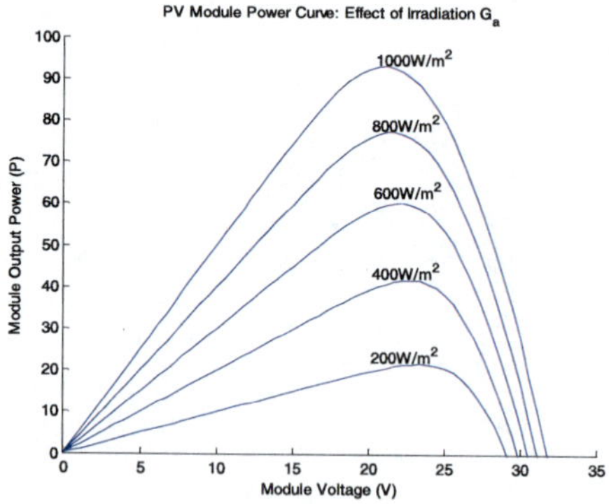

Fig. 24 Power Curve TC = 25°C

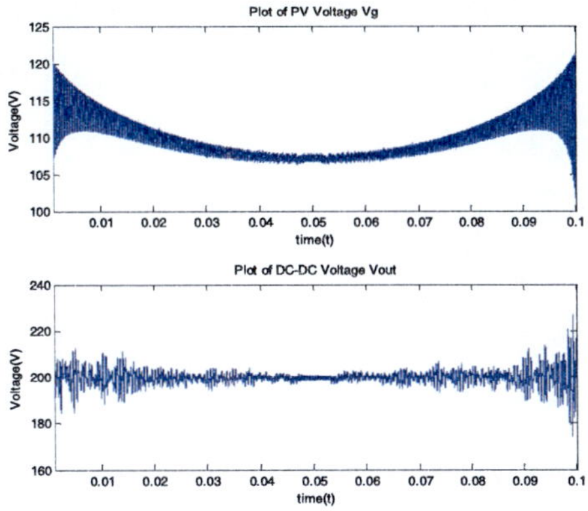

Fig. 25 Plot of PV Voltage and DC/DC Converter Voltage

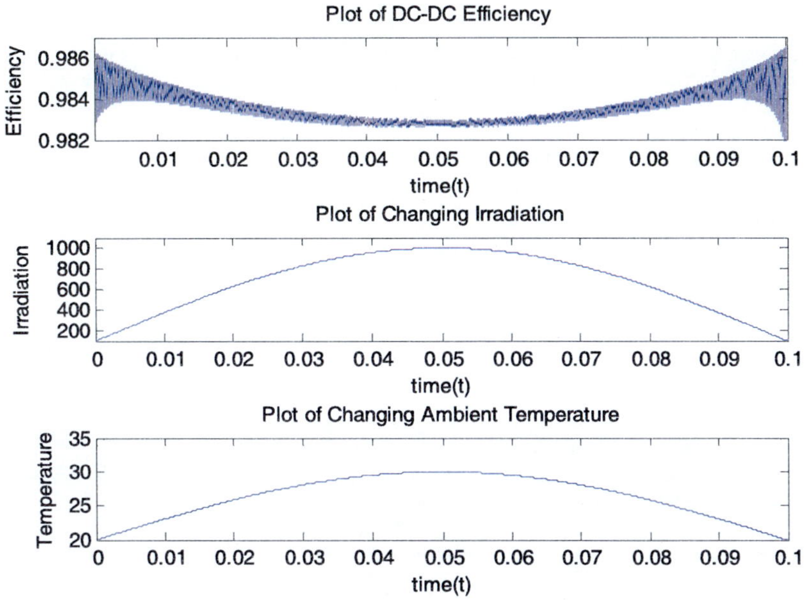

Fig. 26 Plot of DC/DC Efficiency, Irradiation, and Ambient Temperature

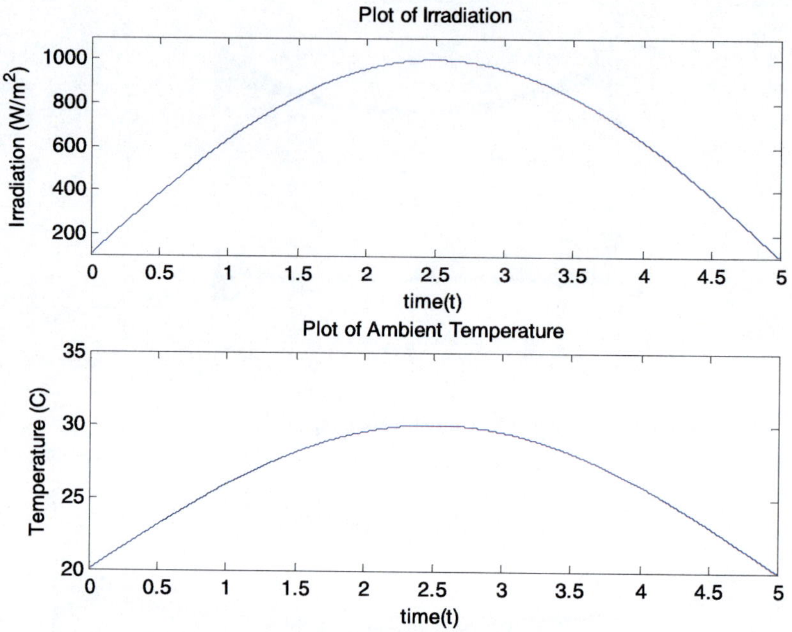

Fig. 27 Plot of Irradiation and Ambient Temperature

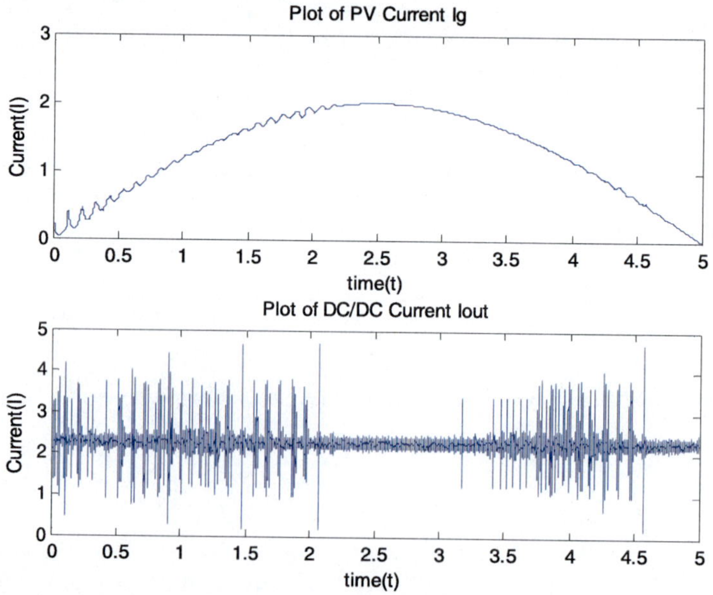

Fig. 28 Plot of PV Current and DC/DC Output Current

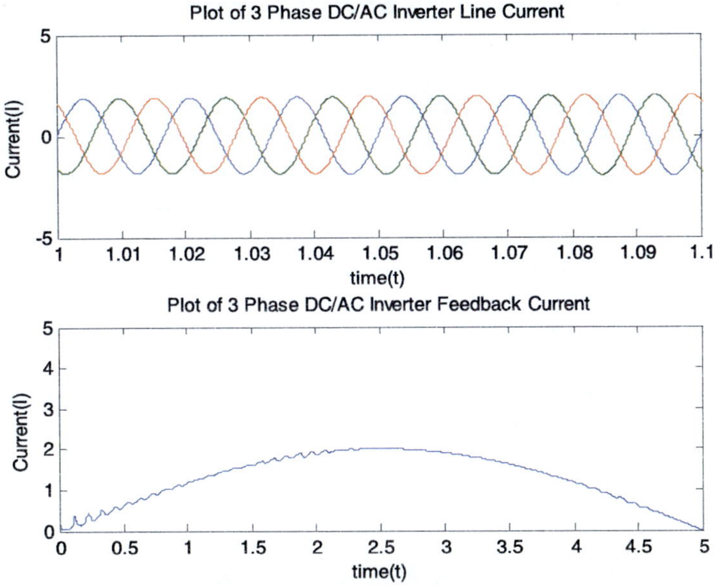

Fig. 29 Plot of Inverter Line Current and Feedback current

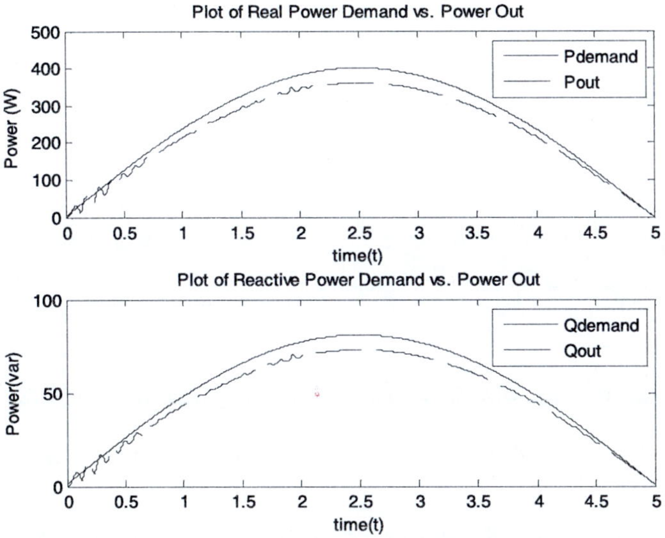

Fig. 30 Plot of P,Q demand vs P,Q out

4 Hybrid Micro-grid Set Up with PV Arrays and PEM Fuel Cell

There are many different scenarios that can take place when a renewable energy power source (i.e. Photovoltaic and Fuel Cell) is connected to the micro grid. The power source can supply power to the grid (light loading), require power from the grid (Heavy Loading), or be separated from the grid (islanding mode). The first scenario that we are going to address is when the micro grid may have been disconnected from the power grid because of problems on either side. In this case, the micro grid is in island mode. This test will consist of two power sources, PV and FC, working simultaneously to provide power to one load. Since the PV source is "free energy source" it will apply all of its available power to the load and the remaining power required will come from the Fuel Cell. The amount of power that the PV can supply is directly related to the amount of irradiation that the PV receives. The following equations are used to calculate V_S, δ, feedback current, and the real and reactive power of the grid. [13,22].

$$V_S = \left[\frac{Z^2}{E^2}(P^2 + Q^2) + E^2 + 2PZcos(\theta_z) + 2QZsin(\theta_z)\right]^{1/2} \tag{45}$$

$$\delta = \theta_z - \cos^{-1}\left(\frac{ZP}{EV_S}\cos(\theta_z)\right) \tag{46}$$

$$I_t = \frac{P_L}{V_S cos\theta} \tag{47}$$

$$I_L = I_t \cos(\theta + \delta) \tag{48}$$

Where, P_L is the real power of the load, and

$$\theta = \cos^{-1}\left[\frac{P_L}{\sqrt{P_L^2 + Q_L^2}}\right], Z = \sqrt{R^2 + X^2}, \text{ and } \theta_z = \tan^{-1}\frac{X}{R}.$$

4.1 *Micro-grid Connected to Power Grid (Islanding Mode)*

To show a simplified representation of the islanding mode a flow diagram is developed as in fig. 31.

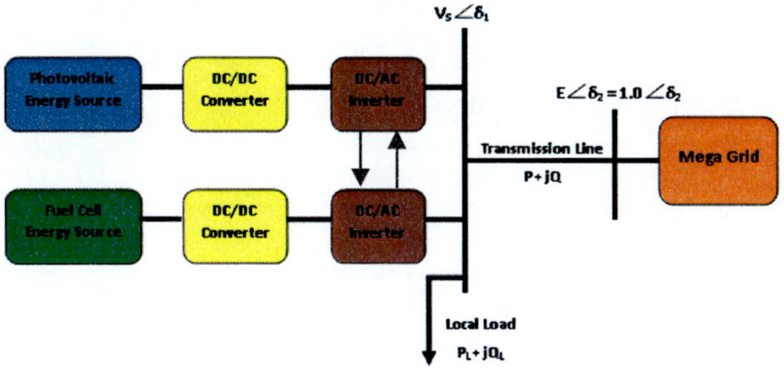

Fig. 31 Islanding Mode of a Fuel Cell and PV Power Source ($E\angle\delta_2 = 1.0\angle 0$ p.u.)

The system is modeled in Simulink® using previously discussed PV and FC models. The PV system used in this simulation has a generating capacity of 450W and the FC has a generating capacity of 1kW. Both systems were placed into a subsystem to feed the same load. An intelligent controller is developed to deliver the appropriate P,Q power from PV array based on the irradiation. The remaining P,Q is supplied by the fuel cell. This is a load following scheme with the micro grid follows the load and the power grid sees a constant load at the point of coupling. The power demand in this case is modeled by a static load (dynamic response will be shown in the following section). Both subsystems for the PV and FC calculate and output V_S and δ_1. E and δ_2 will be equal to V_S and δ_2 which is $1\angle 0$ p.u.). The irradiation and ambient temperature used by the PV are as in fig. 27.

Fig. 32 shows the FC subsystem, PV Array Subsystem, intelligent controller, and an infinite bus subsystem to connect the two outputs together. The intelligent controller measures the irradiation at the time and calculates the amount of power that can be supplied to the load. Once that is determined the fuel cell will feed the remainder of the load demand. The infinite bus subsystem takes the two outputs and connects them in parallel which averages the voltage, given a balanced system, and adds the currents.

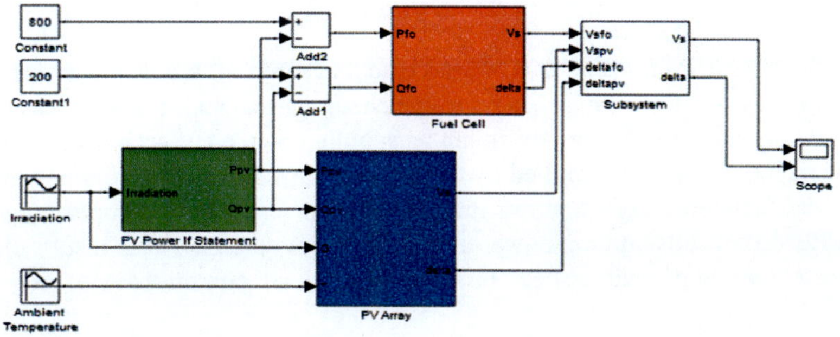

Fig. 32 Island Mode Micro-grid System

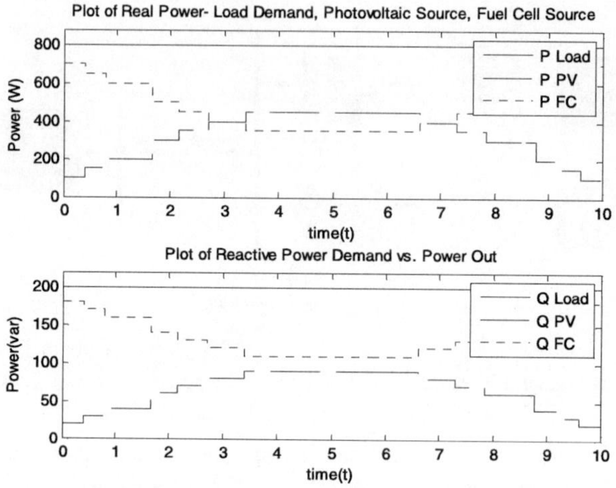

Fig. 33 Plot of Real and Reactive power demand and generation

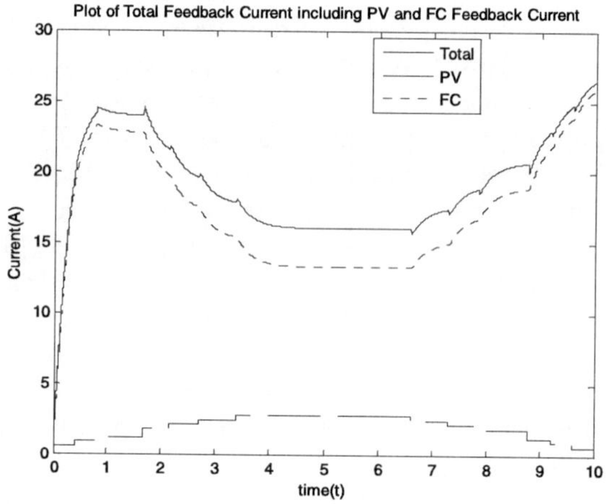

Fig. 34 Total Feedback Current Including PV and FC Feedback Currents

Then the output values for the power and feedback current are observed for various simulated conditions. Fig. 33 shows the changing real and reactive power. As the irradiation increases, more power can be supplied by the PV array from the PV and FC. As more power is supplied by the PV, less power has to be supplied to the load by the fuel cell. Fig. 34 shows the feedback current required controlling the flow in PEM fuel cells and the power output of the PV arrays. These results clearly indicate two important issues. First the light load conditions are fed in a power following manner with priority given for PV arrays. From fig. 33 it can be seen that the P is around 800W and Q around 200 Vars. Second, the hybrid system smartly controls the feedback current thus providing a stable and effective power requirements in the presence of changing load dynamics. Next study shows the effect of connecting this smart micro grid to the power grid.

4.2 Micro-grid Connected to Power Grid (P, Q Positive or Negative)

A micro grid, or distributive generation system, is a cluster of PV and fuel cell energy sources, their controllers, converters and inverters connected to a power grid system through a transmission line and some type of switch or breaker. The mega-grid may contain a large number of interconnected buses, but for this case study, we will use light and heavy loading. Light loading is when there is a heavy local load and the mega grid is calling for a load demand. To simplify, there is power flowing onto the mega grid as shown below. Heavy Loading is when there is a heavy load on the local load and the mega grid is giving a load demand. To simplify, here there is power flowing from the mega grid.

Both of these cases can be shown dynamically in one simulation. Fig. 35 illustrates both these conditions in a one line block diagram for priority control of the power 'P'. For this simulation we will use the same equations developed earlier. The mega-grid was modeled with both renewable energy sources, along with the intelligent controller used to control the power demand for each of the energy sources, and feedback current using δ_1 and δ_2. Changes were made to the 3-phase inverter in each model to successfully model the V_S and δ_1 based on line power and load power and changing values of grid voltage. The Simulink® model for the overall architecture is shown in fig. 36. Parameters include changing irradiation and ambient temperature, changing real and reactive power values, and a changing grid voltage E and δ_2 values. The V_s and δ_1 are calculated depending on the changing E and δ_2. This is used as a reference voltage to be sent back to the VSI inside the Fuel Cell and PV Array. This allows the inverter to accommodate for the power flowing to or from the mega grid. From there the FC and PV work together to produce the appropriate power demand. The feedback current going to the PV and Fuel Cell is in fig. 37. Note the changing E and δ_2 start out in island mode with no power flowing onto or from the mega-grid. Then, the mega grid cycles from demanding power to supplying power a couple of times to ensure that the power sources act in the manner in which they should. Fig. 38 shows the changes in the power (P and Q) from the grid with load. This shows dynamic load changes with the value of P and Q changes with respect to time as shown in fig. 38.

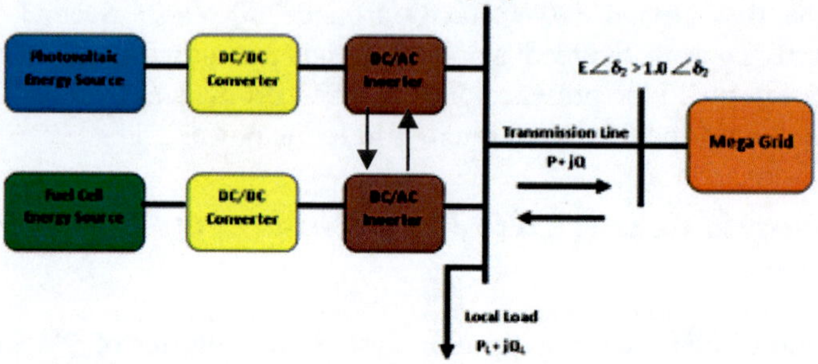

Fig. 35 Light Loading of a Fuel Cell and PV Power Source ($E\angle\delta_2 > 1.0\angle 0$ p.u.)

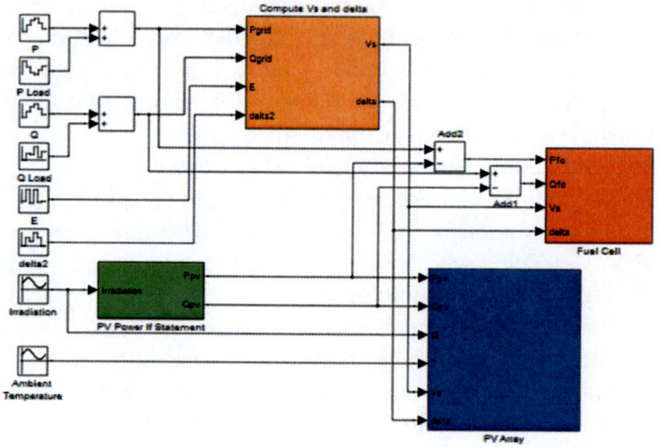

Fig. 36 Simulink Grid-Connected Micro-grid

The PV handles more power when the irradiation is at its highest point. This is proven as the output results shown in fig. 39 and fig. 40. It is noted from fig. 40 that the power grid responds to the initial load changes, and then the micro grid picks up the load changes to keep the grid power at its scheduled value. The simulation results show that the micro grid can follow the load power changes in less than 0.5 s. As irradiation increases, the load handled by the PV array increases and so does the feedback current. The efficiency starts out unstable due to the stabilizing of the system, but steadies out right around one which equates to an almost 100% efficiency. Fig. 41 shows the FC share when the P and Q demand changes. It can be seen that as the irradiation increases, the FC does not have to handle as much power. Just like the PV array, the fuel cell produces 5% extra power to handle transmission line losses. The system takes almost one second to stabilize, so the inverter efficiency varies. It does however become almost one for the remainder of the simulation. Initially the feedback current for the fuel cell becomes lower as more power is handled by the PV and increases as irradiation becomes smaller.

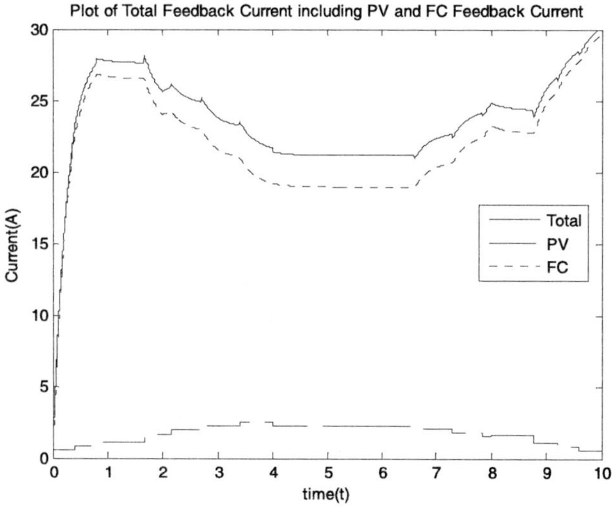

Fig. 37 Total Feedback Current Including PV and FC Feedback Currents

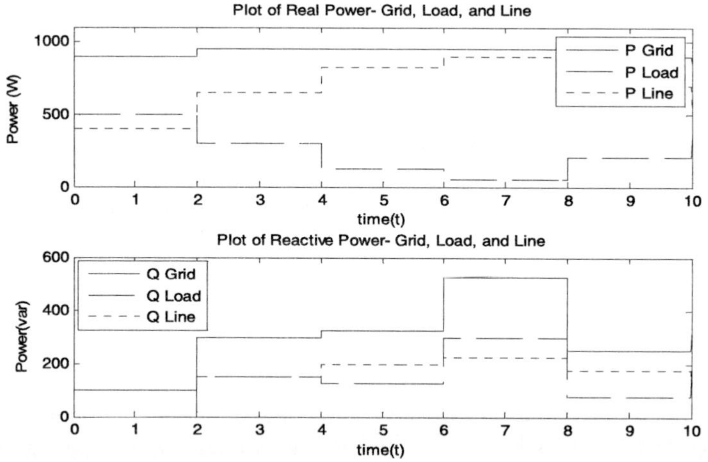

Fig. 38 Plot of Real and Reactive Power (Grid, Load, and Line)

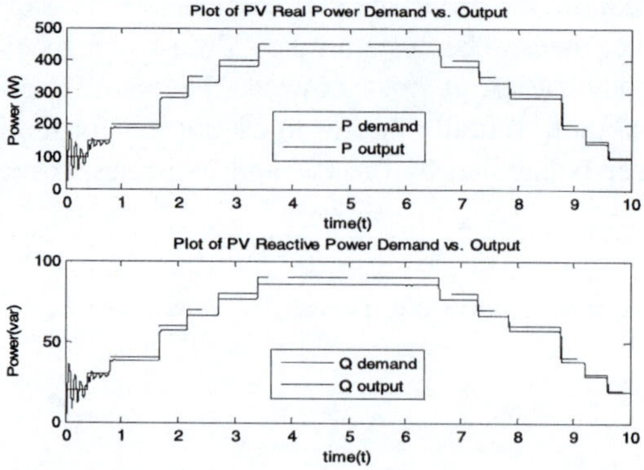

Fig. 39 Plot of Real and Reactive Power Demand vs. Output Handled by the PV

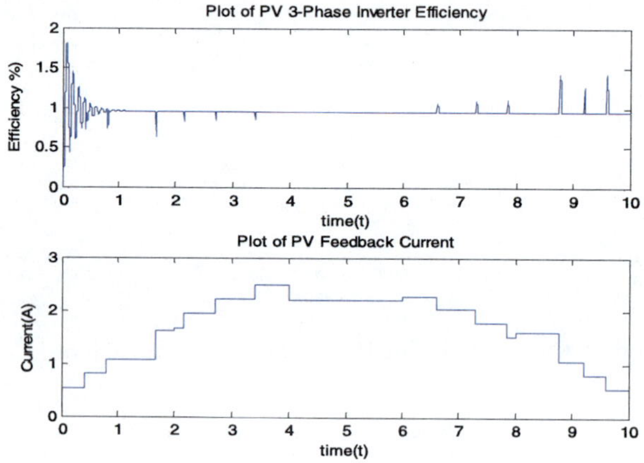

Fig. 40 Plot of PV 3-Phase Inverter Efficiency and Feedback Current

4.3 *Impact of the Micro Grid as a Customer Driven Micro System*

The impact of the micro grid on the power system as a load following customer driven micro system can be seen from above simulation and real-time testing. It could be noted that with the help of intelligent and local controllers, the priority PV micro system can be developed. This is extremely useful when developing a customer driven micro grid. First, the proposed method can be used for giving priority for intermittent renewable energy resources such as PV system. Second, the customer priority can be developed if we have similar such micro grids in the system. This coordination can be done with the help of smart controllers that indicate priority based on a global MPPT to extract maximum energy from the renewable energy resource. Finally, in the event of faulty modes of operation, shut down of individual micro-sources can be initiated if the system is configured correctly.

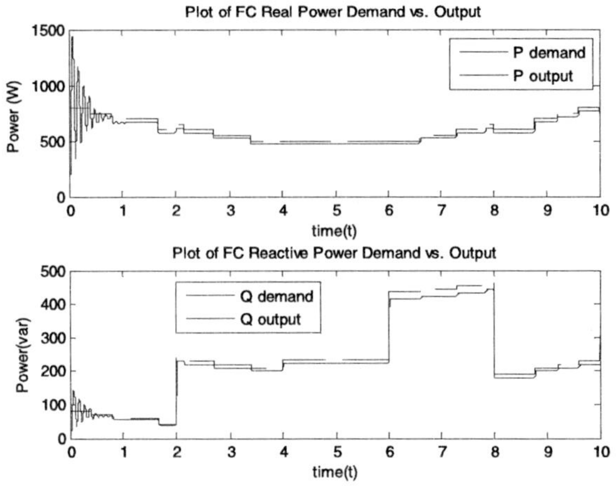

Fig. 41 Plot of FC Real and Reactive Power Demand vs. Output Handled by the FC

5 Conclusions

This chapter has presented an overview of PV and FC systems, the mathematical formulation and the modeling of all components involved. The model consist of mathematical design and development of FC, PV module and arrays, power conditioning DC/DC converter and the inverter set up. Further, a PV charge controller based on Maximum Power Point Tracking (MPPT) has been developed for the control of PV arrays. For FC a feedback controller has been designed. These designs are then integrated with an inverter controller that controls the voltage at a power system bus. Further, an intelligent controller provides a priority tracking scheme in order to develop a load following customer driven micro grid. The developed designs are modeled using Matlab® and Simulink®. For validating the models, real-life systems are used and the comparisons are evaluated. The complete system is used as a micro grid set up connected to a power grid. Testing and evaluations of the micro grid is then conducted using off-grid and on-grid studies and the results show that the micro grid model could emulate real-life scenarios.

Acknowledgement. The first author would like to express since gratitude to National Science Foundation (NSF) for the support in part for the completion of this work provided thought NSF grant # ECS 0748238.

References

[1] US EPA. Inventory of U.S. Greenhouse Gas Emissions and Sinks: 1990-1998, Rep. EPA 236-R-00-01. US EPA, Washington, DC (2000), http://www.epa.gov/globalwarming

[2] Ackermann, T., Andersson, G., Soder, L.: Distributed generation: a definition. Electric Power Systems Research 57, 195–204 (2001)

[3] Simons, G., Sethi, P., Davis, R., DeGroat, K., Comwell, D., Jenkins, B.: The Role of Renewable Distributed Generation in California's Electricity System. In: IEEE PES Summer Meeting, vol. 1, pp. 546–547 (2001)
[4] Fracz, R., Dmowski, A.: Monitoring and Control of Distributed Power Generation Systems such as Photovoltaic System. In: X International PhD Workshop, pp. 313–316 (2008), http://mechatronika.polsl.pl/owd/pdf/313.pdf
[5] Lasseter, R.H.: Microgrids and Distributed Generation. Journal of Energy Engineering 133(3), 144–149 (2007)
[6] Abu-Sharkh, S.: Can Microgrids Make a Major Contribution to UK Energy Supply. Renewable and Sustainable Energy Reviews 10, 78–127 (2006)
[7] El-Sharkh, M.Y., Rahman, A., Alam, M.S.: Neutral Network-based Control of Active and Reactive Power of a Stand-alone PEM Fuel Cell Power Plant. Journal of Power Sources 135, 88–94 (2004)
[8] Hatziadoniu, C.J., Lobo, A.A., Pourboghrat, F., Daneshdoost, M.: A Simplified Dynamic Model of Grid-Connected Fuel-Cell Generators. IEEE Trans. On Power Delivery 17(2), 467–473 (2002)
[9] Amphlett, J.C., Agbossou, K., Laperriere, A., Laurencelle, F., Bose, T.K.: Dynamic Behavior of a PEM Fuel Cell Stack for Stationary Applications. International Journal of Hydrogen Energy 26(6), 625–629 (2001)
[10] Strategic Plan for Distributed Energy Resources, US Department of Energy (September 2000),
http://www.eere.energy.gov/de/pdfs/de_program_plan.pdf (retrieved January 26, 2009)
[11] Gonzalez-Longatt, F.M.: Model of Photovoltaic Module In Matlab. In: Latin American Congress of Students of Electrical Engineering (2005)
[12] Neamen, D.A.: Semiconductor Physics and Devices. Mcgraw-Hill, New York (2003)
[13] El-Sharkh, M.Y., Rahman, A., Alam, M.S., Sakla, A.A., Byrne, P.C., Thomas, T.: Analysis of Active and Reactive Power Control of a Stand-alone PEM Fuel Cell Power Plant. IEEE Trans. Power Systems 19(4), 2022–2028 (2004)
[14] Uzunoglu, M., Alam, M.S.: Dynamic Modeling, Design, and Simulation of a Combined PEM Fuel Cell and Ultracapacitor System for Stand-alone Residential Applications. IEEE Trans. on Energy Conv. 21(3), 767–775 (2006)
[15] Hauer, K.: Analysis tool for fuel cell vehicle hardware and software (controls) with an application to fuel economy comparisons of alternative system designs. Ph.D. Dissertation, Department of Transportation Technology and Policy, University of California, Davis (2001)
[16] Mohan, Undeland, Robbins: Power Electronics – Converters, Applications, and Design, 2nd edn. John Wiley and Sons Inc., Chichester (1995)
[17] Hansen, A.D., Sorensen, P., Hansen, L.H., Binder, H.: Models for a Stand-Alone PV System. Wind Energy and Atmospheric Physics Department (2008)
[18] DenHerder, T.: Design and Simulation of Photovoltaic Super System Using Simulink. California Polytechnic State University (2006)
[19] Tsai, H., et al.: Development of Generalized Photovoltaic Model Using MATLAB/SIMULINK. In: Proceedings of the World Congress on Engineering and Computer Science, San Francisco, CA, October 22-24 (2008)

[20] Jain, S., Agarwal, V.: New current control based MPPT technique for single stage grid connected PV systems. Energy Conversion & Management 48, 625–644 (2007)
[21] Chenni, R., et al.: A detailed modeling method for photovoltaic cells. Energy 32, 1724–1730 (2007)
[22] Wang, Nehrir, Gao: Control of a PEMFC Distribution Generation System. IEEE Trans. on Energy Conversion 21(2), 586–595 (2006)

Optimal Allocation of Wind Turbines in Active Distribution Networks by Using Multi-Period Optimal Power Flow and Genetic Algorithms

P. Siano, P. Chen, Z. Chen, and A. Piccolo

Abstract. In order to achieve an effective reduction of greenhouse gas emissions, the future electrical distribution networks will need to accommodate higher amount of renewable energy based distributed generation such as Wind Turbines.

This will require a re-evaluation and most likely a revision of traditional methodologies, so that they can be used for the planning and management of future electrical distribution networks. Such networks evolve from the current passive systems to active networks and smart grids, managed through systems based on Information Communication Technology.

This chapter proposes a hybrid optimization method that aims at maximizing the Net Present Value related to the investment made by Wind Turbines developers in an active distribution network. The proposed method combines a Genetic Algorithm with a multi-period optimal power flow.

The method, integrating active management schemes such as coordinated voltage control, energy curtailment and power factor control is demonstrated on a 69-bus 11 kV radial distribution network.

1 Introduction

Today, the environmental concerns and energy crisis lead the international politics towards new energy development that calls for the reduction of pollutant emissions and the increase of energy efficiency. Consequently, Distributed Generation (DG) based on renewable sources has become the focus of new energy development (see European Directive 2001/77/EC and 2003/54/EC).

P. Siano · A. Piccolo
Department of Industrial Engineering, University of Salerno, Fisciano (SA), Italy
e-mail: psiano@unisa.it, apiccolo@unisa.it

P. Chen · Z. Chen
Department of Energy Technology, Aalborg University, Pontoppidanstraede 101, Aalborg, Denmark
e-mail: pch@iet.aau.dk, zch@iet.aau.dk

L. Wang (Ed.): *Modeling and Control*, Green Energy and Technology, pp. 249–268.
springerlink.com

In the meantime, distribution network operators (DNOs) are now forced to deal with an increasing number of problems, largely related to increased loads, new environmental policies and economic pressures of the market. Distribution networks should be operated to meet the following three main objectives:

- environmental sustainability
- security and quality of supply
- low cost.

These objectives are essential elements of a modern smart grid (SG). First, a SG should allow a simple connection (plug & play) of renewable energy based DG through common standards of equipment and communication systems. Second, a SG, by means of self-healing systems, should be able to reduce the periods of disservice in an automatic or semi-automatic mode, and thus improve the quality of power supply. Finally, a SG should take advantage of a competitive electricity market and optimize the dispatch of generation and consumption so that the total system operating cost is minimized without degrading the system security.

At the distribution level, a smart grid can be achieved through the evolution of active distribution networks that employ various active network management schemes (ANM). Such ANM schemes intend to make better use of available energy resources [4, 8, 11, 31, 34, 35, 38, 40, 47].

ANM is expected to emerge as the preferred solution to the connection and operation of DG in the near-future. In the initial stage, ANM will allow monitoring and remote control of DG to facilitate its integration in the system. In the intermediate stage, ANM will permit accommodating significant amount of DG once local and global services and trading issues have been defined. Ultimately, the full active power management will rely on network management regime and use real-time communication and remote control to meet the majority of the network service requirements [11].

ANM schemes can be designed by using automatic controllers and data obtained from information communication technologies (ICT). Due to the liberalization of the electricity market, the distribution network has a full connection with the customer, the supplier and the transmission system operator. Such a connection enables the exchange of not only energy but also information. The information exchange can further contribute to the optimization of the market operation and system security. Therefore, the exchange of information is crucial for the energy sector to carry out the proper power system management strategies.

Recent progress in the field of ICT can facilitate such information exchange and accelerate the transformation of distribution networks from passive systems to active ones. Thus, the ICT is considered a strategic instrument to upgrade the existing infrastructure.

Consequently, with the increase in complexity of future distribution systems based on ICTs, it is necessary to develop innovative methodologies, models and measuring techniques that are able to cope with these new contexts.

Furthermore, the operation and planning of a SG involves not only the DNOs, but also DG developers and energy consumers. Among other issues, this chapter focuses on the technical challenges faced by both the DNOs and DG developers when considering the grid integration of large amounts of DG [21, 25, 13, 14, 15, 30].

These challenges are partly caused by the mismatch between the location of energy resources and the capability of local networks to accommodate new generation. Particularly, the location of wind turbines (WTs) is determined by the local wind resources and geographical conditions. However, the current capacity of the network to which the WTs will be connected may not be sufficient to deliver the generated wind power. As a result, network reinforcement needs to be planned by the DNOs and the costs of these reinforcements should be partly paid by WT developers.

Since such network reinforcement usually calls for high capital investment, DNOs and WT developers would like to explore less costly means that can improve the capability of the network to accommodate new generation. One way is to make the best use of the existing network by encouraging development at the most suitable locations [13, 14]. In order to do this, DNOs and WTs developers require a reliable and repeatable method of quantifying the capacity of new DG that may be connected to distribution networks without the need for reinforcement.

The challenge of identifying the best network location and capacity for DG has attracted significant research effort, albeit referred to by several terms: optimal 'capacity evaluation' [12, 13, 14, 15, 30], 'DG placement' [27], or 'capacity allocation' [20, 43, 44]. These optimization problems apply different numerical algorithms with various objectives and constraints. For example, genetic algorithms are used to find the optimal location of DG [2, 22, 23]. Several other algorithms are adopted to handle optimization problems with discrete variables [10, 43]. Other approaches require network locations of interest to be pre-specified with algorithms guiding capacity growth within network constraints [13, 14, 15, 30, 33].

Nevertheless, as values associated with WTs are time- and location-dependent, methods that simply consider one specific power value at a specific moment are not able to account for time dependence. Therefore, WTs optimal allocation should consider their capability of delivering power at the right time and WTs should be located in the right place to be able to deliver energy while satisfying network constraints. Simulating load and generation variations during a year and computing the WTs delivered energy allows including the time dimension, when compared with methods that simply consider the power at one specific point in time [33]. In order to account for the time dependence of load and generation, some approaches focused on the concept of energy from DG [9, 21]. In [9], an optimization method is developed to evaluate annual energy in order to measure the risk of unserved energy for each planning option. However, active management schemes are not considered in either method to exploit higher energy from DG. In [21], an optimization method is proposed to maximize the total energy reaped from a given area and to minimize DG connection costs and system losses.

This chapter proposes a hybrid optimization method that aims at maximizing the Net Present Value (NPV) related to the investment made by WTs developers in a distribution network endowed with active management schemes. The hybrid optimization method combines the Genetic Algorithm (GA) and the multi-period optimal power flow (OPF) algorithm.

The GA is suitable for selecting the optimal site and number of WTs among some selected WTs types allowing WTs developers to maximize the NPV. The evaluation of the wind energy production is based on a multi-period OPF algorithm, which takes into account distribution network constraints. Such a multi-period OPF, derived from the OPF methods of [13, 14, 15] and [39] considers the time-varying characteristics of the load demand and wind power generation. The method also integrates active management schemes such as coordinated voltage control, energy curtailment and power factor control. The analyses are demonstrated on a 69-bus 11 kV radial distribution network. Sections 2 describes the active management schemes adopted in the multi-period OPF which is described in Section 3. Section 4 describes the GA features. Sections 5 and 6 present and comment some case studies. A discussion on the presented results is given in Sections 7 while conclusions are drawn in Section 8.

2 Assessing Maximum Wind Energy Exploitation in Active Distribution Networks by Means of Multi-period OPF

Active management represents an alternative approach to enable national targets for renewable energy and increase the penetration of WTs into the existing distribution networks [24, 32]. It has, indeed, the potential to maximize DG penetration level while minimizing DG-related network reinforcements [1, 5, 6, 26, 28, 29, 36, 37, 41, 46]. In [41] it is demonstrated that networks endowed with active management schemes can potentially accommodate up to three times as much generation.

Active management can be realized, for instance, through generation dispatch, transformer tap adjustment or reactive power compensators. In [24], WTs generation curtailment during low demand, reactive power management using a reactive compensator, and area-based on-load tap-changer coordinated voltage control have been used in the active management.

In [28, 29], a multi-period steady-state analysis for maximizing the capacity of wind generation through an OPF-based technique with active management features has been proposed. However, since wind capacity rather than wind energy is maximized, WTs allocation does not allows maximum wind energy exploitation. Moreover, short-circuit level is computed with a simplified approach.

The more advanced and emerging concept of active management is based on real time measurements of the distribution network parameters and employs real time control of generators, tap-changing transformers, reactive power compensators and communication among the generators and voltage control devices [24].

The multi-period OPF proposed here improves the methods proposed in [39] and [28, 29] by accounting for load and generation time interdependences and by focusing on the concept of energy from WTs. The proposed method allows, in fact, finding the optimal WTs capacities allocation in order to maximise wind energy exploitation under different active management schemes, briefly described in the following subsection.

2.1 Coordinated On-Load Tap-Changer Voltage Control

Traditional control strategies of on-load tap-changers (OLTCs) are either based on the voltage regulation at a single busbar or voltage drop compensation on a particular line [24]. Such voltage control strategies are based on local measurements and are suitable for traditional distribution systems with unidirectional power flow. However, these strategies may cause problems in distribution networks with bi-directional power flows. On the other hand, the area-based control strategy of OLTCs is based on measurements from various locations of the network. In this way, the voltage regulation of OLTCs can be based on the voltage information of the bus that has the most severe over voltage problem [24]. Consequently, the maximum wind energy penetration level may be increased by the implementation of the control strategy.

2.2 Energy Curtailment

In order to alleviate the over voltage problem, it may be necessary to curtail a certain amount of wind energy injected into the network [24]. Although the output wind energy is reduced, the WT developer may still gain more profits due to the possibility of installing more WTs [36].

In the proposed method, wind energy may be curtailed during certain periods in order to alleviate any voltage or thermal constraint violation. For example, for a specific period, there are different possible combinations of load demand and wind power. Wind energy is curtailed during periods of minimum demand and high wind power generation. The same strategy is applied to each of the periods analyzed.

In the method, energy curtailment is implemented in each period by introducing a negative generation variable to represent the curtailed energy from each WT. For a given period, the maximum energy that can be curtailed from a given WT is set to a fraction of the potential energy that the WT could have produced without energy curtailment.

2.3 Coordinated Generator Reactive Power Control

The recent grid codes of many countries, such as Denmark, Germany, Italy, Ireland and the UK, require that WTs should provide reactive power control capabilities and that network operators may specify power factor or reactive power generation requirement for grid-connected WTs [42].

In practice, a grid-connected WT needs to fulfill the specific requirement depending on the regulation of the country. For example, in the Danish grid code for grid-connected WTs, reactive power generation is confined to a control band with respect to active power generation (with a power factor between 1.00 and 0.995 lagging). The German grid code specifies different reactive power limits according to voltage value at interconnection (with a power factor ranging between 1.00 and 0.925 lagging). The Irish grid code requires a power factor between 0.835 leading and 0.835 lagging when the active power output level is below 50% of the rated capacity. In Italy and the UK, the power factor at a WT's terminal should be between 0.95 leading and 0.95 lagging.

Although it is important to fulfill the grid code when connecting a WT, this paper intends to illustrate the concept of the proposed method, but not to design a WT that fulfills a specific requirement.

WTs, especially those with power electronic controllers, are able to provide necessary reactive power support to the grid. The reactive power generation can be dispatched centrally by the DNOs [45]. In other words, power factors of WTs can be controlled so that wind energy penetration level in the network is maximized. The proposed control scheme requires WTs to generate reactive power during load peak hours and low generation, and to absorb reactive power during load off-peak hours and high generation.

3 Multi-Period Optimal Power Flow

The optimization method aims to find the optimal locations and capacities of WTs so that the wind energy exploitation in the network is maximized. Such an objective is subject to a number of technical constraints imposed by regulations including bus voltage limits, line/transformer thermal limits, and system short-circuit levels. By fulfilling these constraints, the network reinforcement due to the connection of WTs may be avoided. In addition, such a method can be used to investigate the impact of the foregoing active management strategies on the maximum wind energy penetration level in the network.

The proposed approach, based on the nonlinear programming formulation of the multi-period OPF described in [39] and [28, 29], has been modified in order to maximize the wind energy exploitation and to include active management schemes, the time-varying characteristics of the load demand and wind power generation, and the system short-circuit constraints.

The multi-period OPF is formulated as:

$$\begin{aligned} &\text{maximize E} = \sum_{j=1}^{N_j} \sum_{g=1}^{N_G} E_g^j(P_g,\ \mathbf{x_j}) \\ &\text{subject to} \\ &\mathbf{h}(\mathbf{x_j}) = 0, \\ &\mathbf{g}(\mathbf{x_j}) \le 0 \end{aligned} \tag{1}$$

where $E_g^j(P_g,\ \mathbf{x_j})$ is the wind energy generated during the time period j by the gth WT with rated capacity P_g, N_j is the total number of periods in a year corresponding to different combinations of load demand and wind power generation; N_G is the number of WTs (indexed by g).

The vector $\mathbf{x_j}$ consists of a set of controllable quantities and dependent variables during each period j. The optimization variables include the capacity of each WT, and for each period j: the secondary voltage of the OLTC, the power factor angle and the curtailed energy of each WT, and the import/export power at the interconnection to the external network.

The equality constraints $\mathbf{h}(\mathbf{x_j})$ represent the static load flow equations such as Kirchhoff current law $\forall j \in J$ and $\forall b \in B$, where J is the set of periods (indexed by j), B is the set of buses (indexed by b) and Kirchhoff voltage law, $\forall j \in J$ and $\forall l \in L$, where L is the set of lines (indexed by l).

The inequality constraints $\mathbf{g}(\mathbf{x_j})$ are listed in the following.

-Capacity constraints for the interconnection to external network (slack bus) $\forall j \in J$, $\forall x \in X$:

$$\begin{aligned} P_x^- \le P_{x,j} \le P_x^+ \\ Q_x^- \le Q_{x,j} \le Q_x^+ \end{aligned} \tag{2}$$

where X is the set of external sources (indexed by x), $P_{x,j}$ and $Q_{x,j}$ are the active and reactive power outputs of x, respectively and P_x^-, P_x^+ and Q_x^-, Q_x^+ are the min/max active and reactive power outputs of x, respectively.

-Capacity constraints for the WTs: maximum capacity that may be installed at each site $\forall j \in J$, $\forall g \in G$:

$$\begin{aligned} P_g^- \le P_{g,j} \le P_g^+ \\ Q_g^- \le Q_{g,j} \le Q_g^+ \end{aligned} \tag{3}$$

where G is the set of WTs (indexed by g), $P_{g,j}$ and $Q_{g,j}$ are the active and reactive power outputs of g, respectively and P_g^-, P_g^+ and Q_g^-, Q_g^+ are the min/max active and reactive power output of g, respectively.

-Voltage level constraints $\forall j \in J$, $\forall b \in B$:

$$V_b^- \le V_{b,j} \le V_b^+ \tag{4}$$

where $V_{b,j}$ is the voltage at bus b, V_b^+ and V_b^- are the max/min voltage at bus b, respectively.

Flow constraints for lines and transformers $\forall j \in J$, $\forall l \in L$:

$$\sqrt{\left(f_{l,j}^P\right)^2 + \left(f_{l,j}^Q\right)^2} \le f_l^+ \tag{5}$$

where $f_{l,j}^{P}$ and $f_{l,j}^{Q}$ represent the active and reactive power injection onto l respectively and f_l^+ the maximum power flow on l.

- Short-circuit level constraint: the requirement of not exceeding the design short-circuit capacity in typical radial networks, fed by a MV/LV substation and with wind generation, should be satisfied as it could constrain new generation capacity. WTs connected to the distribution network may contribute to the short-circuit level at the distribution substation. The upstream grid provides the dominant contribution to the short circuit capacity, which rapidly diminishes downstream the network due to the series impedance of the lines. The short-circuit requirement normally needs to be checked at the MV (or LV) busbars of the substation [2]. Therefore, given the typical radial arrangement of distribution networks, the maximum short-circuit level will be obtained when considering a three-phase short-circuit at the low-voltage side of the substation.

The magnitude of the expected short-circuit current $|I_{cc}|$ at the low-voltage side of the substation, calculated from the phasor sum of the maximum short-circuit currents from the upstream grid, through the step-down transformer, and from the WTs connected to the distribution network, is, therefore, limited by the design short-circuit capacity $I_{cc}^{\max}$.

$$|I_{cc}| \le I_{cc}^{\max} \tag{6}$$

The grid contribution is calculated according to IEC 60909 [16, 17, 18, 19] and the contribution of WTs is computed according to the method proposed in [2].

The additional constraints derived from the active management schemes are: coordinated on-load tap-changer voltage constraint, curtailed energy, WTs power factor angles.

- Curtailed energy constraint $\forall j \in J$:

$$CE_g^j \le CE_{g\max}^j \tag{7}$$

where CE_g^j represents the amount of curtailed energy from generator g during period j and $CE_{g\max}^j = C_f^j \times E_g^{j-\max}$ the maximum permitted curtailed energy from generator g during j, where C_f^j is the curtailment index, varying in the range [0,1] and $E_g^{j-\max}$ is the maximum energy that generator g could have produced during j without curtailment.

- Coordinated on-load tap-changer voltage constraint $\forall j \in J$:

$$V_{OLTC}^{-} < V_{j,OLTC} < V_{OLTC}^{+} \tag{8}$$

where $V_{j,OLTC}$ is the secondary voltage of the OLTC during j, V_{OLTC}^{-} and V_{OLTC}^{+} are the (max/min) voltage of the OLTC, respectively.

- Coordinated generator reactive power constraints, $\forall j \in J$, $\forall g \in G$:

$$\phi_g^- < \phi_{g,j} < \phi_g^+ \tag{9}$$

where $\phi_{g,j}$ is the power factor angle of g during j, ϕ_g^- and ϕ_g^+ are the (max/min) power factor angle of g, respectively.

4 Genetic Algorithm - Multi-Period OPF Hybrid Optimization Method for Optimal Allocation of Wind Turbines

The GA is used in order to select the types and number of WTs to be allocated at each candidate bus. The GA randomly generates the initial population of solutions (individuals) by defining a set of vectors. Each vector, or called a chromosome, has a size $N_e = N_C \times N_T$, where N_C is the number of candidate locations and N_T is the number of defined WT types. This is demonstrated in Fig. 1.

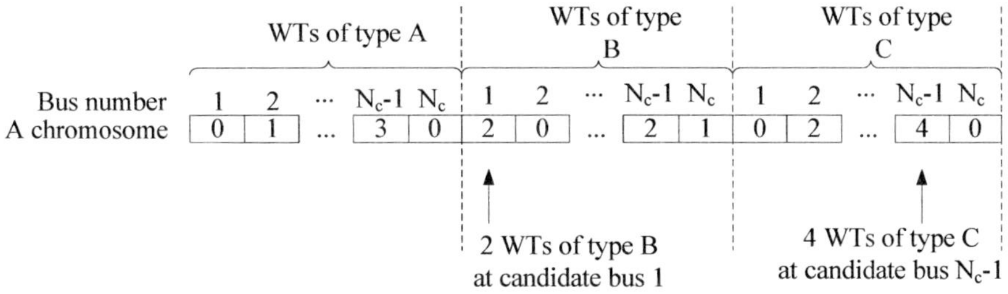

Fig. 1 Schematic of the GA chromosome.

As shown in Fig. 1, a chromosome consists of a vector of integers, each of which represents the number of WTs of a given type to be allocated at a candidate bus. For instance, WTs of type A is associated with the first part of the vector with the size of N_c, which is the number of the candidate locations. Each element of this vector is an integer representing the number of WTs of type A connected to the corresponding bus. As such, the locations and types of WTs are expressed as a string of integers.

At each generation of the GA, a new set of improved individuals is created by selecting individuals according to their fitness; the selection mechanism used here is the normalized geometric ranking scheme. After the new population is selected, genetic operators are applied to selected individuals for a discrete number of times. These genetic operators are simple crossover and binary mutation. A simple crossover randomly selects a cut-point dividing each parent into two segments. Then, two segments from different parents are combined to form a new child (individual). A binary mutation changes each of the bits of the parent based on the probability of mutation. An elitism mechanism is also adopted to ensure the best member of the population is not lost. The iteration process continues until one of the stopping criteria is reached.

For each chromosome of the GA, specifying the number and location of WTs, the maximum wind energy generation over a year is evaluated by a multi-period OPF algorithm nested in the GA algorithm.

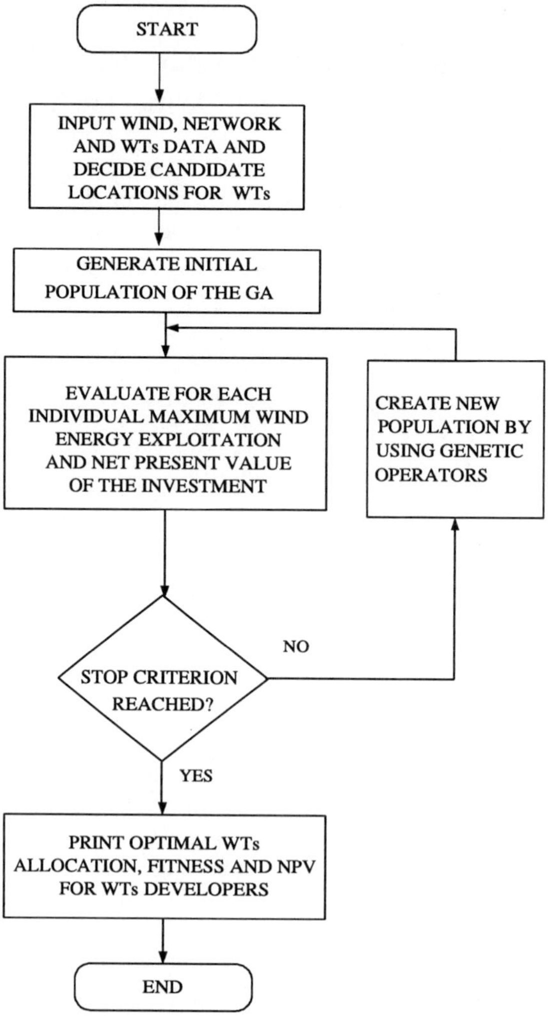

Fig. 2 Flow chart of the Hybrid optimization method.

In order to analyze the profitability of the WTs investment, the GA objective function is the NPV, i.e. the difference between the sum of the discounted cash flows which are expected from the investment and the amount which is initially invested:

$$NPV = -C_0 + \sum_{k=1}^{n} \frac{C_k}{(1+i)^k} \tag{10}$$

where: k the year of the cash flow, i the discount rate (the rate of return that could be earned on an investment in the financial markets with similar risk), C_k the net cash flow (the amount of cash, inflow minus outflow) at year k, C_0 is the investment, n is the lifetime of WTs.

Consequently, this hybrid method will deliver the best locations as well as the best WT types in the end.

The flow chart of the foregoing hybrid optimization method is shown in Fig. 2.

The proposed method has been implemented in Matlab® and is based on MATPOWER suite [48] and demonstrated through the study system described in the following section.

5 System Description

This section describes the distribution system and data that are used to demonstrate the hybrid optimization approach proposed in the previous section for the optimal allocation of WTs.

The following analyses are based on a 69-bus 11 kV radial distribution system whose data are given in [7]. The four feeders are supplied by two identical 6 MVA 33/11 kV transformers. Fig. 3 shows the distribution system and the potential WT locations, selected to demonstrate the capabilities of the method.

5.1 Modelling of Time-Varying Load and Wind Power Generation

For the modeling of time-varying load and wind power generation, real data from the local distribution network in Nordjylland in Denmark have been used and processed. In order to account for the seasonal, weekly and daily variation of load, the measured data are grouped by summer/winter, weekday/weekend, and 24 hours. In order to account for the seasonal and daily variation of wind power generation, the measured data are grouped by summer/winter and 24 hours. In particular, the 365 days of the year have been divided into 153 winter days and 212 summer days and, for each week into 5 weekdays and 2 weekend days. As a result, there are 96 groups for load and 48 groups for wind power generation.

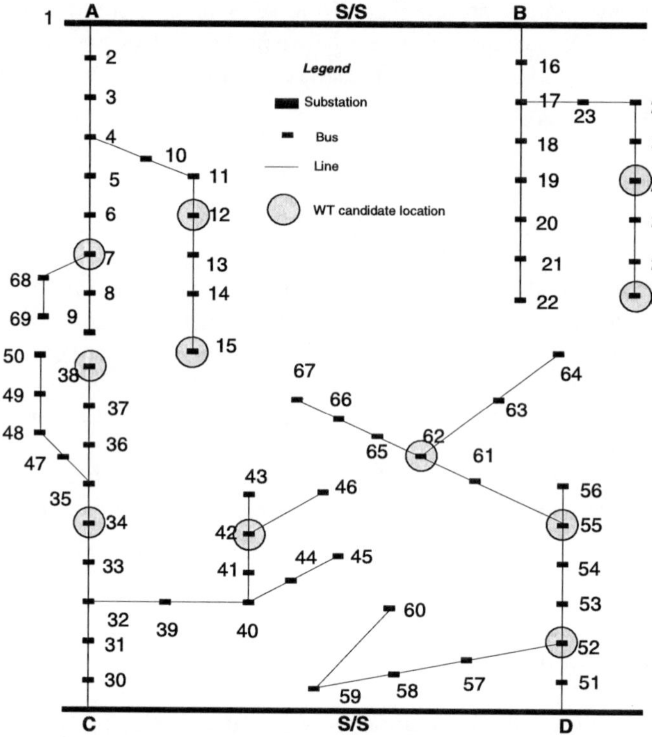

Fig. 3 69-bus network indicating potential locations for WTs.

From each group of load, e.g. 12 o'clock in a summer weekday, a load duration curve is obtained and then discretized into four states. A similar approach is applied to the wind power generation, but discretized into ten states. The discretization is demonstrated in Fig. 4. As a result, for one group of load, there are 4 load states with corresponding 10 wind power generation states.

Each type of day consists of 24 hours each of which can have 40 (10x4) different combinations of load-generation, therefore a total of 3840 load flows ($3840 = 4_{day\_types} \times 24_{hours} \times 40_{load-generation}$), with different load-generation combinations have been analyzed in the MP-OPF. In order to create the multi-period interdependency, at each iteration of the MP-OPF, to a unique set of WTs capacity variables correspond 3840 sets of power flow variables.

The maximum load level of each bus given in [7] is scaled down for the use in the 69-bus network: the corresponding maximum loading levels are summarized in Table 1.

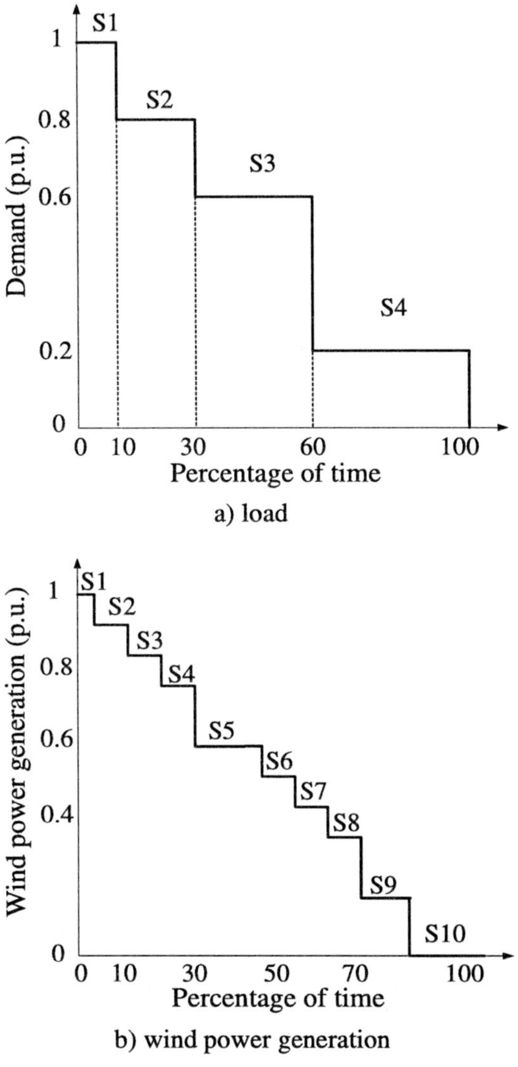

Fig. 4 Discrete states of a) load and b) wind power generation at a given hour.

Table 1 Maximum Network Loading

Active Power [MW]	Reactive Power [MVar]
2.90	1.99

5.2 Short-Circuit Calculations

The short-circuit capacity of the grid is assumed to be 1000 MVA. The grid is connected to the transformers through a series inductive impedance of 8%. Losses for the substation's transformers are not considered. The designed short-circuit capacity for the 11 kV network is assumed to be 200 MVA. As recommended by the IEC Standard [16, 17, 18, 19], the R/X ratio for the corresponding equivalent impedance is assumed to be 0.1. The voltage factor used to consider the variations of the system voltage is assumed to be 1.1 [2]. The WTs are connected to the 11 kV network through a 0.69/11 kV transformer with a series impedance of 4% and a rated resistive component of the short-circuit voltage of 1.2% [2]. In order to evaluate the WTs' contribution to the short-circuit level at each bus, the equivalent short-circuit impedances of WTs and transformers have been computed for each bus according to the installed WTs capacities and to the parallel connections of WTs.

5.3 Network Operation Constraints

Voltage limits are taken to be ±6% of nominal and feeder thermal limits are 5.1 MVA (270 A/phase). The substation power exports to the upstream grid are limited to the capacity of the transformers (12 MVA).

In order to demonstrate the capabilities of the energy curtailment scheme, it has been assumed that wind energy can be curtailed only during the period in which the combination of minimum demand and high wind power occurs. It has been assumed that energy output from each WT operating at the 10th or 9th wind power generation states could be curtailed until it reaches the 8th wind power generation state (a curtailment index C_f^j equal to 0.62 for the 10th wind power generation state and a curtailment index C_f^j equal to 0.86 for the 9th wind power generation state). Energy curtailment during such periods should alleviate over-voltage or overcurrent problems.

Power factor is assumed to vary between 0.9 leading and 0.9 lagging when the coordinated generators reactive power control option is considered. Otherwise, WTs are assumed to operate with a fixed power factor of 0.95 lagging (absorbing reactive power).

The short-circuit limit constraint of 200 MVA has been assumed accordingly to the designed short-circuit capacity for the network.

6 Case Study

The hybrid optimization algorithm was applied on the above-described distribution system.

It is assumed that WTs of three different capacities are chosen by the WT developers. These capacities are 225 kW, 660 kW and 900 kW.

Maximum three WTs of each type are allowed at a given location. This requirement may be set by the available land for building WTs. For another distribution network with a different load level, WTs with different capacities may be considered.

Consequently, GA is used to search for the optimal number of WTs of each type at the candidate locations. It is also assumed that the power factor is the same for all WTs connected to the same bus.

The multi-period OPF has been applied for evaluating its annual maximum wind energy exploitation considering the following active management options simultaneously: coordinated OLTC voltage control, energy curtailment and WTs reactive power control.

The basic parameters of the GA are summarized as follows. The total control variables are 33 (= 3×11), corresponding to the number of three types of WTs at the eleven candidate locations. The population size of each generation is 20. The initial population is generated at random between zero and three.

The GA stops if any of the following conditions is reached:

1) the maximum generation number exceeds 150,

2) there is no improvement in the objective function for 50 consecutive generations, and 3) the cumulative change in the fitness function value over 5 generations is less than $1e^{-6}$.

Sensitivity analyses have been carried out to consider different values for the GA parameters such as stop criteria, population size and genetic operators. From these analyses, it was shown that the used values guarantee the convergence of the algorithm to a satisfactory solution in this case.

In order to evaluate the effectiveness of the proposed hybrid method, different scenarios for the wind speed distributions in each bus of the network have been assumed.

Table 2 Weibull wind speed distributions parameters

Weibull distribution	Scale Parameter	Shape Parameter
WD1	6	3
WD2	10	3
WD3	18	3

Table 3 Capital costs (Cc) associated to the three candidate WTs

WT Type	Rated output electric power [kW]	Capital cost [€/kW]	Total capital cost [M€]
A	225	1400	0.315
B	660	1200	0.792
C	900	900	0.810

In particular, three different Weibull wind speed distributions have been considered, namely WD1, WD2 and WD3. The corresponding wind power generation have been derived and discretized on the basis of power curves of commercial WTs. The Weibull wind speed distributions parameters are shown in Table 2. Table 3 lists the capital costs (C_c) associated to the three candidate WTs.

Table 4 NPV for each WT considering the different wind distributions

WT Type	NPV [€]		
	WD1	WD2	WD3
A	98 320	745 100	1 307 100
B	-264 570	1 964 100	3 958 000
C	-248 600	2 407 500	5 432 200

Table 5 Optimal numbers of WTs at different locations found by the GA and corresponding wind speed distribution

Bus no.	Weibull distribution	225 kW	660 kW	900 kW	Capacity (kW)
7	WD3	2	1	3	3810
12	WD3	2	1	3	3810
15	WD3	3	0	0	675
29	WD2	2	2	0	1770
26	WD2	1	1	0	885
34	WD3	2	0	2	2250
38	WD3	2	0	2	2250
42	WD3	2	2	3	4470
55	WD1	3	0	0	675
52	WD1	3	0	0	675
62	WD1	1	0	0	225
Total		**23**	**7**	**13**	**21495**

Table 4 presents the NPV for a single WT of type A, B, and C in correspondence of different wind distributions WD1, WD2 and WD2. The negative value of the NPV indicates that the initial investment is higher than the total benefits obtained from selling wind power over N years.

It is observed that, under the wind speed distribution of WD3, the investment in a single WT of type C produces higher profits than the investment in one type A WT plus one type B WT. On the contrary, under the wind speed distribution of WD2, the investment in one type A WT plus one type B WT is more beneficial than the investment in one single type C WT.

Moreover, under the wind speed distribution of WD2, the investment in three WTs of type A produces higher profits than the investment in one type B WT. In contrast, under the wind speed distribution of WD3, the investment in one type B WT is more beneficial than the investment in three type A WTs. However, under the wind speed distribution of WD1, only the investment in a type A WT is profitable.

Table 5 lists the optimal numbers of WTs at different locations found by the GA and the corresponding wind speed distribution assumed at each location.

As shown in Table 6 a total WTs capacity of 21.495 MW is installed allowing delivering 126,880 MWh/year, with a curtailed wind energy of 5319 MWh/year.

The total capital cost of the investment is of 23.319 M€ and the NPV equals 108.810 M€.

Table 6 Results of WTs obtained from the hybrid optimisation method

Total capacity [MW]	Delivered wind energy [MWh]	Total capital cost [M€]	NPV [M€]	Curtailed wind energy [MWh]
21.495	126 880	23.319	108.820	5319

7 Discussion

The proposed optimization method combines the GA and the multi-period OPF and considers the time-varying characteristics of the load demand and wind power generation. The proposed method allows the WTs developers to optimally allocate a chosen number and types of WTs among a large number of potential combinations in an active distribution network. Furthermore, the method can be used to evaluate the feasibility of a project in WTs before carrying out investments. Simulation results evidenced that the proposed hybrid method, which maximizes the NPV related to the investment in WTs, is suitable for selecting the optimal site and number of WTs among selected WTs types.

Different active management schemes have been considered in the proposed optimization formulation as they are able to enhance the total amount of wind energy exploitation and thus offer more economic benefits to both WTs developers and DNOs. In fact, active management is expected to provide higher profits to the WTs developers by allowing them connecting more WTs to the network. The active management is also an effective and indispensable strategy for DNOs to integrate and operate WTs in distribution networks and to defer network investments caused by annual load growth and/or DG connections.

Nevertheless, practical implementation of active management schemes requires additional commercial arrangements and financial evaluations. New market rules should be implemented to offer economic benefits to DNOs in order to drive them to provide the active management service to WTs developers. On the other hand, new revenue mechanisms should be developed so that WTs developers and DNOs share the benefits as well as the costs of active management.

Further simulations with larger networks, not presented here, have demonstrated the scalability of the proposed method and its applicability to larger networks [30]. The method is also able to cope with a larger number of control variables [30], [39]. Although this will lead to an increase in the computing time, this is not a constraint as the method is intended for long-term planning studies. Different load profiles (by considering a mix of industrial, commercial and residential customers) for each node can be easily introduced in the method.

The main drawback of the proposed hybrid optimization algorithm is that the simulation time is very long. This is due to the evaluation of the fitness function which each time calls for a lengthy multi-period OPF. However, this drawback can be tolerated as simulation time is not a major concern for long-term system planning. In addition, a more powerful computer may improve the simulation speed to a certain extent.

8 Conclusion

This chapter has described a hybrid optimization method that can help WTs developers to plan investments in an active distribution network. The hybrid optimization method, combining the GA and the multi-period OPF, maximizes the NPV related to the investment in WTs and allows evaluating the economic benefits deriving from active management schemes. Simulation results on a 69-bus 11 kV radial distribution network confirmed the effectiveness of the proposed method in selecting the optimal site and number of WTs among different WTs types.

References

1. Ault, G.W., Currie, R.A.F., McDonald, J.R.: Active Power Flow Management Solutions for Maximising DG Connection Capacity. In: Proc. of IEEE Power Eng. Society General Meeting, pp. 1–5 (2006)
2. Boutsika, T.N., Papathanassiou, S.A.: Shortcircuit calculations in networks with distributed generation. Electric Power Systems Research 78, 1181–1191 (2008)
3. Celli, G., Ghiani, E., Mocci, S., Pilo, F.: A multiobjective evolutionary algorithm for the sizing and siting of distributed generation. IEEE Trans. Power Systems 20(2), 750–757 (2005a)
4. Celli, G., Pilo, F., Pisano, G., Soma, G.G.: Optimal participation of a microgrid to the energy market with an intelligent EMS. In: Proc. of the 7th International Power Engineering Conference, vol. 2, pp. 663–668 (2005b)
5. Currie, R.A.F., Ault, G.W., Foote, C.E.T., Burt, G.M., McDonald, J.R.: Fundamental research challenges for active management of distribution networks with high levels of renewable generation. In: Proc. of 39th Int. Universities Power Eng. Conf., vol. 2, pp. 1024–1028 (2004)
6. Currie, R.A.F., Ault, G.W., Foote, C.E.T., McDonald, J.R.: Active power-flow management utilising operating margins for the increased connection of distributed generation. IET Generation, Transmission & Distribution 1(1), 197–202 (2007)
7. Das, D.: A Fuzzy Multiobjective Approach for Network Reconfiguration of Distribution Systems. IEEE Trans. Power Delivery 21(1), 202–209 (2006)
8. Djapic, P., Ramsay, C., Pudjianto, D., Strbac, G., Mutale, J., Jenkins, N., Allan, R.: Taking an active approach. IEEE Pow. & Energy Magazine 5(4), 68–77 (2007)
9. Dugan, R., Wacławiak, M.: Using energy as a measure of risk in distribution planning. In: Proc. of 19th Int. Conf. on Electricity Distribution, pp. 1–4 (2007)
10. El-Khaltam, W., Bhattacharya, K., Hegazy, Y., Salama, M.M.A.: Optimal investment planning for distributed generation in a competitive electricity market. IEEE Trans. Power Systems 19(3), 1674–1684 (2004)

11. European Commission, EUR 22040: European Technology Platform SmartGrids, Office for Official Publications of the European Communities (2006)
12. Harrison, G.P., Wallace, A.R.: Optimal power flow evaluation of distribution network capacity for the connection of distributed generation. IEE Proceedings Generation, Transmission & Distribution 152(1), 115–122 (2005)
13. Harrison, G., Piccolo, A., Siano, P., Wallace, A.R.: Exploring the Trade-offs Between Incentives for Distributed Generation Developers and DNOs. IEEE Trans. on Power Systems 22, 821–828 (2007b)
14. Harrison, G.P., Piccolo, A., Siano, P., Wallace, A.R.: Hybrid GA and OPF evaluation of network capacity for distributed generation connections. Electrical Power Systems Research 78, 392–398 (2008)
15. Harrison, G.P., Piccolo, A., Siano, P., Wallace, A.R.: Distributed Generation Capacity Evaluation Using Combined Genetic Algorithm and OPF. International Journal of Emerging Electric Power Systems 8, 1–13 (2007a)
16. IEC 60909-1: Short-circuit currents in three-phase a.c. systems—Part 1:factors for the calculation of short-circuit currents according to IEC 60909-0 (2002)
17. IEC 60909-2: Electrical equipment—Data for short-circuit current calculations in accordance with IEC 909 (1992)
18. IEC 60909-3: Short-circuit currents in three-phase a.c. systems—Part 3: currents during two separate simultaneous line-to-earth short circuits and partial short-circuit currents flowing through earth (2003)
19. IEC 60909-4 : Short-circuit currents in three-phase a.c. systems—Part 4:examples for the calculation of short-circuit currents (2000)
20. Keane, A., O'Malley, M.: Optimal Allocation of Embedded Generation on Distribution Networks. IEEE Trans. Power Systems 20(3), 1640–1646 (2005)
21. Keane, A., O'Malley, M.: Optimal Utilization of Distribution Networks for Energy Harvesting. IEEE Trans. on Power Systems 22(1), 467–475 (2007)
22. Kim, K.H., Lee, Y.J., Rhee, S.B., Lee, S.K., You, S.K.: Dispersed generator placement using fuzzy-GA in distribution systems. In: Proc. of IEEE PES Summer Meeting, Chicago, USA, pp. 1148–1153 (2002)
23. Kuri, B., Redfern, M., Li, F.: Optimization of rating and positioning of dispersed generation with minimum network disruption. In: Proc. of IEEE Power Eng. Soc. Gen. Meeting, Denver, CO, pp. 2074–2078 (2004)
24. Liew, S.N., Strbac, G.: Maximising penetration of wind generation in existing distribution networks. In: Proc. of IEE Proc. Generation, Transmission and Distribution, vol. 149, pp. 256–262 (2002)
25. Masters, C.L.: Voltage rise: The big issue when connecting embedded generation to long 11 kV overhead lines. Power Eng. J. 16(1), 5–12 (2002)
26. Mutale, J.: Benefits of Active Management of Distribution Networks with Distributed Generation. In: Proc. of Power Systems Conf. and Exp., pp. 601–606 (2006)
27. Nara, K., Hayashi, Y., Ikeda, K., Ashizawa, T.: Application of tabu search to optimal placement of distributed generators. In: Proc. of IEEE PES Winter Meeting, pp. 918–923 (2001)
28. Ochoa, L.F., Dent, C.J., Harrison, G.P.: Maximisation of intermittent distributed generation in active networks. In: Proc. of IET-CIRED Seminar SmartGrids for Distribution, pp. 1–4 (2008)
29. Ochoa, L.F., Keane, A., Dent, C., Harrison, G.P.: Applying active network management schemes to an Irish distribution network for wind power maximisation. In: Proc. of Int. Conf. on Electricity Distribution, pp. 1–4 (2009)

30. Piccolo, A., Siano, P.: Evaluating the Impact of Network Investment Deferral on Distributed Generation Expansion. IEEE Trans. Power Systems 24(3), 1559–1567 (2009)
31. Prica, M., Ilic, M.D.: Optimal Distribution Service Pricing for Investment Planning. In: Proc. of IEEE PES General Meeting, Tampa (Florida, USA), pp. 1–7 (2007)
32. Pudjianto, D., Castro, M., Djapic, P., Stojkovska, B., Strbac, G., Allan, R.N.: Transmission Investment and Pricing in Systems with Significant Penetration of Wind Generation. In: Proc. of IEEE Power Engineering Society General Meeting, pp. 1–3 (2007)
33. Rau, N.S., Wan, Y.H.: Optimum location of resources in distributed planning. IEEE Trans. Power Systems 9(4), 2014–2020 (1994)
34. Roberts, V., Collinson, A., Beddoes, A.: Active networks for the accommodation of dispersed generation. In: Proc. of 2003 CIRED Conf., pp. 1–6 (2003)
35. Samotyj, M., Howe, B.: Creating Tomorrow's Intelligent Electric Power Delivery System. In: Proc. of 2005 CIRED Conf., pp. 1–5 (2005)
36. Schwaegerl, C., Bollen, M.H.J., Karoui, K., Yagmur, A.: Voltage control in distribution systems as a limitation of the hosting capacity for distributed energy resources. In: Proc. of 18th Int. Conf. on Electricity Distribution, pp. 1–5 (2005)
37. Shafiu, A., Bopp, T., Chilvers, I., Strbac, G.: Active management and protection of distribution networks with distributed generation. In: Proc. of IEEE Power Eng. Society General Meeting, vol. 1, pp. 1098–1103 (2004)
38. Shafiu, A., Jenkins, N., Strbac, G.: Measurement location for state estimation of distribution networks with generation. IEE Proc. Gen., Trans. and Distr. 152, 240–246 (2005)
39. Siano, P., Ochoa, L.F., Harrison, G.P., Piccolo, A.: Assessing the strategic benefits of distributed generation ownership for DNOs. IET Generation, Transmission & Distribution 3(3), 225–236 (2009)
40. SUSTELNET. Review of technical options and constraints for integration of Distributed Generation in electricity networks (2003), `http://www.sustelnet.net`
41. Strbac, G., Jenkins, N., Hird, M., Djapic, P., Nicholson, G.: Integration of operation of embedded generation and distribution networks. Technical Report (2002)
42. Tsili, M., Papathanassiou, S.: A review of grid code technical requirements for wind farms. IET Renewable Power Generation 3(3), 308–332 (2009)
43. Vovos, P.N., Harrison, G.P., Wallace, A.R., Bialek, J.W.: Optimal Power Flow as a tool for fault level constrained network capacity analysis. IEEE Trans. Power Systems 20(2), 734–741 (2005)
44. Vovos, P.N., Bialek, J.W.: Direct incorporation of fault level constraints in optimal power flow as a tool for network capacity analysis. IEEE Trans. Power Systems 20(4), 2125–2134 (2005)
45. Vovos, P.N., Kiprakis, A.E., Wallace, A.R., Harrison, G.P.: Centralised and Distributed Voltage Control: Impact on Distributed Generation Penetration. IEEE Trans. Power Systems 22(1), 476–483 (2007)
46. Zhang, J., Cheng, H., Wang, C., Xia, Y., Shen, X., Yu, J.: Quantitive assessment of active management of distribution network with distributed generation. In: Proc. of Third Int. Conf. on Electric Utility Deregulation and Restructuring and Power Technologies, pp. 2519–2524 (2008)
47. Zhang, J., Cheng, H., Wanga, C.: Technical and economic impacts of active management on distribution network. Electrical Power and Energy Systems 31, 130–138 (2009)
48. Zimmerman, R.D., Gan, D.: MATPOWER – A MATLAB Power System Simulation Package, User's Manual, School of Electrical Engineering, Cornell University (1997)

Planning the Grid for Winds of Change

Yunzhi Cheng and Mandhir Sahni

Abstract. With larger wind generation penetration and concentration on the grid, the planning/operation requirements associated with the wind generation have evolved drastically over the last decade or so. While the old challenges such as wind resource forecasting are still under study and of significance, the new trend/attitude of treating the wind generation as the conventional generation from the perspective of the fair market principle of the ISO (Independent System Operator) brings many new challenges associated with wind generation integration. Thus the Voltage Ride-though (VRT) capability, reactive power support, frequency control and sub-synchronous oscillation become key topics from a power system planning standpoint. This chapter briefly introduces all these new study aspects including relevant discussions on WGR model Equivalencing, requirement and assessment of VRT and reactive power, inertial and frequency control of the wind generation, all from power system planning perspective.

1 Introduction/Background

With the wind generation technology maturing over the last decade or so, the penetration of wind generation resources (WGR) in transmission system networks across the world has been on the rise consistently. With close to 76,000 MW of installed capacity as of 2006, wind generation can now be effectively termed as "utility scale". With the large amounts of wind penetration comes the ability of wind to impact generation and transmission planning for regional reliability and/or utility organizations across the world. Figure 1 depicts the United States' national fuel mix as of 2009 as per Edison Electric Institute (EEI) [1]. Figure 2 depicts the growth of installed wind capacity in the U.S over the last decade or so [2]. Figure 3 depicts a map of announced wind projects by state totaling over 48 GW with expected completion dates out till 2015 [3]. Figure 4 depicts the wind generation interconnection queues by reliability organization in GW [3].

Yunzhi Cheng · Mandhir Sahni
PwrSolutions Inc, 2777 N Stemmons Fwy Suite 1520 Dallas TX USA 75207

L. Wang (Ed.): *Modeling and Control*, Green Energy and Technology, pp. 269–309.
springerlink.com

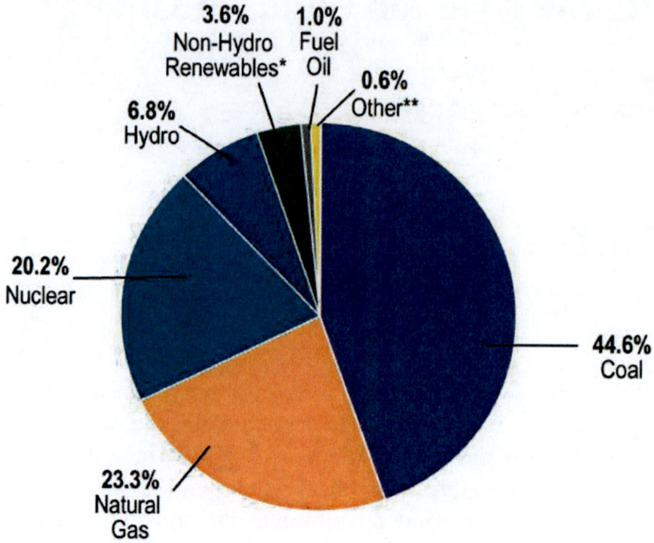

Fig. 1 U.S National Fuel Mix as of 2009

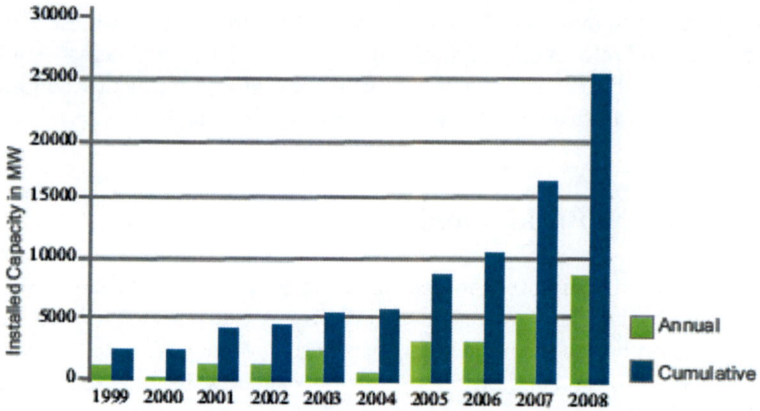

Fig. 2 Growth in U.S Wind Energy Installations

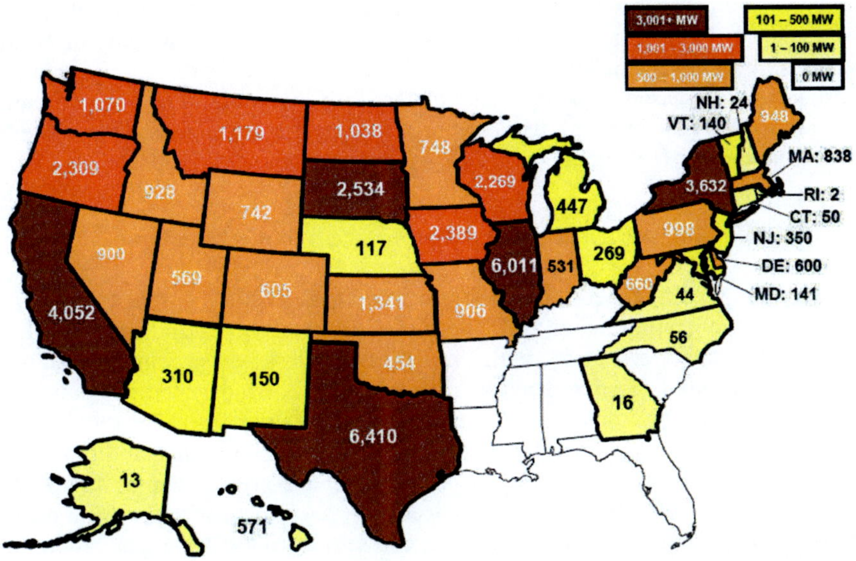

Fig. 3 Wind Capacity in Development by State

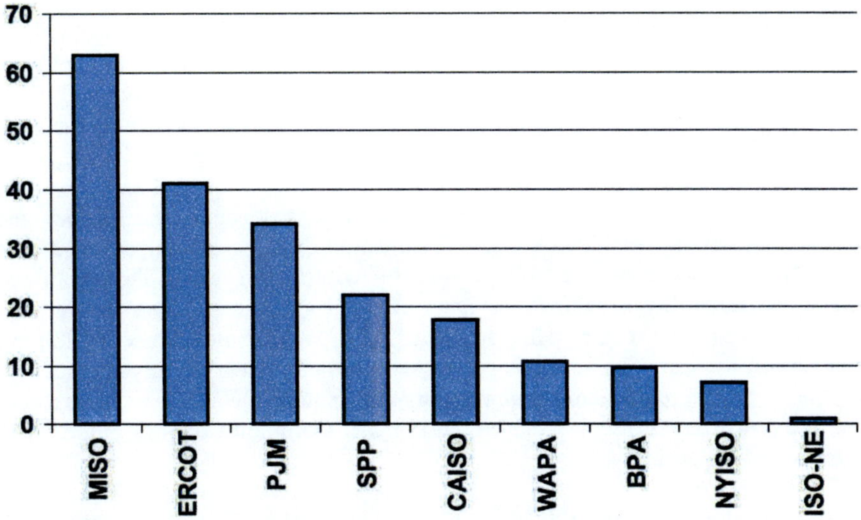

Fig. 4 Wind Interconnection Queues (GW) by Reliability Region

While the presence of the aforementioned large amounts of wind generation on the transmission system does go a long way in achieving regional renewable portfolio standard (RPS) goals, the grid integration of such large amounts of wind presents some unique challenges. Some of the key aspects that have presented themselves as increased challenges associated with integrating the large amount (and continuously increasing) wind generation in different transmission system networks across the U.S and the world in general are:

- **Scheduling sufficient generation resources to meet demand**

The challenges associated this aspect are many fold. Historical trends are indicative of wind generation being available during off-peak hours and months. Furthermore, uncertainty associated with availability of wind power coupled with the inaccuracies in wind forecast make it difficult for the wind generating units to either store or schedule their primary energy source. Regional Reliability Organizations and/or Balancing Authorities, in order to fulfill their responsibility of ensuring reliable delivery of electric power to the entire grid within its footprint, have to ensure that the aforementioned uncertainty associated with the wind energy must be accounted for in the form of reserves from controllable generation resources in the system.

- **Maintaining adequate dynamic voltage support (reactive power capability) to respond to voltage events**

Generation resources utilizing traditional synchronous generation technology employ a standard Automatic Voltage Regulator (AVR) to ensure automatic voltage control. In the case of wind units, while the older wind units rely on external switching devices (capacitor banks, Static Var Compensator (SVC) etc) for reactive power support the newer wind units utilize the in-built power electronic devices to perform the same function. Apart from the other issues raised by external switching or power electronic devices (harmonics, resonance etc), proper and timely operation of these devices has to be ensured by the system operator during any voltage events to ensure voltage stability within its system. To that effect, most utility/reliability organizations have effected specific steady state and dynamic reactive power requirements associated with WGRs interconnecting to their transmission system. The reactive power requirements are typically included in the generation interconnection requirements and require the WGR to meet a prespecified power factor requirement at the Point of Common Coupling (PCC) with the transmission grid.

- **Frequency Control & Primary Frequency Response**

The issues associated with maintaining frequency control in the wake of increased wind penetration are again many fold. First of all, wind generation units being variable energy sources are utilized to meet the Balancing energy need in the system. With increasing penetration of wind, a greater percentage of the energy sources utilized for providing balancing energy is being derived from wind energy sources. Given the uncertainties associated with the wind energy output, the risk associated with not meeting the balancing energy need and maintaining frequency within acceptable limits is getting higher.

Additionally, unlike the VRT and Reactive capability requirements, the ability to provide primary frequency response is a more recent grid interconnection requirements associated with WGRs connecting to the transmission grid. Figure 5 provides an illustration of typical frequency response associated with a generation unit trip [4]. Under normal operating conditions, the balance between the generation and load results in the system frequency being at nominal value (i.e. 60 Hz). However, the loss of a generation unit results in the aforementioned load-generation balance being perturbed thereby causing the system frequency to drop. The initial decay of the system frequency (from point A to C) is arrested by a combination of the system inertial response and primary frequency response. It is important to understand that while the primary governor response would assist in the recovery of frequency close to nominal (from point C to B) , supplemental or secondary frequency control in the form of Automatic Generation Control (AGC) or Load Frequency Controller action is needed to restore frequency back to nominal value.

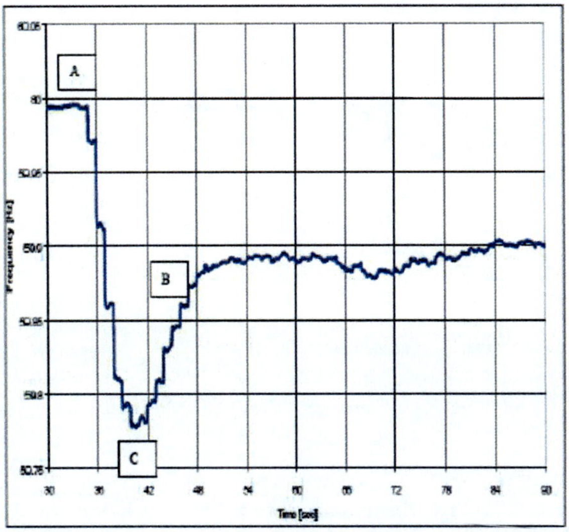

Fig. 5 Typical Frequency Response associated with generation unit trip depicting NERC defined A, B & C points

The primary frequency response requirement for the interconnecting WGRs is predicated from the fact that increasing concentration of WGRs (which effectively replace conventional generation in the unit commitment stack) tend to reduce the available system inertia available to respond to low-frequency events. Effectively, the frequency nadir associated with a low frequency event is exaggerated during high wind conditions thereby necessitating the need for fast acting primary frequency response on WGRs connected to the grid during the event to ensure timely and adequate frequency recovery. To that effect, numerous utilities require interconnecting WGRs to demonstrate the ability to provide primary frequency response with characteristics similar to the 5% droop characteristics associated with conventional generators.

- **Voltage Ride Through (VRT) Capability**

Based on the technology associated with the wind turbine(s) utilized for the wind generation units, various wind generation units have varying capabilities to remain online during a fault/outage or a dip in voltage. The ability of a WGR to ride through voltage dips on the transmission system for a pre-specified period of time without tripping i.e. VRT capability has become one of the key grid interconnection requirements when connecting to transmission grids across the U.S. The issue has gained additional emphasis with the increasing penetration of wind generation across the country. Figures 6 and 7 provide pictorial and tabulated versions of the summary of VRT standards across different national grid codes respectively [5].

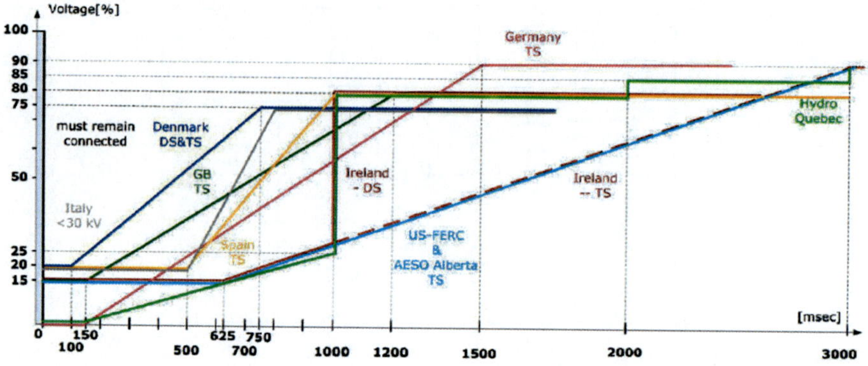

Fig. 6 Summary of Voltage Ride-Through (VRT) Standards across various national grid codes

Unlike the assessments discussed above, the Sub-Synchronous Resonance (SSR) & Sub-Synchronous Torsional Interaction (SSTI) assessments are not typically part of the grid interconnection requirements. However, with the advent of series compensated transmission lines coupled with the large concentration of wind units such assessments are becoming part of the WGR interconnection process. The ideal example of the same is the Competitive Renewable Energy Zones (CREZ) Transmission Development in Electric Reliability Council of Texas (ERCOT) wherein the presence of series compensated lines in close proximity to wind

Country	Voltage Level	Fault ride-through capability				
		Fault duration	Voltage drop level	Recovery time	Voltage profile	Reactive current injection
Denmark	DS	100 msec	25%U_r	1 sec	2, 3-ph	no
	TS	100 msec	25%U_r	1 sec	1, 2, 3-ph	no
Ireland	DS/TS	625 msec	15%U_r	3 sec	1, 2, 3- ph	no
Germany	DS/TS	150 msec	0%U_r	1.5 sec	generic	Up to 100%
Great Britain	DS/TS	140 msec	15%U_r	1.2 sec	generic	no
Spain	TS	500 msec	20%U_r	1 sec	generic	Up to 100%
Italy	> 35 kV	500 msec	20%U_r	0.3 sec	generic	no
USA	TS	625 msec	15%U_r	2.3 sec	generic	no
Ontario	TS	625 msec	15%Ur	-	-	no
Quebec	TS	150 msec	0%U_r	0.18 sec	Positive-sequence	no

Fig. 7 Summary of Voltage Ride-Through (VRT) Standards across various national grid codes

collection stations has necessitated the screening for SSR & SSTI phenomenon associated with WGRs. The categories of sub-synchronous phenomenon relevant to the discussion at hand are:

- **Torsional Interaction (TI):** TI usually occurs when the generators connected to a series compensated transmission system have one or more natural frequencies which are the synchronous frequency complements of one or more of the spring-mass natural frequencies. TI can lead to growing shaft torque oscillations at one of the natural mechanical frequencies of the turbine-generator spring-mass system. Additionally SSTI can also manifest itself in terms of interactions between conventional generator and HVDC and/or conventional generators and power electronic devices such as wind farms.
- **Induction Generator Effect (IGE):** Unlike the TI, the IGE is purely an electrical phenomenon that results from the generator appearing as a negative resistance circuit at prevailing sub-synchronous frequencies due to the flow of sub-synchronous currents in the armature circuit of a sub-synchronous machine. IGE could result in current and voltage large enough to be damaging for the generator and power system equipment. In the eventuality, that the sub-synchronous currents in the armature circuit of the synchronous machine are at a frequency corresponding to one of the turbine-generator spring-mass modes, large oscillatory shaft torque could be experienced.
- **Torque Amplification (TA):** TA refers to a phenomenon that usually occurs following the clearing of a fault on a series compensated line, resulting in high energy storage in the series capacitor banks being discharged through the generators in the form of a current which has a frequency corresponding to one of the natural frequencies of the

turbine-generator spring-mass system. Unlike the TI and IGE, the growth rate associated with TA oscillations is high and can cause extensive damage in a very short period of time.

Apart from the classic SSR and/or SSTI issues, another concern associated with interconnecting WGRs on weak transmission grids is the presence of Sub-Synchronous Control Instability (SSCI). The power electronic devices and fast control loops associated with modern day WGR technologies are designed to operate on transmission grids with short circuit ratios between 2 to 5. However, with wind resource availability typically corresponding to areas of the transmission grid that are very weak, short circuit ratios in the order of 0 to 2 are not uncommon. The low short circuit ratios associated with the weak transmission grid coupled with the fast control loops and high over-all gain of the WGRs can potentially result in marginally stable or unstable responses of the control loops. While the detailed discussion presented in the ensuing sections of this chapter focuses on the more common WGR grid interconnection requirements (VRT capability, reactive power capability & frequency response assessment), the authors felt it appropriate to outline some lesser observed and more complex issues associated with grid integration of wind resources.

2 Voltage Ride-Through Capability

2.1 Conceptual Discussion and Background

The ability of a Wind Generation Resource (WGR) to ride through voltage dips on the transmission system (especially the Point of Interconnection i.e. POI) has gained significant importance over the last decade. The relatively negligible concentration of WGRs in the power system network in the past allowed the WGRs to disconnect from the network during and/or immediately following system disturbances with little or no system impact. However, the ever increasing penetration of WGRs (as a percent of the over-all system resource mix), has resulted in the need for WGRs to stay connected and ride through voltage dips on the transmission system for certain pre-specified durations. An ideal illustration of the impact of increasing WGR penetration on the need to possess VRT capability exists in the form of the Electric Reliability Council of Texas (ERCOT) grid or the "Texas Connection". With close to 9.5 GW of installed wind capacity as of December 2010 [6], even 30% of the installed wind capacity being dispatched on any given windy day/hour amounts close to 3.2 GW. In the eventuality that most WGRs in such a system do no possess the ability to ride through voltage dips on the transmission system, the grid would stand to loose about 3.2 GW of generation due to a major transmission system disturbance. Under such a situation, the amount of generation lost or disconnected from the transmission grid would tend to exceed the responsive reserves available to the system to arrest the resulting frequency decline thereby resulting in serious system reliability concerns.

To that effect, almost all transmission system and/or grid operators have outlined specific Voltage Ride-Through (VRT) requirements as part of their Planning/Operating Guides and/or grid codes. Some of the most notable VRT guidelines and/or standards are those associated with the E.ON Grid Code, Red Eléctrica de España (REE) Grid Code (the Spanish transmission system operator), the German and Danish power system network, the Western Electricity Coordinating Council (WECC) VRT standard and the Federal Energy Regulatory Commission (FERC) 661-a standard for WGR VRT requirements. [7] - [11] outline the WGR VRT requirements for the aforementioned power system networks as outlined in their interconnection procedures to integrate WGRs into the respective networks. In United States, while most regional reliability organizations have stipulated individual WGR VRT requirements, the VRT requirements outlined in FERC Order 661-A has served as the primary guideline. FERC Order 661-A calls for WGRs to ride through a 3- Phase fault on the high side of the wind farm station transformer (typically the POI of the wind farm with the transmission grid) for a duration of no more 9 cycles depending on the primary fault clearing time of the fault interrupting circuit breakers at the POI.

The advent of such VRT grid code requirements associated with WGRs has resulted in active research and development directed to the enhancement of wind turbine technology to aid turbines to ride through such voltage dips on the transmission system. While the Doubly Fed Induction Generator (DFIG) and full converter machines have been the focus of research and development associated with VRT capability lately, there are considerable amounts of Fixed Speed Induction Generator (FSIG/Type I) and Rotor Current Controlled (RCC/Type II) wind turbines installed in various power system networks across the world. Additionally, due to the inherent design of Type I and Type II machines, the LVRT capability of these turbine types assumes more significance. Some of the notable technologies employed to improve the VRT capability of Type I and/or II turbines include the utilization of central dynamic VAR (DVAR) devices such as Static VAR Compensators (SVC) and/or STATCOM to provide dynamic reactive power compensation during the voltage dip period, the use of Energy Capacitor System (ECS) to improve the LVRT capability of FSIG machines and the employment of dynamic braking resistors to improve the VRT capability for a FSIG machine just to name a few [12] - [14].

The DFIG technology on the other hand has been the focus of extensive research and development in terms of enhancement of VRT capability. Some of the key technology enhancements associated with equipping DFIG machines with advanced VRT capability include [15] - [20]:

- Efficient control strategies utilizing a combination of passive (typically a crowbar arrangement) and active LVRT compensators resulting in reduced electric torque oscillation, rotor instantaneous power and transient response of DFIG during a voltage dip thereby enhancing the VRT capability
- Utilization of series converters on the stator of the DFIG machine to prevent over-current conditions on the rotor side
- Limiting the fluctuation of the DC link voltage associated with the rotor-side converter in a DFIG machine

- Integration of short-term Energy Storage Device (ESD) in a DFIG machine design to smoothen fast wind-induced power variations and to strengthen the DC bus during the transient period to enhance VRT capability

While the background associated with VRT requirements for various power system grid codes and the technology aspects associated with VRT capability enhancement for different turbine types is useful, the focus of discussion in this section is the assessment of the VRT capability from a transmission planning perspective. From a transmission system planning perspective, all of the aforementioned standards and/or codes require the interconnecting WGR to demonstrate the ability to meet the VRT requirements at the POI by means of dynamic simulation studies. Additionally, most electric utilities, as per the grid code requirements, are required to test each interconnecting WGR for compliance with the host grid VRT requirements. The discussion presented in the ensuing subsections of this section focus on the modeling and simulation aspects associated with the VRT assessment of WGRs from a transmission planning standpoint.

2.2 *Key Modeling Requirements*

There are 2 main categories of models that are required for setting up and performing accurate dynamic simulations associated with WGR grid integration and/or interconnection studies, namely:

- Steady State Model
- Dynamic Model

The steady state model includes accurate and relevant modeling of the turbines, pad-mount transformers, wind farm collection system, station transformers and any other power system equipment that may affect the performance of the wind farm from a power flow and/or voltage standpoint. The steady state model is a key ingredient to setting up a acceptable "initial conditions" for the dynamic simulation. The dynamic model on the other hand primarily corresponds to the dynamic model of the turbines comprising the wind farm unless there are other external dynamic devices (such as SVCs and/or STATCOMS to name a few) in which case models for the same would also be accounted for under dynamic modeling. The dynamic model for wind turbines are typically compiled source codes for most commercial power system simulation software due to intellectual property issues. While access to the source code for such turbine models has been difficult in the past, there are methods to adequately test the turbine dynamic model to ensure acceptable dynamic initialization and behavior of the model. The key aspects associated with developing and/or ensuring acceptable steady state and dynamic models associated with a WGR for the purposes of grid integration VRT studies are presented in this sub-section.

2.2.1 Steady State Modeling

The following aspects need to be taken into account during the development of detailed steady state models to represent WGR facilities irrespective of the software environment being utilized by utility transmission planners and/or power system engineers:

- Accurate modeling of reactive capability of various wind turbine types
- Accurate modeling of reactive operation/control scheme for various wind turbine types
- Ability to distinguish between reactive capability and control amongst various types of wind turbines
    - Type of Control
    - Range of Control
    - Point of Control
- Station Transformer Load Tap Changer (LTC) Settings
    - Tap Settings & Step
    - Location of the tap changer
    - Side controlled by LTC – Numerous WGR facilities locate the LTC on the high side controlling the medium voltage side to provide the facility the ability to control the high voltages on the turbine side especially in the presence of sub-station capacitor banks
- Medium Voltage Capacitor/Reactor Banks
    - Size
    - Type – Fixed/Discrete/Continuous
    - Control Mechanism – Manual/Automated Switching
    - Control Point & Settings – Avoid conflicts with LTC and/or turbine control
- Accurate Detailed Collection System
    - It is necessary to account for the detailed collector system from a steady state and dynamic modeling perspective
    - The increasing size of WGR facilities (with campuses as large as 800 MW in the ERCOT system for example) further necessitate the need to account for the reactive power consumption and/or charging originating from collector system model
    - It is important to take into account the reactive control and operational schemes associated with various wind turbine types while developing detailed collector system models from a steady state standpoint. This fact is further re-iterated by the fact that the collection system of an operational 800 MW WGR campus (alluded to above) was found to be producing 64 MVAR when turbines were not producing any output. In other words, the shunt charging originating from the cable segments associated with collection systems of large WGR campuses do have significant modeling impact in terms of grid studies and cannot be ignored and/or assumed.

  - It is important to take into account the impedance associated with the collector system for WGR facilities when assessing VRT capabilities. The impedance between the point of common coupling with the system and the individual wind turbines does impact the ability of a wind turbine to ride through system disturbances for varying amounts of time.
- Accurate impedances (per unit length) for all major under-ground (U/G) cables and/or over-head (O/H) lines utilized for collection system
- Short Circuit Standpoint
  - The representation of wind turbines by means of short circuit models associated with conventional generation units is necessitated due to the absence of dedicated models for wind turbines in commercial power system software for short circuit analysis. To that effect, wind turbine models are represented by means of a constant voltage source behind an impedance which typically represents the "source impedance" (Z_{source}) in case of conventional generation models. However, in case of wind turbine models, this impedance is reflective of the "equivalent impedance" behind the turbine terminal.
  - Accurate reflection of source impedance for various turbines which is usually significantly different from the traditional transient (Xd') & sub-transient (Xd'') reactance for conventional machines. This is especially true for DFIG and/or full-conversion machines wherein the effective impedance up to the machine terminals also includes the impedance of the associated power electronic devices
  - Grounding scheme
  - Pad-Mount Transformers including impedance, connection configuration and grounding scheme
  - Station Transformers – Yg-Yg-Delta
- In more cases than not, there is a need to aggregate the collection system associated with the WGR when performing grid level interconnection studies. The aforementioned need arises due to numerous reasons with the most common one being the complexity of the resulting network if all wind farms in the network are modeled in detail to the last turbine and the resulting computational burden for dynamic simulations.
- The aggregation model for grouping N generators into 1 equivalent generator:
  - $R_{source\text{-}turbine\text{-}eq} = R_{source\text{-}turbine}/N$ (1)
  - $X_{source\text{-}turbine\text{-}eq} = X_{source\text{-}turbine}/ N$ (2)
- The aggregation model for grouping N pad-mount transfomers into 1 equivalent pad-mount transformer:

- $R_{PM\text{-}EQ} = R_{tran}/N$ (3)
- $X_{PM\text{-}EQ} = X_{tran}/N$ (4)

- The most popular aggregation method of the median voltage line/cable collector system is based on equal power loss principle:
  - For each line/cable segment of the median voltage collector system:
    - R_i is the resistance of the line/cable segment
    - X_i is the reactance of the line/cable segment
    - B_i is the suseptance of the line/cable segment
    - Count the downstream turbine number, N_i
    - i is the line/cable segment index, M is the total number of segments
  - The equivalent impedance of the voltage line/cable collector system can be derived by:
    - $$R_{eq} = \frac{\sum_{i=1}^{M} R_i * N_i^2}{N^2} \quad (5)$$
    - $$X_{eq} = \frac{\sum_{i=1}^{M} X_i * N_i^2}{N^2} \quad (6)$$
  - The equivalent impedance can be derived by:

$$B_{eq} = \sum_{i=1}^{M} B_i \quad (7)$$

- Care should be taken by utility engineers to ensure that the aggregated model associated with the wind farm has been validated with its detailed counterpart from the following standpoints:
  - Steady State Standpoint
    - Active and Reactive Power Losses typically assessed at POI
  - Short Circuit Standpoint
    - 3-Phase and S-L-G fault duties at POI
  - Dynamic Response standpoint
    - Terminal Voltage
    - Active & Reactive Power Response
    - Frequency Response
  - For the dynamic response, it is important to note that the voltage diversity across the wind farm campus arising due to the collection system impedance needs to be accounted for and retained in the aggregated model in order to ensure that the aggregated model adequately

and accurately represents the detail model especially in terms of VRT assessment. [21] discusses and approach developed by the authors of this chapter that lends itself well to achieving this while being applicable in the practical environment. This aspect assumes greater significance in case of VRT assessments since the voltage diversity across the wind farm could result in only a section of the farm tripping as opposed to a single machine aggregated model wherein the entire wind farm is modeled as being online or tripped.

The aforementioned aspects serve as key guidelines in ensuring the set-up and utilization of an accurate steady state model associated with WGRs for grid interconnection based VRT assessments.

2.2.2 Dynamic Modeling

The focus of the discussion presented in this section is not on the development of dynamic models associated with various wind turbine technologies but to equip power system planners with guidelines to assess the acceptable dynamic initialization and behavior of commercial grade turbine dynamic models prior to performing dynamic simulations similar to those required for VRT assessment.

Prior to the addition of any wind turbine model to the study dynamic data set or testing the dynamic behavior and/or flat start associated with a wind turbine model, the dynamic behavior of the model should be validated on a simple test system preferably containing only conventional unit dynamic models. This is done in order to prevent interaction between dynamic models associated with different wind turbine types and/or vendors.

The addition of the wind turbine model should be followed by the test of an acceptable flat start of the resulting dynamic data set utilizing a 10-sec no disturbance simulation. The following quantities should be monitored to assess the worst deviation as part of the wind flat start assessment:

- Conventional (Non-wind Generation Resources)
    - Electrical Active and Reactive Power Output
    - Terminal Voltage
    - Machine Rotor Angle
- Wind Generation Resources
    - Electrical Active and Reactive Power Output
    - Terminal Voltage
- System Frequency Deviation

Care should be taken to ensure that the addition of dynamic data associated with the wind turbine model does not have adverse impact on the dynamic behavior of the conventional machines reflected in terms of unacceptable deviations on the rotor angles of the conventional machines. The rotor angle of wind turbine model has been generally found to be an unsatisfactory indicator of dynamic initialization of the model and should not be utilized as one of the parameters.

The dynamic behavior of the wind turbine model should further be validated by simulating a 3-Phase normally cleared dynamic event at the POI associated with the WGR utilizing the wind turbine model. Typically, a 3-Phase fault lasting 6 cycles and subsequently cleared without tripping any transmission system elements should be simulated to assess the dynamic response of the wind turbine model by monitoring the following quantities:

- Pelec & Qelec
- Eterm
- System Frequency Deviation & Voltage at POI

Some general guidelines to bear in mind during wind turbine model validation:

- Certain wind turbine models do not lend themselves well to aggregation beyond a certain number and the dynamic performance of the resulting aggregated model may suffer.
- Observe the Pelec associated with the wind turbine model to ensure timely dissipation of the energy stored in the rotor following the clearing of the fault during a dynamic event simulation.
- The Xsource associated with a wind turbine model has profound impact on the wind turbine model behavior and should be one of the aspects investigated when a model exhibits unacceptable dynamic response.
- Certain wind turbine models have limitations on time step of integration below and/or above which they do not exhibit acceptable dynamic behavior. It is important to verify the same prior to running dynamic simulation.
- Wind turbine models for different machines do impact each others behavior when included in the same dynamic data set and hence the model validation exercise associated with a particular wind turbine model should be restricted to having only that wind turbine model in the sample dynamic data set.

All new versions of wind turbine models or new wind turbine models should be subjected to the above described model validation prior to including the models in the study dynamic data set.

2.3 Individual WGR VRT Capability Assessment

This section presents a discussion on the procedure associated with the assessment of the VRT capability for an individual WGR and its compliance with the host grid code requirements. The PSS/E sample ("SAVNW") system from Siemens PTI has been utilized as the test system to assess the behavior of the wind farm from a transient response standpoint [22]. The sample ("SAVNW") system comprises of 23 buses, 34 branches (including lines and transformers), and 6 generators. The load level associated with the sample ("SAVNW") system is 3,200MW. Figure 8 depicts a one-line schematic associated with the sample "savnw" case utilized as the test system.

The "SAVNW" case is modified to incrementally add the WGR under study. The 300MW generator at bus #3018 in the original "savnw" system is replaced by a wind farm with the same capacity comprising of the turbines under test for VRT compliance. The turbine type and/or vendor have not been dwelt upon since the approach is independent of the turbine type and/or vendor. The 35kV collector system equivalent and substation transformer (35kV/230kV) associated with the 300 MW WGR have also been modeled. The data associated with the WGR incrementally added to the "savnw" dataset is depicted in Table 1 below.

Table 1 WGR Collection System, Station Transformer & Capacitor Data

Devices	Parameters	Value
Substation Transformer (35kV/230kV)	MVA Base	333
	R (p.u. on MVA Base)	0.002
	X (p.u. on MVA Base)	0.085
35kV Collector System Equivalent Circuit	R (p.u. on 100MVA Base)	0.014
	X(p.u. on 100MVA Base)	0.026
	B (p.u. on 100MVA Base)	0.0618
35kV Capacitor (only applied for the wind farm with fixed speed wind turbine generators)	MVAR	48

The resulting dynamic dataset, following the inclusion of the test WGR, was tested for a 10-sec no-disturbance flat start to assess acceptable dynamic initialization of the WGR in the study dataset. Maximum deviations associated with the following power system quantities were observed for the duration of the 10-sec flat start and were observed to be within acceptable limits:

- Rotor Angles of conventional units
- Terminal Voltages
- Active & Reactive Power
- Frequency

It is important to note that the dynamic behavior of the turbine being utilized for the test wind farm has been exhaustively tested utilizing the procedures outlined above prior to being utilized for dynamic simulation. However, discussion associated with the details of the same are beyond the purview of this chapter.

The ERCOT VRT requirement curve has been utilized to illustrate the procedure associated with the assessment of the VRT capability of the test wind farm in terms of compliance. Figure 9 depicts the VRT boundary associated with the ERCOT VRT requirements for WGRs as provided in ERCOT Operating Guides Section 3.1.4.6.1[23].

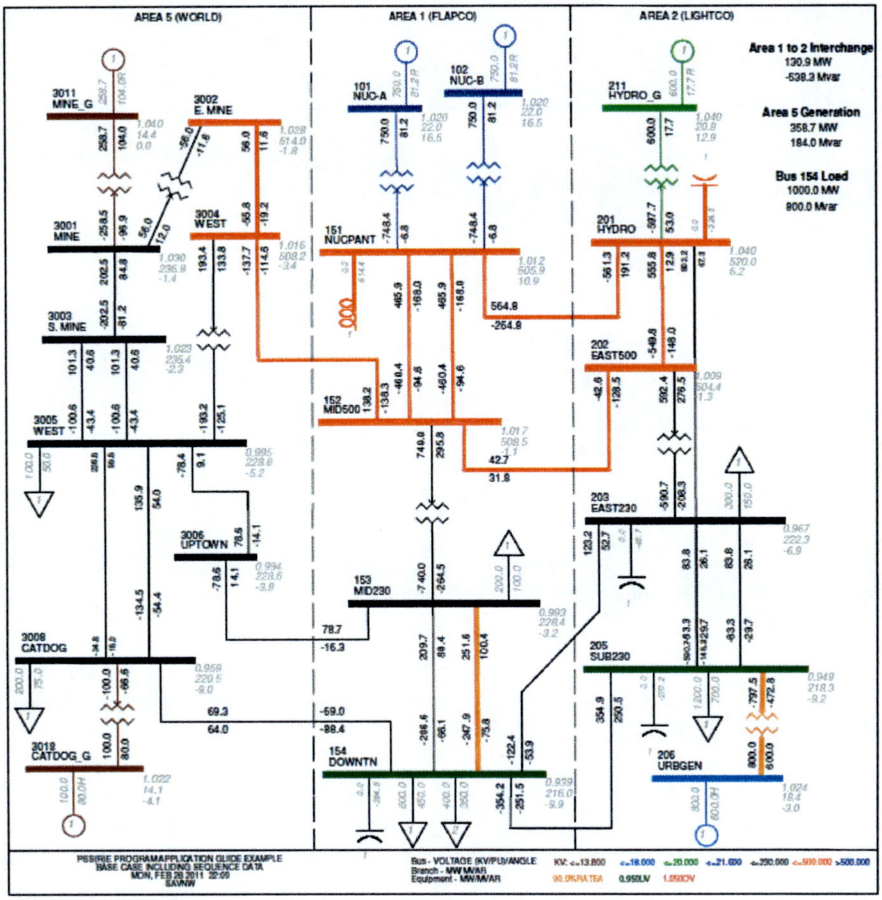

Fig. 8 Sample "savnw" transmission system one-line schematic

Section 3.1.4.6.1 of the ERCOT Operating Guide states that Wind-powered Generation Resources (WGRs) are required to set generator voltage relays to remain in-service during all transmission faults (no more than nine (9) cycles) in accordance with Voltage Ride-Through Boundaries For Wind-powered Generation Resources. Faults on individual phases with delayed clearing (zone 2) may result in phase voltages outside this boundary but if the phase voltages remain inside this boundary then generator voltage relays are required to be set to remain connected and recover within the voltage recovery boundary.

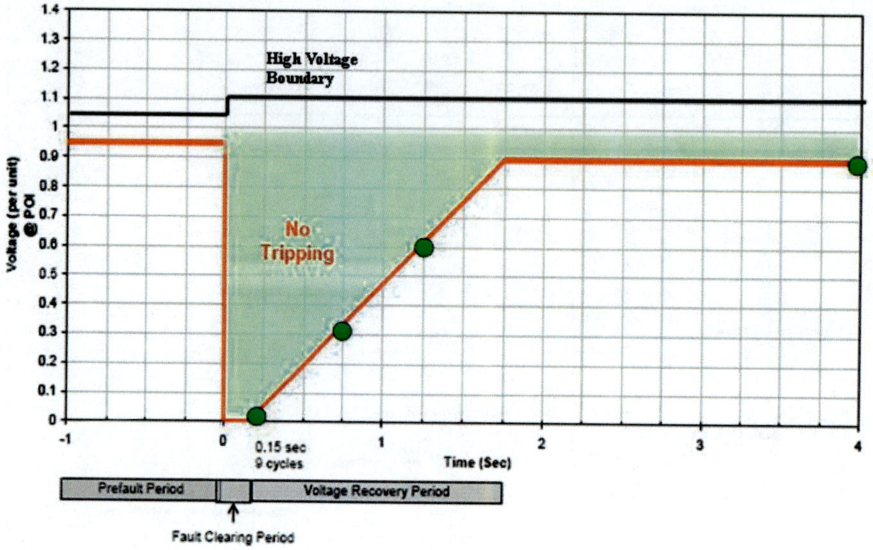

Fig. 9 ERCOT VRT Boundary, ERCOT Operating Guides

The ability of the test WGR added to the sample system to comply with the ERCOT VRT standard as depicted in Figure 9 has been tested utilizing the approach mentioned below.

A 3-Phase fault at the POI substation with varying fault admittances and fault durations is executed to assess the ability of the test wind farm to stay connected at different locations on the voltage recovery boundary as depicted in Figure 9. The varying levels of fault admittances were utilized to obtain the varying voltage levels depicted in the voltage boundary in Figure 9.

The following voltage and fault duration combinations have been utilized to assess the ability of the proposed project to comply with the VRT boundary depicted in Figure 9:

- Low Voltage Ride Through Assessment
  - Voltage Dip Level: 0 pu; Voltage Dip Duration: 0.15s
  - Voltage Dip Level: 0.3 pu; Voltage Dip Duration: 0.68s
  - Voltage Dip Level: 0.6 pu; Voltage Dip Duration: 1.22s
  - Voltage Dip Level: 0.9 pu; Voltage Dip Duration: 4s
- High Voltage Ride Through Assessment
  - Over-Voltage Level: 1.1 pu; Over-Voltage Duration: 10 sec flat start

Figure 10. depicts the terminal voltage response associated with the test wind farm for various fault durations and fault admittances outlined above. As can be seen from Figure 10, the test wind farm does seem to possess the ability to ride through varying levels and durations of voltage dips within the VRT boundary depicted in Figure 9. It is important to note that the sample "savnw" dynamic dataset does not include dynamic behavior of the load models and hence Fault Induced Delayed Voltage Recovery (FIDVR), if any, has not been captured in the voltage response depicted in Figure 10. However, the approach outlined above lends itself well to the inclusion of the dynamic load models, if available.

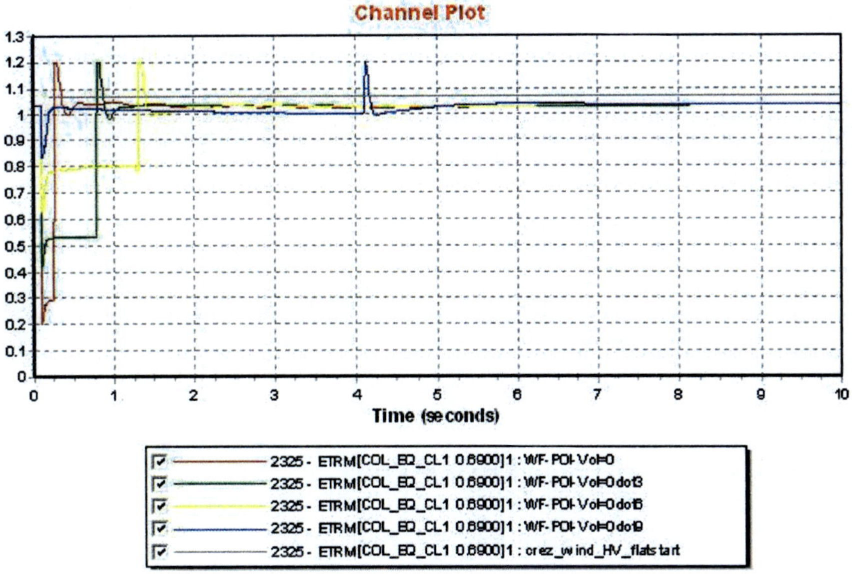

Fig. 10 Terminal Voltage Response of test wind farm, VRT Performance Assessment under voltage dips of varying magnitude and duration

To further illustrate the approach and its application, the test wind farm was simulated under 2 conditions namely:

- Test wind farm comprising of Type I turbines possessing limited VRT capability
- Test wind farm comprising of Type IV turbines possessing zero-voltage ride through capability

Figure 11 depicts the terminal voltage response of the test wind farm when comprising of the Type I (FSIG) turbines and the Type IV (full-converter) turbines. As can be seen from Figure 11, the advanced VRT capability of the Type IV turbines allows it to ride-through the voltage dip created by a normally clearing 3-Phase fault whereas the Type I turbine trips off due to the lack of adequate VRT capability.

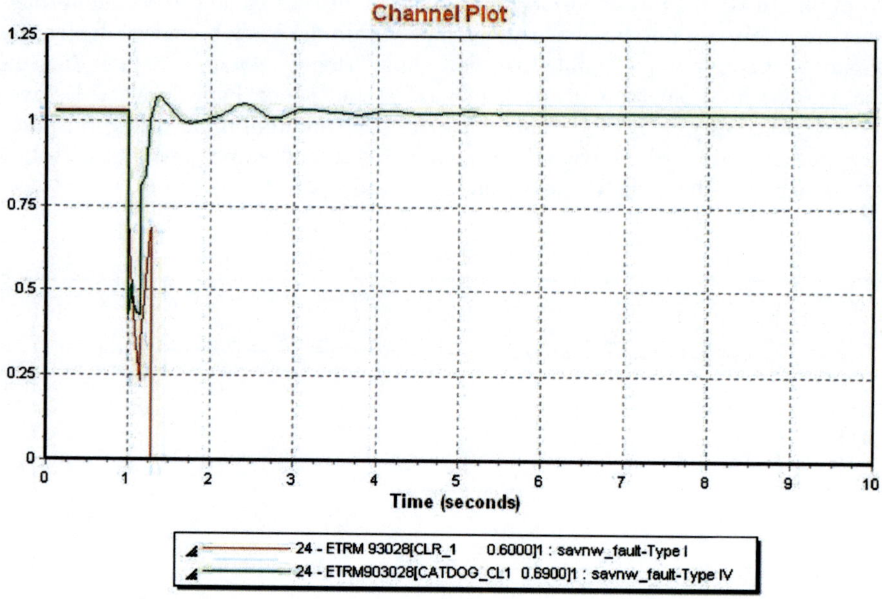

Fig. 11 Comparative Assessment of VRT Performance of Type I Vs Type IV turbines

3 Reactive Power Support Requirements

3.1 Conceptual Discussion and Background

Apart from VRT capability, another key requirement associated with the integration of wind generation technology into the transmission grid has been the need for reactive power support. As in the case of VRT capability requirements, the increasing penetration of WGRs has resulted in the regional reliability organizations and/or utilities mandating a certain level of reactive power support requirements from interconnecting wind farms. While the potential reactive power support that could be obtained from WGRs was relatively limited during low penetration levels, the same is not the case with transmission systems possessing 10-12% penetration of WGRs. While utilities could rely on tripping wind farms that consumed VARs during periods of temporary voltage dips to prevent the situation from getting exacerbated, the same is not possible today. On the contrary, utilities and regional reliability organizations expect WGRs to contribute to the grid support during periods of voltage dips and/or reactive power deficiency.

To that effect, numerous transmission system grids require WGRs to exhibit the ability to meet a pre-specified reactive power capability irrespective of the active power output. Additionally, the WGRs interconnecting to the transmission grid need to meet the reactive power and/or power factor requirements from a steady state and dynamic standpoint. While numerous technology based advancements have been made by turbine vendors to offer varying packages that provide different levels of reactive power compensation, transmission planners and utility engineers have limited information associated with testing these capabilities in the planning environment. The discussion presented in this section focuses on certain techniques associated with assessing the ability of WGRs to meet pre-specified reactive power and/or power factor requirements outlined by the regional reliability organization and/or local utility.

3.2 Steady State Assessment

It goes without saying that all aspects outlined in Section 3 with respect to detailed and accurate steady state modeling of WGRs is a key requirement to performing such assessments. Following the development of a detailed WGR steady state model taking into account all aspects outlined in Section 3, the following aspects need to be accounted for in the steady state reactive capability and/or power factor assessment of WGRs:

- Assessment of reactive capability of the WGR campus at varying active power output levels
- Accurate modeling of turbine reactive capability and variation in reactive power capability owing to:
  - Terminal Voltage
  - Active Power Output
  - A combination of the two factors above
- Adequate and accurate modeling of detailed collection system to account for the VAR losses therein
- Accurate modeling of the station capacitor banks that may assist in providing additional reactive power or may account for the losses in the collection system thereby allowing the campus to fully leverage capability of the turbines
- Controls associated with the stations transformer which may allow regulation of the medium side voltage to fully leverage the capability of the turbines

It is also important to note that while a lot of the aforementioned aspects may play a decisive role in enhancing the steady state reactive capability of the WGR campus, the very same aspects may have little to no impact on the dynamic reactive capability of the campus. Additional discussion on this is presented in the ensuing sub-sections.

$$P_{out} = P_{gen} - P_{loss} \tag{8}$$

$$Q_{out} = Q_{gen} + Q_{shunt} - Q_{loss} \tag{9}$$

Where:

P_{out}, Q_{out}: Active Power and Reactive Power injection to the POI of the wind farm

P_{gen}, Q_{gen}: Active Power and Reactive Power from the wind generators

P_{loss}, Q_{loss}: Active Power and Reactive Power losses of the wind farm including pad-mount transformers, median voltage collector system, substation transformers, and high-voltage lines

Q_{shunt}: Reative Power Generation/Consumption from wind farm shunt devices

3.2.1 Case Study

A 500 MW WGR campus connecting to the transmission grid at 138kV has been utilized to illustrate the steady state reactive capability assessment approach. The 500 MW campus is comprised of Type II, III and IV turbines each comprising of 1 of the 3 phases of the WGR campus. The details associated with each phase of the 500 MW WGR campus are as follows:

- Phase I
    - Type II Turbine
    - Number of Turbines: 70
    - Turbine Reactive Capability: Turbine side capacitor banks to maintain Unity Power Factor
    - Medium Voltage (34.5kV) capacitor banks
    - Sub-station Transformer (Tap Changer Type): Load Tap Changer (LTC)
- Phase II
    - Type III Turbine
    - Number of turbines: 145
    - Turbine Reactive Capability: +/-0.95 pf capability at nominal voltage at turbine terminals
    - Medium Voltage (34.5kV) capacitor banks
    - Sub-station Transformer (Tap Changer Type): De-Energized Tap Changer (DETC)
- Phase III
    - Type IV Turbines
    - Number of Turbines: 70
    - Turbine Reactive Capability: +/- 0.9 pf capability at nominal voltage
    - No Medium Voltage Capacitor Banks
    - Sub-station Transformer (Tap Changer Type): De-energized Tap Changer DETC

The following scenarios have been assessed to gauge the reactive power and/or power factor capability associated with the WGR under varying operational conditions:

- Scenario #1: Reactive power capability of the WGR campus at 1.0 per unit voltage at POI with varying active power output levels associated with the campus
- Scenario #2: Full Buck capability of the WGR campus with 1.05 pu voltage at the POI and full boost capability of WGR campus with 0.95 pu voltage at the POI bus with varying active power output levels

Figures 12 and 13 depict the results of the steady state reactive capability assessment of the WGR campus for the 2 scenarios mentioned above. The "cone curve" outlining the levels associated with +/- 0.95 pf levels at various active power output levels have also been identified in Figures 12 and 13. As per the local utility, the following requirements govern acceptable reactive power capability for the interconnecting WGR:

- The ability to demonstrate +/- 0.95 power factor capability at all active power output levels with the help of external VAR devices if needed at the following POI voltage levels:
    - 1.0 per unit
    - 1.05 per unit for full buck and 0.95 per unit for full boost
- The ability to maintain enough reactive capability at all active power levels so as to never fall inside the "cone curve"

As can be seen from Figures 12 and 13, the WGR campus does meet the ability to maintain 0.95 pf capability lag/lead at all active power levels except for at 100% output at 1.0 per unit voltage at the POI. According to Figure 20, the WGR campus falls short of meeting the 0.95 lag capability at 100% output when maintaining 1.0 per unit voltage at POI. However, apart from that specific condition, the WGR campus does meet the ability to perform 0.95 power factor lag or lead at all active power output levels at various voltages outlined in the scenario definitions. It would be important to remember that the following aspects govern the over-all reactive capability of the WGR campus:

- No load tap settings for turbine pad-mount transformers
- LTC and/or DETC settings for station transformers
- Station capacitor banks for each phase of the WGR campus
- Individual turbine reactive capability

3.2.2 Key Observations

Based on the results of the illustrative case study presented above, the following key observations can be made with regards to the assessment of the WGR steady state reactive capability from a transmission planning standpoint:

- The modeling aspects required for accurate steady state representation of the WGR model as outlined above play a key role in the assessment

- The steady state reactive capability assessment for WGR campuses is impacted by and should take into account the following:
  - o Assessment across varying active power output levels and POI voltage profiles
  - o Ability to accurately model turbine reactive capability and its dependency on the following:
    - Active Power Output Level
    - Turbine Terminal Voltage
  - o No load tap settings for turbine pad-mount transformers
  - o LTC and/or DETC settings for station transformers
  - o Station capacitor banks for each phase of the WGR campus
  - o Accurate modeling of the detailed collection system for the WGR campus

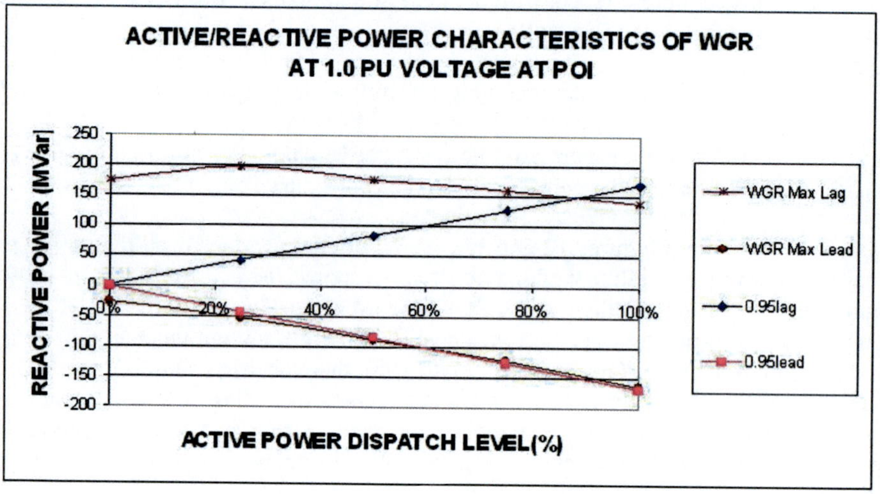

Fig. 12 Steady State Lag/Lead Reactive Capability Assessment of WGR Campus, 1.0 per unit voltage at POI

3.3 Dynamic Assessment

As outlined earlier, the following aspects need to be taken into account when developing a dynamic model associated with a WGR campus for the purposes of dynamic reactive capability assessment:

- Comprehensive validation of acceptable dynamic initialization and behavior of the dynamic model for the turbine utilizing the techniques discussed above
- Development of detailed steady state model of the WGR campus taking into account all aspects outlined in the previous sub-section

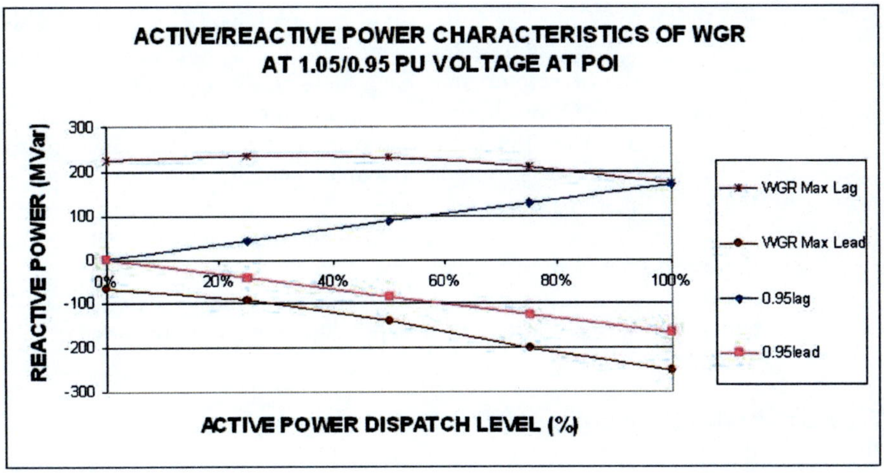

Fig. 13 Steady State Lag/Lead Reactive Capability Assessment of WGR Campus, 1.05/0.95 per unit voltage at POI

- Inclusion of the dynamic model associated with the WGR under study into the study dynamic dataset
- Ensuring proper dynamic initialization of the resulting dynamic dataset following the addition of the dynamic data for the WGR by means of a 10-sec no disturbance flat start simulation

Following the development and validation of an acceptable dynamic dataset that can be utilized for the dynamic assessment, the following approach characterizes the dynamic reactive capability assessment of the WGR campus:

- Perform dynamic simulations associated with simulating the injection and/or withdrawal of reactive power equaling the requirement amount by means of switched shunt activation during the simulation
    - Initiate dynamic simulation and run normal operating conditions for 1 second
    - Simulate switched shunt action at the POI of the WGR campus to inject or withdraw a pre-determined amount of reactive power
    - The size of the switched shunt should be such that the VAR injected or withdrawn should equal or exceed the reactive capability requirements associated with host grid
    - In case of testing dynamic reactive capability leading, the switched shunt action would comprise of capacitor switching thereby increasing the voltage at POI resulting in the turbine consuming reactive power

- Conversely, in case of testing dynamic reactive capability lagging, the switched shunt action would comprise of reactor switching thereby reducing the voltage at POI resulting in the turbine injecting reactive power
- The turbine must be made to operate in voltage control mode during the dynamic simulation

- Repeat simulations at varying levels of active power dispatch to assess the ability of the WGR campus to meet the reactive capability requirements set forth by the host grid
- Utilize the dynamic simulation results to assess the ability of the WGR campus to comply with the reactive capability requirements from a dynamic standpoint

The ensuing sub-section presents an illustrative case study to demonstrate the application of the approach on a sample WGR campus.

3.3.1 Case Study

The sample "savnw" dynamic dataset, as described in the previous sections, has been utilized to demonstrate the dynamic reactive capability assessment for WGRs. A 300 M sample WGR model comprising of Type III turbines has been included in the "savnw" dataset including the following aspects of the WGR model:

- Turbine equivalent
- Pad-mount transformer equivalent
- Collection system equivalent
- Station transformer

The Type III turbine model is operated in voltage control mode. The WGR campus is tested to ensure that it has the ability to meet 0.95 power factor lag or lead capability from a dynamic standpoint which would amount close to 95.3 MVAR. Figure 14 depicts the one-line schematic associated with the WGR campus as integrated into the sample "savnw" dataset.

As mentioned earlier, the POI voltage is changed by means of a switched shunt action during the dynamic simulation in order to assess the ability of the WGR to dynamically inject and/or withdraw reactive power to the tune of 0.95 power factor lag or lead. Additionally, this exercise is repeated at different active power levels.

The illustrative case study has been executed for the following scenarios and sub-scenarios:

- Scenario #1 – 0.95 pf lag
  - Tested for 10%, 40%, 70% and 100% active power dispatch levels associated with test WGR campus
- Scenario #1 – 0.95 pf lead
  - Tested for 10%, 40%, 70% and 100% active power dispatch levels associated with test WGR campus

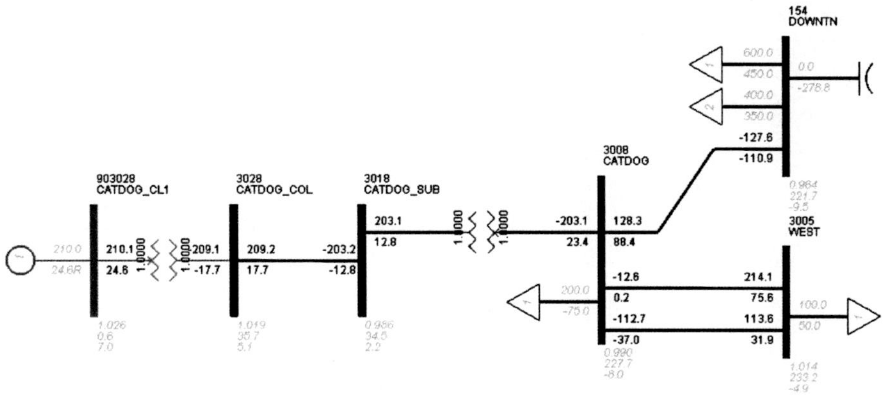

Fig. 14 One-line Schematic of test 300 MW WGR model, Sample "savnw" system

Figures 15 through 18 depict the lag/lead dynamic reactive capability and/or response associated with varying active power output levels for the WGR campus.

Figures 15 through 18 depict the response of the reactive power of the test WGR system over the period of the dynamic simulation. The reactive power response depicted in the figures is on a 100 MVA base. As outlined earlier, for the test 300 MW WGR, 0.95 lag/lead power factor capability would correspond to 95 MVAR which would be about 0.95 per unit on a 100 MVA base. As can be seen from Figures 15 through 18, the test WGR depicts the ability of meeting the 0.95 lag/lead power factor capability at varying active power levels from a dynamic standpoint.

For comparative purposes, the same WGR model was altered to comprise of Type I turbines with minimal dynamic reactive capability. The resulting sample system was simulated for 70% dispatch level of the test WGR in the same fashion as described above to assess the ability of the WGR to meet 0.95 lag/lead reactive capability from a dynamic standpoint. Figure 19 depicts the dynamic reactive power response associated with the test WGR altered to comprise of Type I turbines. As in the case of Figures 15 through 18, the WGR reactive power response is depicted in per unit on a 100 MVA base. As can be seen from Figure 19, the WGR comprising of Type I turbines does not meet the 0.95 lag/lead reactive power capability from a dynamic response standpoint. In such a situation, a typical solution expected to be adopted in order to comply with the dynamic reactive capability requirements set forth by the host grid would be the application of optimally sized DVAR devices such as SVCs and/or STATCOMs.

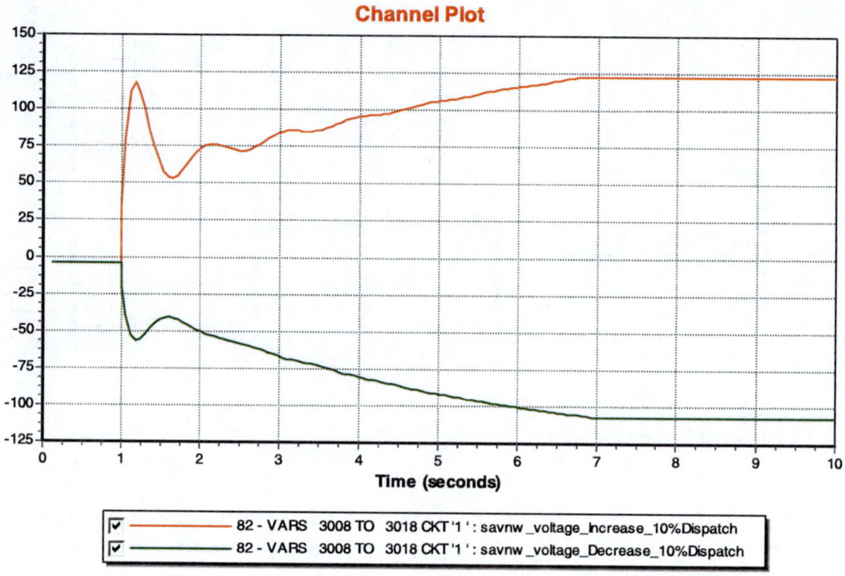

Fig. 15 Dynamic Lag/Lead Reactive Power Capability for Test WGR utilizing Type III turbines, 10% Active Power

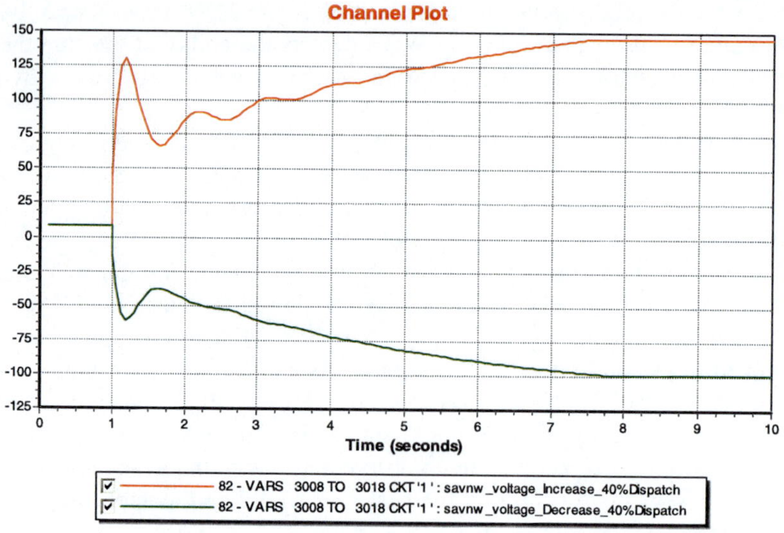

Fig. 16 Dynamic Lag/Lead Reactive Power Capability for Test WGR utilizing Type III turbines, 40% Active Power

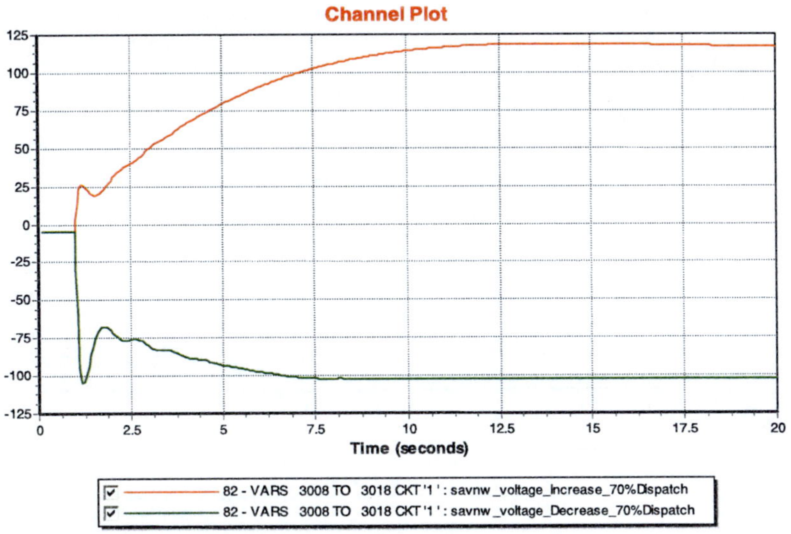

Fig. 17 Dynamic Lag/Lead Reactive Power Capability for Test WGR utilizing Type III turbines, 70% Active Power

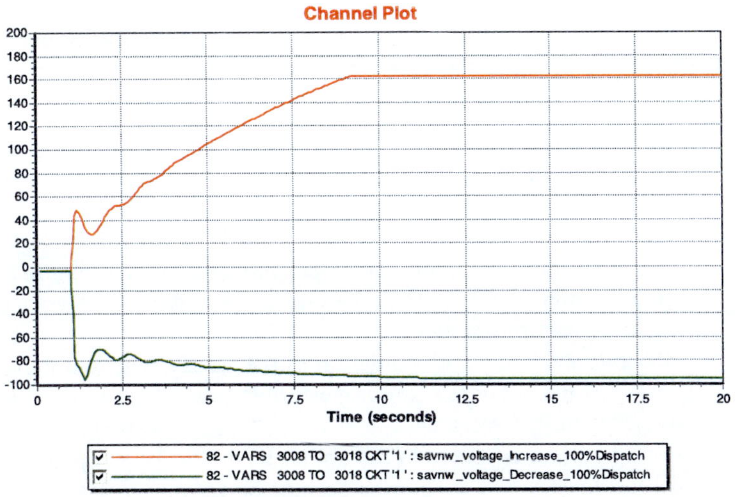

Fig. 18 Dynamic Lag/Lead Reactive Power Capability for Test WGR utilizing Type III turbines, 100% Active Power

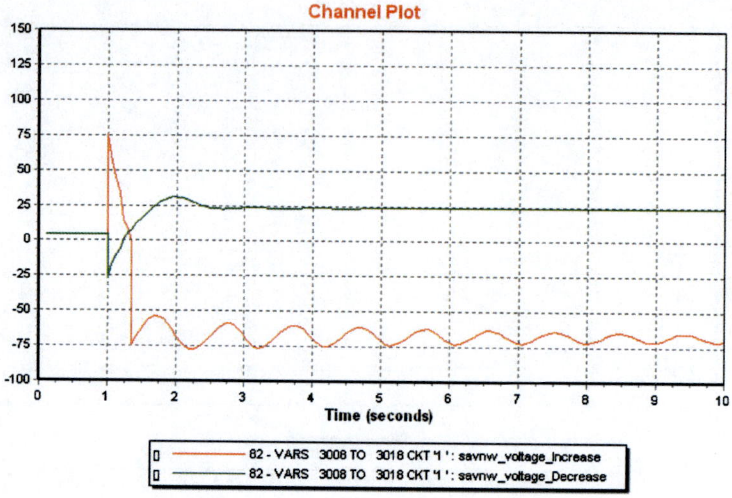

Fig. 19 Dynamic Lag/Lead Reactive Power Capability for Test WGR utilizing Type I turbines, 70% Active Power

3.3.2 Key Observations

Based on the illustrative case study presented above, the following observations can be made with respect to the assessment of dynamic reactive capability for WGRs:

- Reactive devices such as capacitor banks, station transformer LTC and no load tap changers for pad-mount transformers do not play a significant role in determining the reactive capability of the WGR from a dynamic response standpoint
- This is so since the timeframe associated with the dynamic reactive response is much smaller than the typical response time associated with the aforementioned devices
- Accurate modeling and validation of the steady state and dynamic behavior of the WGR are key to performing an adequate dynamic reactive capability assessment
- It is important to ascertain the ability of the WGR campus to dynamically supply and/or absorb the designated amount of reactive power at varying active power levels in order to meet the host grid's reactive power requirements
- The modeling of the dynamic reactive capability of the turbine comprising the WGR is the key aspect in assessing the reactive capability of the WGR campus as a whole
- The reactive capability has to be tested under dynamic simulation environment by means of creating a deficit or surplus of reactive power that the turbine dynamic model can respond to during the simulation timeframe

4 Inertial Response and Frequency Control Requirements

4.1 Conceptual Discussion and Background

The steadily increasing penetration of wind in various power system grids has attracted a lot of attention on the ability of various wind turbine technologies to provide dynamic reactive power and/or voltage support. The increased penetration of wind has also resulted in displacement and/or de-commitment of conventional generation resources thereby reducing the grid inertia derived from them. In such situations, the ability of wind turbine technologies to provide primary frequency control and/or inertial response capabilities is of prime significance.

Frequency response is defined as the automatic corrective response in order to balance demand and supply. The frequency response of a typical power system can be broadly classified into 3 categories based on the response time namely:

- Inertial Frequency Response: This category of frequency response is inherently present in the system due to the rotating masses in the system typically comprising of conventional synchronous generation and motor loads. The Inertial Frequency Response typically responds to system disturbances within seconds of the disturbance initiation to arrest the frequency deviation.
- Primary Frequency Response (PFR): This category of frequency response can be defined as the increase and/or decrease in active power output in proportion to the system frequency deviations. This type of response typically derived from synchronous generation units acting on primary governor response. The timeframe associated with the primary frequency response is typically within 12 to 14 seconds. However, in case of a system disturbance all governor units will respond to the deviation based on the governor droop characteristics. While this action reduces the frequency deviation, supplemental control is necessary to adjust the load reference set-point thereby restoring the steady state frequency to its nominal value.
- Secondary Frequency Response: This category of frequency response corresponds to Automatic Generation Control (AGC) action typically deployed to regulate the frequency back to 60Hz after the deployment of the PFR. The objective of AGC, apart from restoring the frequency to its nominal value, is to ensure the maintenance of power transfer between control areas at scheduled values by adjusting the output of select generation units.

The increasing penetration of WGR on the power system grid has come at the cost of displacing conventional synchronous generation from the "order –of – merit" stack. The economic benefits of this displacement have been vastly quantified in terms of reduced over-all production costs and increased cost savings owing to the lower price wind generation. However, the price that system operators and utilities tend to pay for this displacement in terms of lost system inertia has rarely if ever quantified until recently. The impact of the reduced system inertia is more exaggerated due to the coincidence of high wind penetration periods on the power system network with light load conditions. Due to high wind availability during light load conditions, the system inertia under such conditions is at its minimum. The wind generation resources cannot provide near as much inertia to the system due to the decoupling of the machines from the system by power electronic devices. This situation is obviously more exacerbated in case of Type III and Type IV turbines.

Inertial Frequency Response is defined as "The power delivered by the Interconnection in response to any change in frequency due to the rotating mass of machines synchronously connected to the Bulk Power System (BPS), including both-load and generation" [3]. System frequency drops whenever there is shortage of generation to supply demand and frequency increases whenever there is excess of energy. Sudden loss of supply or demand will result in frequency deviation from the nominal frequency. The rate of change in frequency due to imbalance depends on the system inertia. System inertia is directly proportional to synchronously rotating mass in the system (includes synchronous generation and motor load). The general equation for calculating rate of change of frequency using system inertia constant (H) is illustrated by:

$$\frac{df}{dt} = \frac{\Delta P}{2H} * f_0 + \frac{D}{2H} * \Delta f \tag{10}$$

Where:

H: system inertia constant on system base (s)
D: Load Damping Constant (pu/Hz)
f_0: frequency at the time of disturbance (Hz)
df/dt: Rate of Change of Frequency (Hz/sec)
ΔP: Power Change
Δf: Change in frequency

As is obvious from the discussion presented above, the inertial response and frequency control requirements and assessments associated with WGRs is gaining significance across various power system networks. Utilities and/or regional reliability organizations have placed increased stress on WGR's ability to provide frequency control as part of the interconnecting and grid integration process. The ensuing sub-sections under this section present a qualitative and quantitative assessment of frequency control and inertial response capabilities associated with various turbine types with special focus on DFIG frequency control capabilities.

R(per unit), the slope of the "droop" curve, is defined as Δf(p.u.)/ ΔP(p.u.)

Where:

$\Delta f(p.u.) = \Delta f(HZ) / 60.0$

$\Delta P(p.u.) = \Delta P(MW) / \text{Unit Capacity}$

In order to understand the concept of a 5% droop characteristic, the following illustration is presented:

For a 600 MW unit that has a governor response of 20 MW for a frequency excursion that settles out at 59.9 HZ:

$R = \Delta f(p.u.) / \Delta P(p.u.) = (0.1/60)/(20/600) = 0.05$ *or 5% droop*

In other words, once the droop is known, the MW response to frequency deviation can be determined by:

$(\Delta P/\Delta f) = (1/R)$, *or* $\Delta P = (1/R) \times \Delta f$

Taking the aforementioned illustration, for the 600 MW unit with 5% droop:

$(\Delta P/600) = (1/0.05) \times (\Delta f/60)$, *or* $\Delta P = 200MW/HZ$

Figure 20 depicts the graphic illustration of a droop characteristic described above:

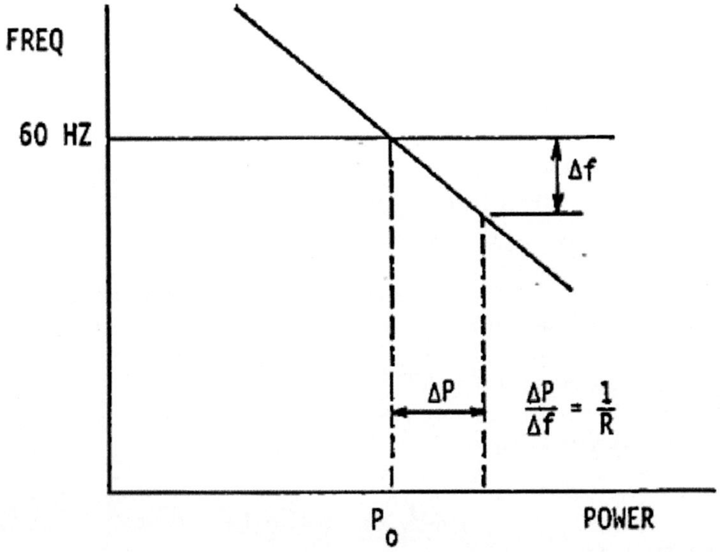

Fig. 20 Typical Steady State Droop Curve Characteristic

The same concept can then be extended to a system containing more than one generator with varying governor droop characteristics. For example, Table 2 depicts a system with 3 generators and associated capacities and droop characteristics.

Table 2 Sample System with 3 Generators – Capacity & Droop Characteristics

UNIT	Pgen	CAPACITY	R (DROOP)	ΔP/Δf
#1	80 MW	300MW	0.100 (10%)	50MW/HZ
#2	120 MW	450MW	0.075 (7.5%)	100MW/HZ
#3	160 MW	600MW	0.050 (5%)	200MW/HZ

Following the addition of 21 MW of load on the aforementioned system, the resulting steady state frequency and amount picked up by each generator can be simply calculated as:

- **Steady State Frequency**

  Unit #1: ΔP1=50 X Δf
  Unit #2: ΔP2=100 X Δf
  Unit #3: ΔP3=200 X Δf
  ΔΣPi=350Δf=21MW,
  Δf=21/350=0.06HZ
  Frequency=60-0.06=59.94HZ

- **MW Amount picked up by each generator**

  ΔP1=50 X 0.06=3MW
  ΔP2=100 X 0.06=6MW
  ΔP3=200 X 0.06=12MW
  Check: ΔΣPi=21MW

The aforementioned illustration also underlines the fact that governor response alone cannot restore the system frequency back to nominal in the absence of supplemental control in the form of AGC.

To that effect, generator droop curve, ranges of frequency and/or active power output and the dead-band region are all key aspects in assessing the need and extent of primary frequency response associated with a generator.

4.2 DFIG Frequency Control and Inertial Response – Qualitative Assessment

As mentioned earlier, newer versions of the DFIG machines have augmented advanced inertial response modeling capabilities to represent incremental features/capabilities associated with their turbines.

The over-all turbine model associated with a DFIG primarily comprises of 3 key device models:

- Generator/Converter Model – injects active and reactive power into the grid/network based on control commands. This model also represents the high/low voltage protection functions associated with the WTG model.
- Electrical Control Model – contains open and closed real/reactive power control/regulation loops and is responsible for sending real and reactive commands to the Generator/Converter model in order to regulate the active and reactive power injected into the grid.
- Turbine & Turbine Control Model – comprises of mechanical controls including pitch, active power, rotor speed/inertia and power order control signals to the converter

In the absence of any optional control functions, under all frequency conditions, the maximum power set point is utilized as input for the pitch compensation and power response rate limit function. However, in the absence of the control function options, the turbine terminal bus frequency signal does not play any role in the calculation of the final power order sent to the converter model to alter the active power output of the model.

The purpose of the inertial response option is to equip the DFIG turbine to provide an inertial response capability for large under-frequency events. System disturbances including loss of generation result in significant system frequency excursions with factors such as depth of excursion, rate of frequency decline and time to return to normal being a function of the dynamic characteristics of the generation connected to the grid. The inertial response option aims to provide the turbines with inertial response capability that could reduce the rate of frequency decline thereby allowing other synchronous generation to increase power output during these events. Some typical aspects characterizing the inertial response option for DFIG turbines are:

- This control is asymmetric in nature only responding to under-frequency events. A dead band filter is utilized to filter only large under-frequency events thereby preventing the control to respond to minor frequency deviations experienced in day-to-day operations.
- Furthermore, the impact of the inertial response option is primarily limited to the transient period immediately following the initiation of the event leading to under-frequency issues. This feature temporarily increases the output of the turbine to within 5-10% of the rated turbine power to allow other slow acting conventional generators enough time to increase their power output in response to the under-frequency event. The duration of the power increase is typically in the order of several seconds.
- The frequency error is a deviation from the normal which is passed through the dead-band filter. The presence of the dead-band filter

prevents the control from responding to either over-frequency events or under-frequency events below a certain threshold.

- The output of the dead-band is further filtered, coordinated, limited and the final output is utilized to obtain the new power order.

4.3 *DFIG Frequency Control and Inertial Response – Dynamic Simulation*

The Type III wind turbine model included in the sample "savnw" system, as discussed in Section 4, has been utilized to perform illustrative simulations to demonstrate the inertial response option associated with DFIG machines.

Figures 21 and 22 depict the active power and frequency response associated with the Type III turbine in the wake of an under-frequency event with and without the inertial response option enabled. PTI sample "savnw" case and dynamic data set were utilized for the simulation. The under-frequency event was created by tripping a 750 MW conventional unit without creating any fault event. This was done in purpose to differentiate between active power response of the Type III turbine stemming from frequency excursions as opposed to terminal voltage changes. The turbine relay model was excluded from the simulation in order to assess the precise nature of the response.

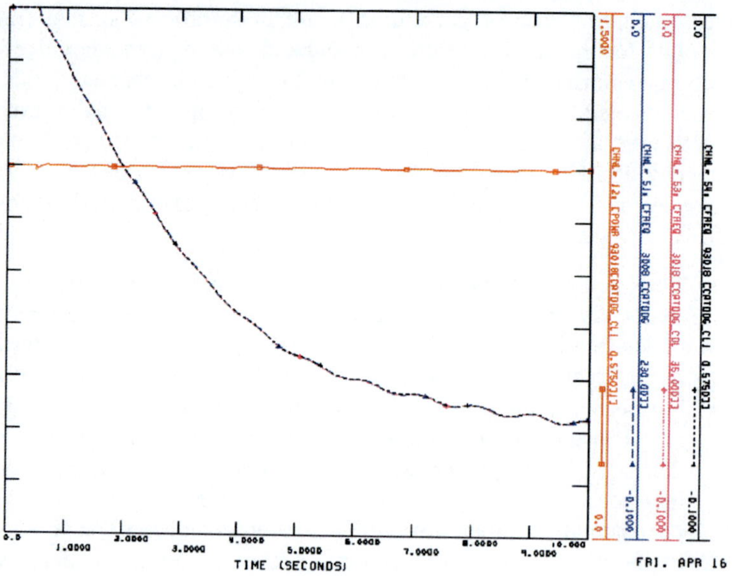

Fig. 21 Pelec & Fter Response of Type III turbine (without inertial response Option) during the Trip Event (SAVNW case, Gen 101 Gen 750MW Trip)

Figure 21 underlines the lack of a frequency response during an under-frequency event for the Type III turbine model when the inertial response option is disabled. Figure 22 depicts the temporary increase in active power owing to the increased power order by virtue of the dead-band filtered, coordinated and limited signal arising from the enabled inertial response option. The under-frequency event was created to be large enough to pass through the dead-band filter associated with the inertial response option.

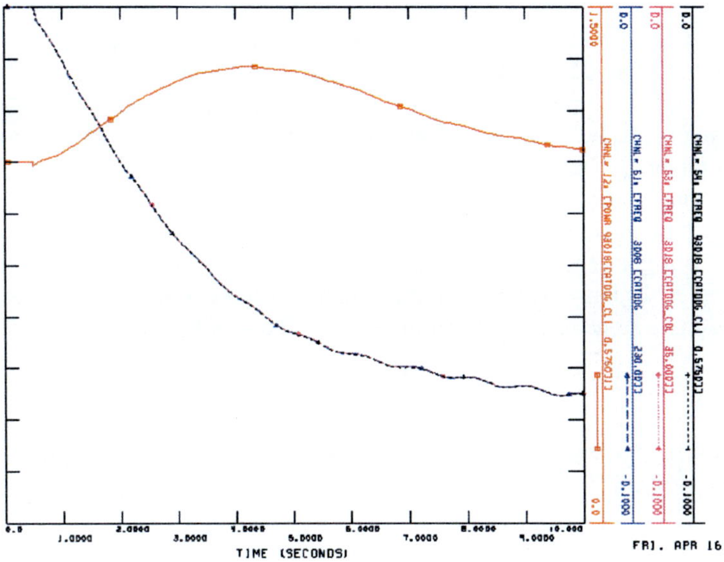

Fig. 22 Pelec & Fter Response of Type III turbine (with inertial response Option) during the Trip Event (SAVNW case, Gen 101 Gen 750MW Trip)

4.4 Frequency Control and Inertial Response – Other Turbine Technologies

The active power and frequency response associated with various wind turbine technologies have also been simulated utilizing the sample "savnw" system and the WGR set-up described in Section 4. All events involved the tripping of generation without any fault and/or line tripping events to observe the primary frequency response while simultaneously observing any major variations in WGR terminal voltage to decipher the reason of the active power response of the turbine.

Figure 23 depicts the active power and terminal frequency response associated with Type I turbines utilizing the same set-up as that utilized for Type III turbines described above. While the first transient active power increase may be attributed to the increase in the terminal voltage of the machine equivalent following the tripping of 700 MW of generation, the Fixed-Speed Induction Generator (FSIG) turbines seem to exhibit a slightly improved frequency response in general when compared to the DFIG turbines.

Figure 24 depicts the active power and terminal frequency response associated with Type IV wind turbine replacing the Type I turbine utilizing the set-up described above. The full-converter based wind turbine does not seem to have any significant transient active power response following the under-frequency excursion.

Figure 25 depicts the active power and terminal frequency response associated with another Type III turbine albeit from a different vendor. This Type III turbine seems to exhibit significantly more profound active power response. The documentation associated with these Type III turbines is indicative of the presence of optional features of active power/frequency regulation with response time of 5% Pr/s utilizing control algorithms implemented with the aid of turbine side PLCs. However, the dynamic data associated with the model provides no information associated with this aspect of the wind turbine. In such a scenario, it is unclear if the active power response demonstrated in Figure 25 is by virtue of the terminal voltage variations alone or a combination of the voltage and frequency excursions.

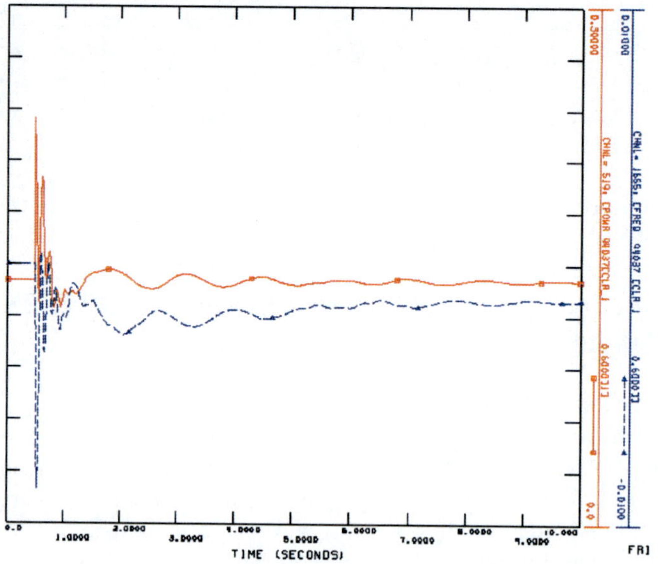

Fig. 23 Pelec & Fter Response of Type I turbine during the Trip Event (SAVNW case, Gen 101 Gen 750MW Trip)

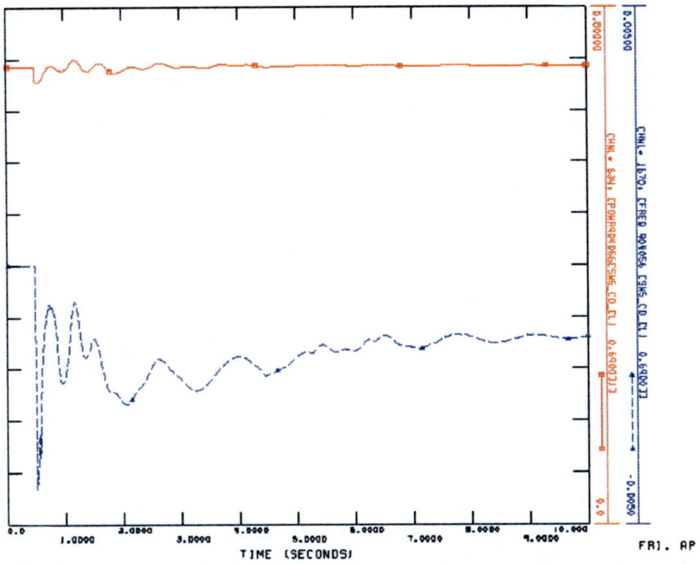

Fig. 24 Pelec & Fter Response of Type IV turbine during the Trip Event (SAVNW case, Gen 101 Gen 750MW Trip)

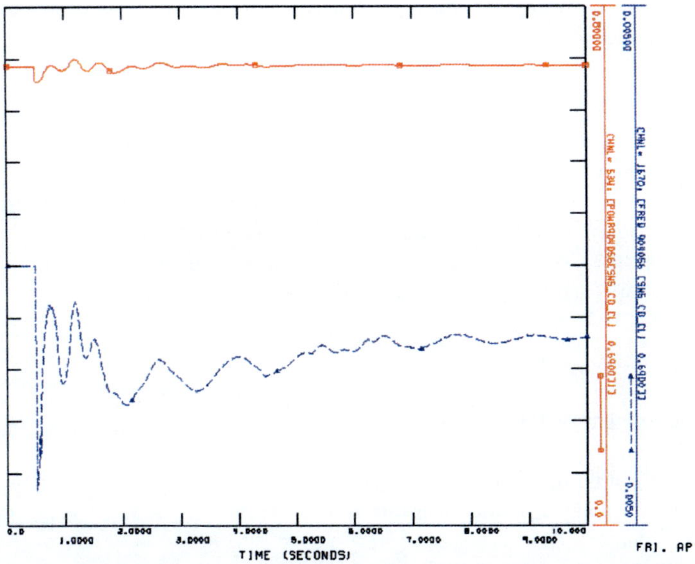

Fig. 25 Pelec & Fter Response of Type III (alternate vendor) turbine during the Trip Event (SAVNW case, Gen 101 Gen 750MW Trip)

4.5 Key Observations

Based on the event simulations and analysis presented in this section, the following observations associated with the active power and/or inertial response capabilities of the wind turbine types assessed can be made:

- The typical inertial response option associated with DFIG machines is an asymmetric option which only comes into play for under-frequency excursions. Furthermore the dead-band filter logic in the inertial response option ensures that the machine provides inertial response only to large under-frequency excursions above a certain threshold and not for minor under-frequency excursions experienced in day-to-day operations.
- The other frequency response controls in DFIG turbines are typically provided through the active power control option which utilizes the terminal bus frequency and a pre-set frequency response curve (resulting in an interpolated factor) to determine a power set point which in turn is utilized to determine new power order for the generator/converter model. However, the frequency response curve in the this option typically corresponds to compliance with the host grid droop requirements albeit within the limits imposed by the turbine rated power and/or available power.
- In general, the simulation results seem to suggest FSIG turbines providing slightly improved inertia response in comparison to DFIG and/or full-converter machines without any additional frequency regulation options enabled.
- Most DFIG machines (irrespective of the vendor type) seem to provide an optional feature to provide frequency regulation and active power control.
- Detailed simulations of individual machine models on the sample "savnw" case do tend to indicate differences in the active power response by virtue of enabling the inertial response features across the same turbine types albeit from different vendors

References

1. U.S Department of Energy, Energy Information Administration, Power Plant Operations Report
2. American Wind Energy Association (2009)
3. 20% Wind Energy Penetration in the U.S. A Technical Analysis of the Energy Resource (October 2007)
4. NERC Frequency Response Initiative (April 2010)
5. Mapping of Grid Faults & Grid Codes, RISO fdf
6. Monthly Status Report to Reliability and Operations Subcommittee (December 2010), ERCOT, http://www.ercot.com/content/meetings/ros/keydocs/2011/0113/07._Planning_Report_December_2010.doc

7. PRC-024-WECC-1-CR - Generator Low Voltage Ride-Through Criterion, `http://www.wecc.biz/Standards/Development/WECC-60/Shared%20Documents/WECC-0060%20PRC-024-WECC-1-CR%20Low%20Voltage%20Ride%20Through%20Criterion%20Draft%201%202-20-08.pdf`
8. Morales, A., Robe, X., Sala, M., Prats, P., Aguerri, C., Torres, E.: Advanced grid requirements for the integration of wind farms into the Spanish transmission system. IET Renewable Power Generation 2(1), 47–59 (2008)
9. Erlich, I., Winter, W., Dittrich, A.: Advanced grid requirements for the integration of wind turbines into the German transmission system. In: IEEE Power Engineering Society General Meeting (2006)
10. Transmission Lines Department (Denmark). Wind turbines con-nected to grids with voltages below 100 kV TechnicalRegulations TF3.2.6 (May 2004), `http://www.wtcertification.dk/Com-mon/Regulation%20for%20Windturbines%20TF%203.2.6.pdf.`
11. Ullah, N.R., Thiringer, T., Karlsson, D.: Voltage and Transient Stability Support by Wind Farms Complying With the E.ON Netz Grid Code. IEEE Transactions on Power Systems 22(4), 1647–1656 (2007)
12. Molinas, M., Suul, J.A., Undeland, T.: Low Voltage Ride Through of Wind Farms With Cage Generators: STATCOM Versus SVC. IEEE Transactions On Power Electronics 23(3), 1104–1117 (2008)
13. Muyeen, S.M., Takahashi, R., Murata, T., Tamura, J., Ali, M.H., Matsumura, Y., Kuwayama, A., Matsumoto, T.: Low voltage ride through capability enhancement of wind turbine generator system during network disturbance. IET Renewable Power Generation 3(1), 65–74 (2009)
14. Causebrook, A., Atkinson, D.J., Jack, A.G.: Fault Ride-Through of Large Wind Farms Using Series Dynamic Braking Resistors. IEEE Transactions On Power Systems 22(3), 966–975 (2007)
15. Ab-del-Baqi, O., Nasiri, A.: A Dynamic LVRT Solution for Doubly Fed Induction Generators. IEEE Transactions On Power Electronics 25(1), 193–196 (2010)
16. Yao, J., Li, H., Liao, Y., Chen, Z.: An Improved Control Strategy of Limiting the DC-Link Voltage Fluctuation for a Doubly Fed Induction Wind Generator. IEEE Transactions On Power Electronics 23(3), 1205–1213 (2008)
17. Flannery, P.S., Venkataramanan, G.: Unbalanced Voltage Sag Ride-Through of a Doubly Fed Induction Generator Wind Turbine With Series Grid-Side Converter. IEEE Transactions On Industry Applications 45(5), 1879–1887 (2009)
18. Abbey, C., Joos, G.: Supercapacitor Energy Storage for Wind Energy Applications. IEEE Transactions On Industry Applications 43(3), 769–776 (2007)
19. Mullane, A., Lightbody, G., Yacamini, R.: Wind-Turbine Fault Ride-Through Enhancement. IEEE Transactions On Power Systems 20(4), 1929–1937 (2005)
20. Conroy, J.F., Watson, R.: Low-voltage ride-through of a full converter wind turbine with permanent magnet generator. IET Renewable Power Generation 3(1), 182–189 (2007)
21. Cheng, Y., Sahni, M., et al.: A Voltage-profile based approach for developing collection system aggre-gated models for Wind Generation Resources for grid VRT studies. IET Renewable Power Generation (2011)
22. PSS/E 31.0 Users Manual, Appendix E, Siemens PTI (2007)
23. ERCOT Operational Guides, Section 3 on Operational Interfaces, pp. 3-24

Topologies and Control Strategies of Multilevel Converters

Arash Khoshkbar Sadigh and S. Masoud Barakati

Abstract. Multilevel converters have been continuously developed in recent years due to the necessity of increase in power level of industrial applications especially high power applications such as high power AC motor drives, active power filters, reactive power compensation, FACTS devices, and renewable energies [1-9]. Multilevel converters include an array of power semiconductors and capacitor voltage sources which generate step-waveform output voltages. The commutation of the switches permits the addition of the capacitors voltages and generates high voltage at the output [8, 10, 11]. The term multilevel starts with the three-level converter introduced by Nabae [12]. By increasing the number of levels in the converter, the output voltage has more steps generating a staircase waveform which has a reduced harmonic distortion [13]. However, a high number of levels increases the control complexity and introduces voltage unbalance problems [10].

The Neutral Point Clamped (NPC) converter, presented in the early 80's [12], is now a standard topology in industry on its 3-level version. However, for a high number of levels, this topology presents some problems, mainly with the clamping diodes and the balance of the dc-link capacitors. An alternative for the NPC converter are the Multicell topologies. Different cells and ways to interconnect them generate several topologies which the most important ones, described in Section II, are the Cascaded Multicell (CM) and the Flying Capacitor Multicell (FCM) with its sub-topology Stacked Multicell (SM) converters [11-14].

The CM converter is the series connection of 2-level H-bridge converter, that several configurations have been proposed for this topology [13]. Since this topology consists of series power conversion cells, the voltage and power levels may be scaled easily. As other alternative topologies, the FCM converter [15, 16], and its

Arash Khoshkbar Sadigh
EECS Department, University of California-Irvine, Irvine, CA, 92617, US
email: a.khoshkbar.sadigh@ieee.org

S. Masoud Barakati
Department of Electrical and Computer Engineering,
University of Sistan and Baluchestan, Zahedan, Iran
email: smbarak@ece.usb.ac.ir

L. Wang (Ed.): *Modeling and Control*, Green Energy and Technology, pp. 311–340.
springerlink.com

derivative, the SM converter [17-19], have many attractive properties for medium voltage applications [15-22].

To control the multilevel converters, there are several modulation methods which can be classified to high and low switching frequency. High switching frequencies based methods have several commutations during one period of the fundamental output voltage. The pulse width modulation (PWM), sinusoidal pulse width modulation (SPWM) and space vector PWM are common methods for the high switching frequency. Low switching frequencies based methods have one or two commutations during one period of the fundamental output voltage, generating a staircase waveform. The multilevel selective harmonic elimination (SHE) and the space vector control (SVC) are common use for the low switching frequency.

The mentioned topologies of multilevel converters as well as their several control methods are discussed in this chapter.

1 Multilevel Converters Topologies

1.1 Cascade Multicell (CM) Converter

The CM converter was introduced in the early 90s [23, 24]. This topology is based on the series connection of units known as cells with three-level output voltage, as shown in Fig. 1. Structure of each cell is based on an isolated voltage source, Fig. 1(a). When only one dc voltage source is available, a bulky and complex multi-secondary input transformer is required, Fig. 1(b). Therefore, the cost and size of the converter is increased.

Since this topology consists of series power conversion cells, the voltage and power levels may be scaled easily and a maximum of $2n+1$ output voltage levels is obtained. The total output voltage, corresponding to the sum of each cell output voltage, is:

$$v_{out} = \sum_{i=1}^{n} v_i \tag{1}$$

where n is the number of cells connected in series.

An additional advantage of this topology is that when an internal fault is detected and the faulty cell is identified, it can be easily isolated through an external switch and replaced by a new operative cell without turning off the converter [25]. However, while the replacement is done, the maximum output voltage in the faulty leg is reduced to:

$$v_f = v_{out}\left(1-\frac{f}{n}\right) \tag{2}$$

where f is the number of faulty cells.

As the 2-cell-5-level CM converter is controlled by phase shifted-SPWM (PS-SPWM) and operated with a modulation index equal to 0.8 ($M = 0.8$), its control strategy, switches states as well as the output voltage are shown in Fig. 2. The

Switch X is on when its state is 1 and is off when its state is 0. The PS-SPWM is a regular phase shifted-SPWM where the phase shift between the carriers of each cell is:

$$\phi = 2\pi / n \tag{3}$$

where n is number of cells. There are other control methods in addition to PS-SPWM which will be discussed in next section. Switches states of the 2-cell-5-level CM converter, using PS-SPWM method, are illustrated in Table 1.

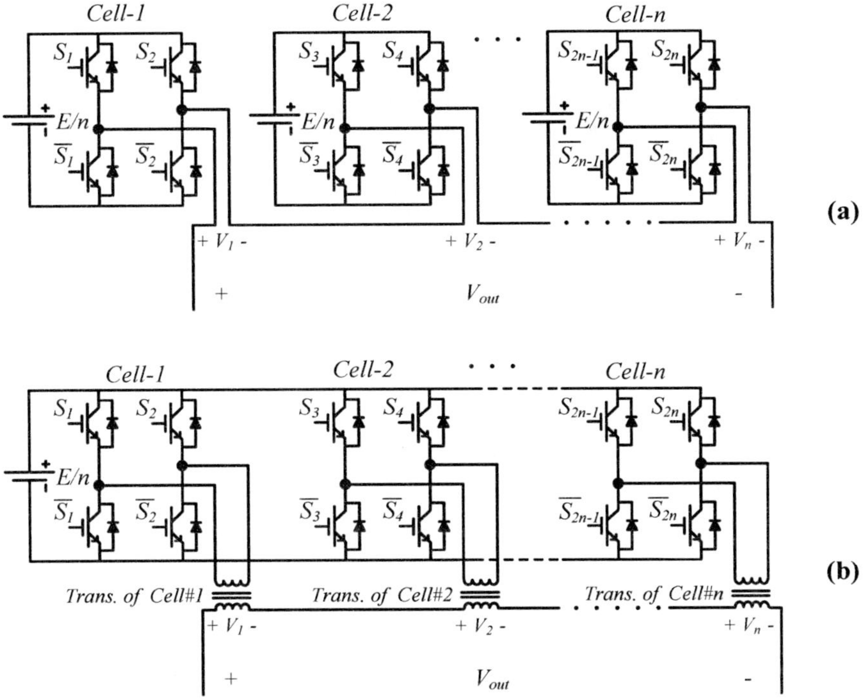

Fig. 1 2n+1 level cascade multicell converter with maximum output voltage value of E: (a) based on isolated dc voltage sources; (b) based on one dc voltage source and isolating transformers.

The voltage of dc sources in different cells of the CM converter topology, as shown in Fig. 1, discussed in the previous are equal and this configuration of CM converter is called symmetrical CM converter. However, it is not essential to have dc voltage sources with same voltage value in all series connected cells. Alternatively, unequal dc voltages may be selected for the power cells. In fact, a proper choice of voltage asymmetry between dc voltage sources of connected cells can produce a different combination of voltage levels and eliminate redundancies. This issue causes to increase the number of voltage levels in the converter output voltage waveform for a given number of cells without necessarily increasing the number of H-bridge cells [13, 26]. This topology of CM converter is asymmetrical

CM converter. Fig. 3 shows two topologies of asymmetrical CM converter, where the dc voltages for the H-bridge cells are not equal. The relationship between the voltage levels and their corresponding switching states in two-cell seven-level asymmetrical CM converter is illustrated in Table 2. There are some drawbacks associated with the asymmetrical CM converters. The modularity advantage of symmetrical CM converter does not exist in asymmetrical CM converter. In addition, switching pattern design becomes much more difficult due to the reduction in redundant switching states [26]. Therefore, the topology of asymmetrical CM converter has limited industrial applications.

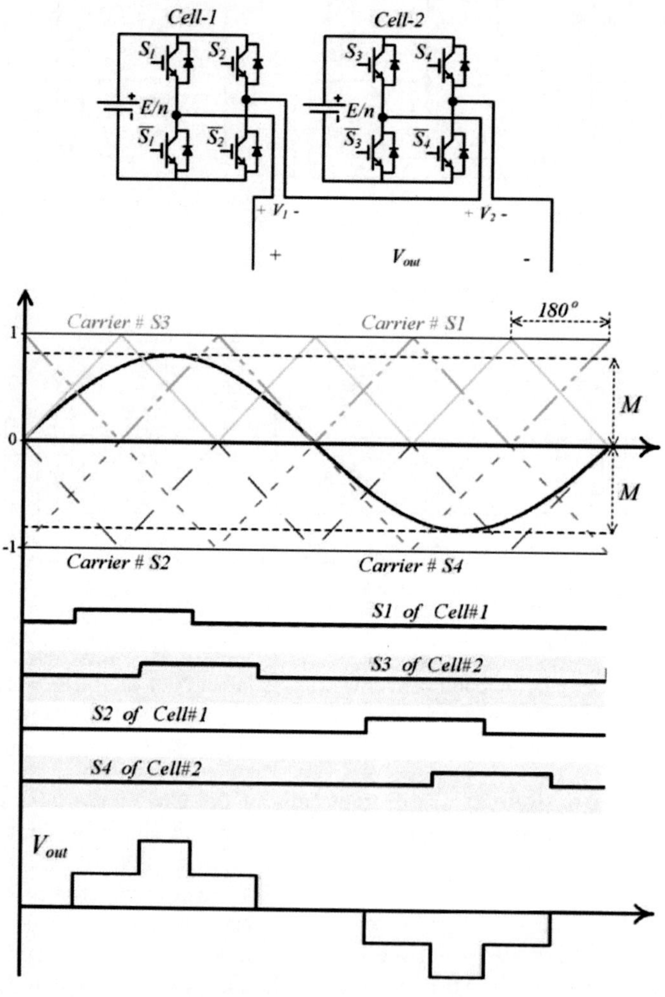

Fig. 2 2-cell-5-level cascade multicell configuration, its control strategy based on PS-SPWM, switches states and output voltage.

1.2 Flying Capacitor Multicell (FCM) Converter

A *2n*-cell FCM converter, as shown in Fig. 4, is composed of *4n* switches forming *2n*-commutation cells controlled with equal duty cycles and phase shifted of $2\pi/2n$ and $2n-1$ flying capacitors with the same capacitance. As a result, the electrical stresses on each switch are reduced and more equally distributed, so that each switch must withstand *2E/2n* volts [27-30].

Table 1 Switches states of 2-cell-5-level cascade multicell converter.

Output Voltage Level	State of switches (S_1, S_2, S_3, S_4)	Number of States
$+\frac{2}{2}E$	(1, 0 , 1 , 0)	1
$+\frac{1}{2}E$	(1, 0 , 1 , 1) , (1, 0 , 0 , 0) (1, 1 , 1 , 0) , (0, 0 , 1 , 0)	4
0	(1, 1 , 1 , 1) , (1, 1 , 0 , 0) (0, 0, 0, 0) , (0, 0 , 1 , 1)	4
$-\frac{1}{2}E$	(0 , 1 , 1 , 1) , (0, 1 , 0 , 0) (1, 1 , 0 , 1) , (0, 0 , 0 , 1)	4
$-\frac{2}{2}E$	(0, 1 , 0 , 1)	1

The output voltage of 2n-cell FCM converter has $2n+1$ level and its frequency spectrum has the harmonics around the $(2n.k.f_{switching})^{th}$ harmonic where k and $f_{switching}$ are the integer number and the switching frequency, respectively. Due to the similar waveform of current in all flying capacitors, they have the same capacitance in order to obtain the same voltage ripple. However, their dc voltage ratings are different and equal to E/n, $2E/n$, ..., $(2n-1)E/n$, so that the energy stored in the capacitor k is [30-32]:

$$U_K = \frac{1}{2}C\left(\frac{KE}{n}\right)^2 \quad (4)$$

The FCM converter and its derivative, the SM converter have many attractive properties for medium voltage applications including an advantage of transformer-less operation and ability to naturally maintain the flying capacitors voltages called natural balancing [31, 32].

The advantage of transformer-less operation causes that the FCM converter does not require a complex input transformer and in the case of internal fault of one cell, the maximum output voltage remains constant, but the number of levels decreases to:

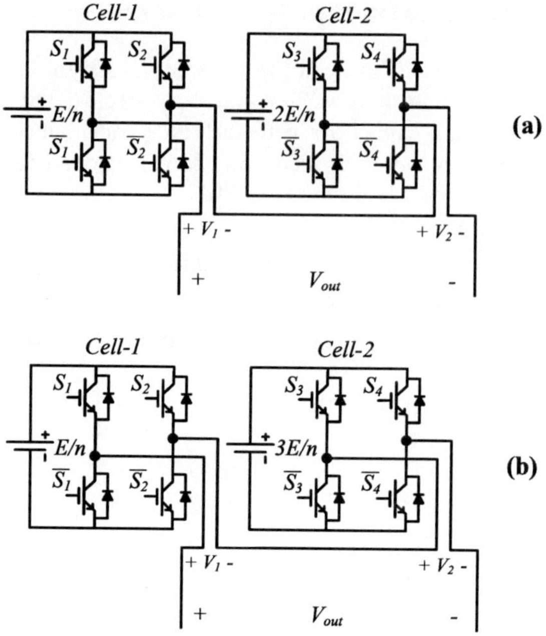

Fig. 3 Two-cell asymmetrical cascade multicell converter configuration: (a) seven-level; (b) nine-level.

Table 2 Switches states of two-cell-seven-level asymmetrical cascade multicell converter.

Output Voltage Level	State of switches (S_1, S_2, S_3, S_4)	Output Voltage of cell-1 (V_1)	Output Voltage of cell-2 (V_2)	Number of States
$+\frac{3}{3}E$	(1, 0, 1, 0)	$+\frac{1}{3}E$	$+\frac{2}{3}E$	1
$+\frac{2}{3}E$	(1, 1, 1, 0), (0, 0, 1, 0)	0	$+\frac{2}{3}E$	2
$+\frac{1}{3}E$	(1, 0, 0, 0), (1, 0, 1, 1)	$+\frac{1}{3}E$	0	2
0	(1, 1, 0, 0), (0, 0, 0, 0) (1, 1, 1, 1), (0, 0, 1, 1)	0	0	4
$-\frac{1}{3}E$	(0, 1, 1, 1), (0, 1, 0, 0)	$-\frac{1}{3}E$	0	2
$-\frac{2}{3}E$	(0, 0, 0, 1), (1, 1, 0, 1)	0	$-\frac{2}{3}E$	2
$-\frac{3}{3}E$	(0, 1, 0, 1)	$-\frac{1}{3}E$	$-\frac{2}{3}E$	1

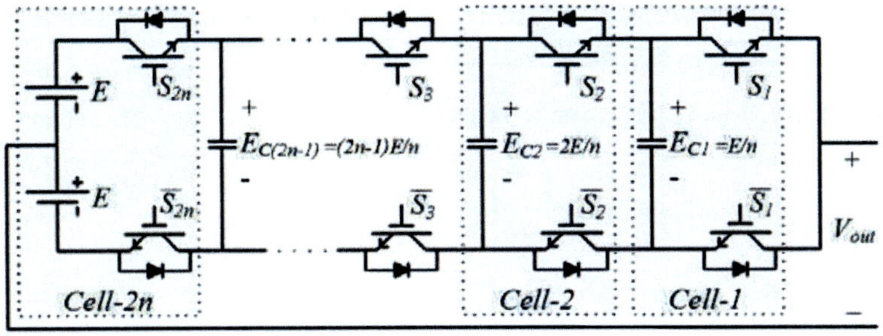

Fig. 4 2n+1 Level flying capacitor multicell converter with maximum output voltage value of *E*.

$$L_f = L - f \tag{5}$$

where f is the number of faulty cells.

As an example, Fig. 5 shows control strategy, switches states, and the output voltage of a 2-cell-5-level FCM converter controlled by the PS-SPWM and operated with a modulation index equal to 0.8 ($M = 0.8$). The Switch *X* is on when its state is 1 and is off when its state is 0. Switches states of the 2-cell-5-level FCM converter, using PS-SPWM method, are illustrated in Table 3.

Natural self-balancing process of the flying capacitors voltages, as one of the advantages of FCM converter, occurs without any feedback control and this process causes that the flying capacitors reach to their different dc voltage ratings equal to E/n, $2E/n$, ..., $(2n-1)E/n$. A necessary self-balancing condition is that the average flying capacitors currents must be zero. As a result, each cell must be controlled with the same duty cycle and a regular phase shifted progression along the cells. Generally, an output RLC filter (balance booster circuit), tuned to the switching frequency, has to be connected across the load in order to accelerate the self-balancing process in the transient states and to reduce the control signal fault effects [16-19, 31]. In this case, the dynamic of the transient depends on the impedance of load at the switching frequency. If the impedance at the switching frequency is high then the natural balancing is very slow and inversely. The output RLC filter is tuned to the switching frequency as follow:

$$\sqrt{L \cdot C} = \frac{1}{2 \cdot \pi \cdot f_{SW}} \tag{6}$$

where, f_{SW} is the switching frequency, *L* and *C* are inductance and capacitance of the output RLC filter, respectively.

Table 3 Switches states of 4-cell-5-level flying capacitor multicell converter

Output Voltage Level	State of switches (S_4, S_3, S_2, S_1)	Number of States
$+\frac{2}{4}E$	(1, 1 , 1 , 1)	1
$+\frac{1}{4}E$	(1, 1 , 1 , 0) , (1, 1 , 0 , 1) (1, 0 , 1 , 1) , (0, 1 , 1 , 1)	4
0	(1, 1 , 0 , 0) , (1, 0 , 0 , 1) (0, 0, 1, 1) , (0, 1 , 1 , 0)	4
$-\frac{1}{4}E$	(0, 0 , 0 , 1) , (0, 0 , 1 , 0) (0, 1 , 0 , 0) , (1, 0 , 0 , 0)	4
$-\frac{2}{4}E$	(0, 0 , 0 , 0)	1

1.3 Stacked Multicell (SM) Converter

An alternative topology based on the FCM converter is the SM converter which stacks two FCM converters together; the upper stack is switched only when a positive output is required and the lower stack is switched only when a negative output is required [31-33]. A $2\times n$ -cell SM converter, as shown in Fig. 6, is composed of *4n* switches forming 2*n*-commutation cells controlled with equal duty cycles, $2n-2$ flying capacitors with the same capacitance and different dc rating voltages equal to $E/2n$, $2E/2n$, …, $(n-1)E/2n$. As a result, the electrical stress on each switch is reduced and more equally distributed, so that each switch must support $E/2n$ volts [27].

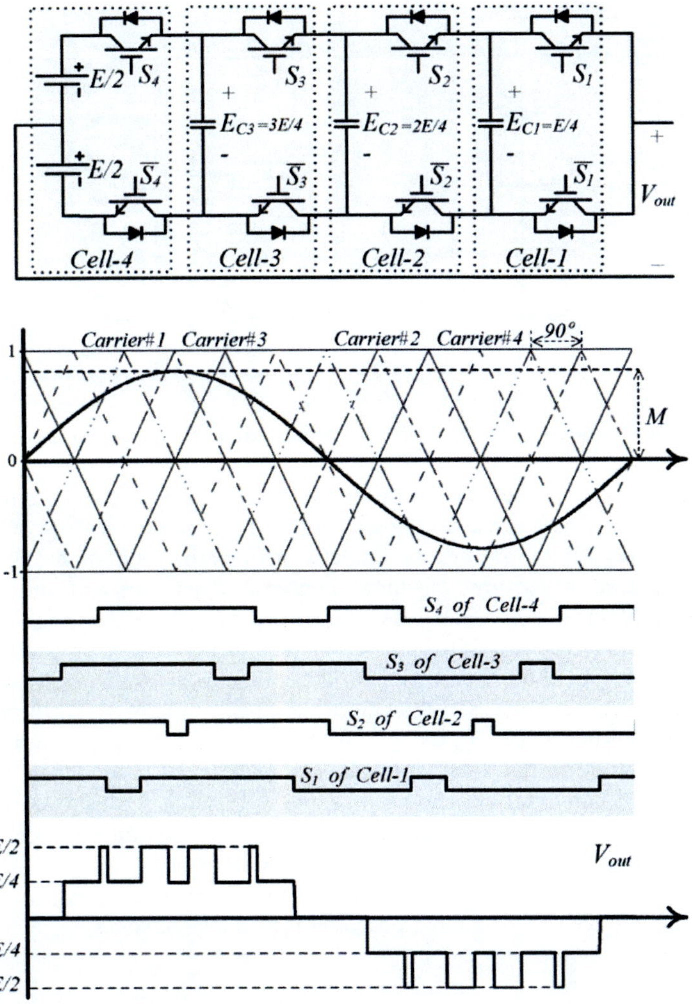

Fig. 5 4-cell-5-level flying capacitor multicell configuration, its control strategy based on PS-SPWM, switches states and output voltage.

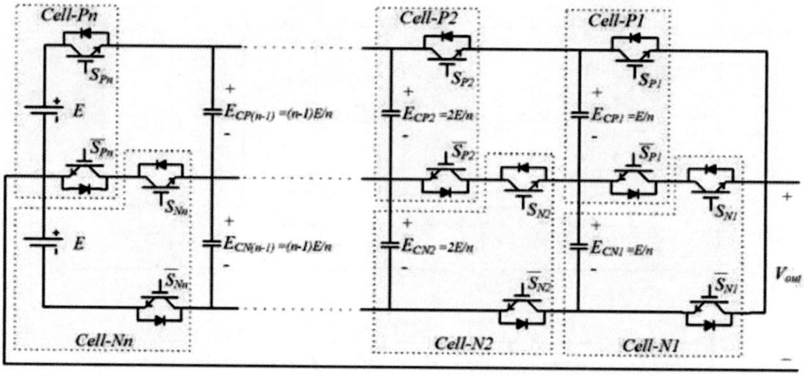

Fig. 6 2n+1 level stacked multicell converter with maximum output voltage value of *E*.

The main advantages of this configuration are that the number of combinations to obtain a desired voltage level is increased and the voltage ratings of capacitors and stored energy in the flying capacitors as well as the semiconductor losses are reduced. However, it requires the same number of capacitors and semiconductors in comparison with the equivalent FCM converter for the same number of output voltage levels [31, 32].

As shown in Fig. 7, the 4-cell-5-level SM converter as like as FCM converter is controlled by a regular PS-SPWM where the phase shift between the carriers of each cell is the same as (14.3); while, *n* is number of cells which work in each half cycle and equals 2 in the 4-cell-5-level SM converter. The switches states of 4-cell-5-level SM converter are shown in Table 4. Similar the FCM converter, the SM converter has the self-balancing advantage and each cell must be controlled with the same duty cycle and a regular phase shifted progression along the cells [21, 30]. In addition, the output RLC filter can be connected across the load in order to accelerate the self-balancing process in the transient states.

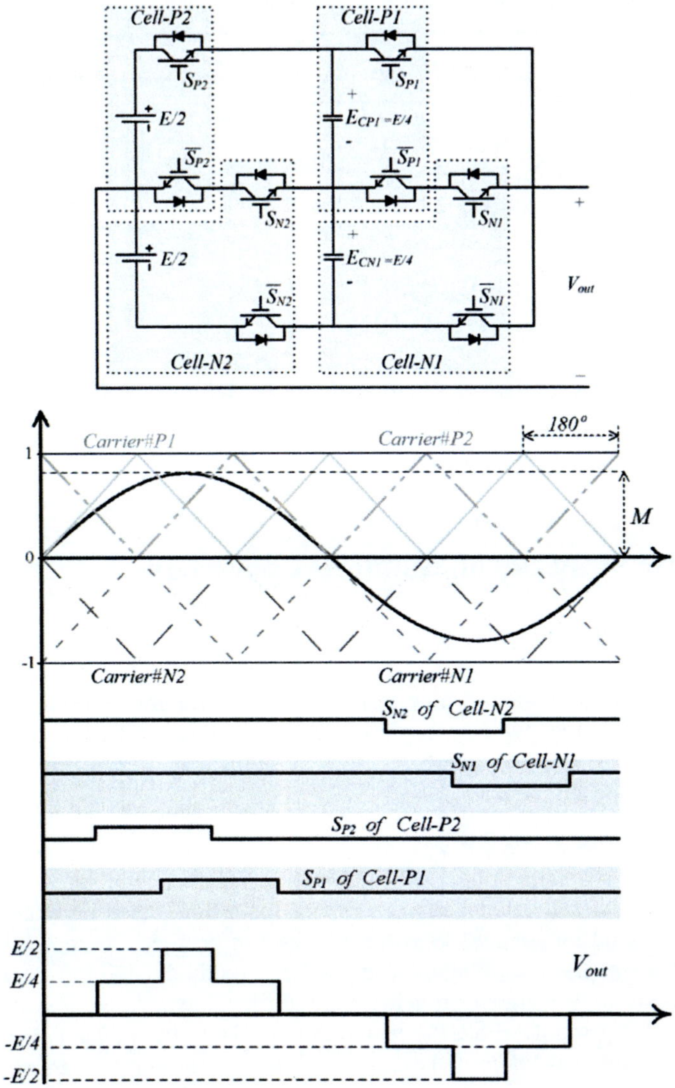

Fig. 7 4-cell-5-level stacked multicell configuration, its control strategy based on PS-SPWM, switches states, and output voltage.

Table 4 Switches states of 4-cell-5-level stacked multicell converter

Output Voltage Level	State of switches $\{(S_{P2}, S_{P1}), (S_{N2}, S_{N1})\}$	Number of States
$+\frac{2}{4}E$	{(1 , 1) , (1 , 1)}	1
$+\frac{1}{4}E$	{(1 , 0) , (1 , 1)} , {(0 , 1) , (1 , 1)}	2
0	{(0 , 0) , (1 , 1)} , {(1 , 0) , (1 , 0)} {(0 , 1) , (0 , 1)}	3
$-\frac{1}{4}E$	{(0 , 0) , (1 , 0)} , {(0 , 0) , (0 , 1)}	2
$-\frac{2}{4}E$	{(0 , 0) , (0 , 0)}	1

2 Control Strategies of Multilevel Converters

2.1 Carrier Based SPWM Algorithms

The carrier-based modulation algorithms for multilevel converters can be generally classified into two categories: phase-shifted and level-shifted modulations.

2.1.1 Phase-Shifted SPWM

Phase-shifted SPWM (PS-SPWM) is a natural extension of traditional SPWM technique [34], specially conceived for FCM and CM converters. Since each FC cell is a two-level converter, and each CM cell is a three-level inverter, the traditional bipolar and unipolar SPWM techniques can be used, respectively. Due to the modularity of these topologies, each cell can be modulated independently using the same reference signal. In general, a multilevel inverter with m voltage levels controlled by bipolar PS-SPWM requires $(m - 1)$ triangular carriers. In the phase-shifted multicarrier modulation, all of $(m - 1)$ triangular carriers have the same frequency (f_{cr}) and the same peak amplitude ($\hat{V}_{cr}$), but there is a phase shift between any two adjacent carrier waves [16-19], given by:

$$\phi_{cr} = 2\pi / (m-1) \tag{7}$$

The reference signal, V_m, is usually a sinusoidal wave with adjustable amplitude, $\hat{V}_m$, and frequency. The gate signals are generated by comparing the modulating sinusoidal reference with the carrier waves [31-33]. Since all the cells are controlled with the same reference and carrier frequency, the switch device usage and the average power handled by each cell is evenly distributed. Another interesting feature is that the total output voltage has a switching frequency with k times of the switching frequency of each cell, where k is an integer number. This multiplicative effect is produced by the phase-shift of the carriers and results in better total harmonic distortion (THD) at the output.

Figs. 8 and 9 show the principle of the PS-SPWM for a five-level CM and FCM converter, respectively. In Fig. 8 the PS-SPWM for CM converter is

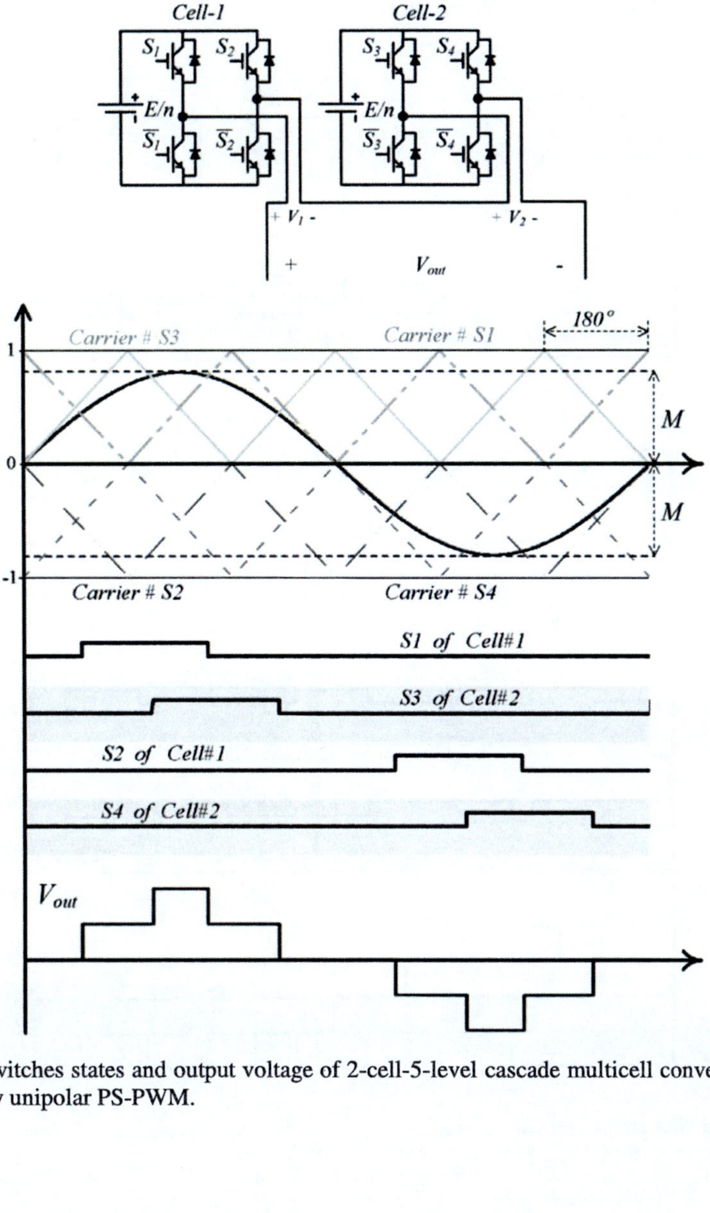

Fig. 8 Switches states and output voltage of 2-cell-5-level cascade multicell converter controlled by unipolar PS-PWM.

unipolar in which the operations of four switches (S_1 , S_2 , S_3 , S_4) are independent from each other. There are two series of carrier waves, one series including the carriers of S_1 and S_3, between 0 and +1, and the other one including the carriers of S_2 and S_4, between 0 and -1. On the other hand, in the CM converter controlled by bipolar PS-SPWM, the pair switch of S_1 and S_2 as well as pair switch of S_3 and S_4 will operate complementary and there will be one series of carrier waves between +1 and -1. In addition, the PS-PWM for FCM converter is bipolar, as shown in Fig. 9, in which the operation of two switches in each cell is complementary and there is one series of carrier waves between +1 and -1.

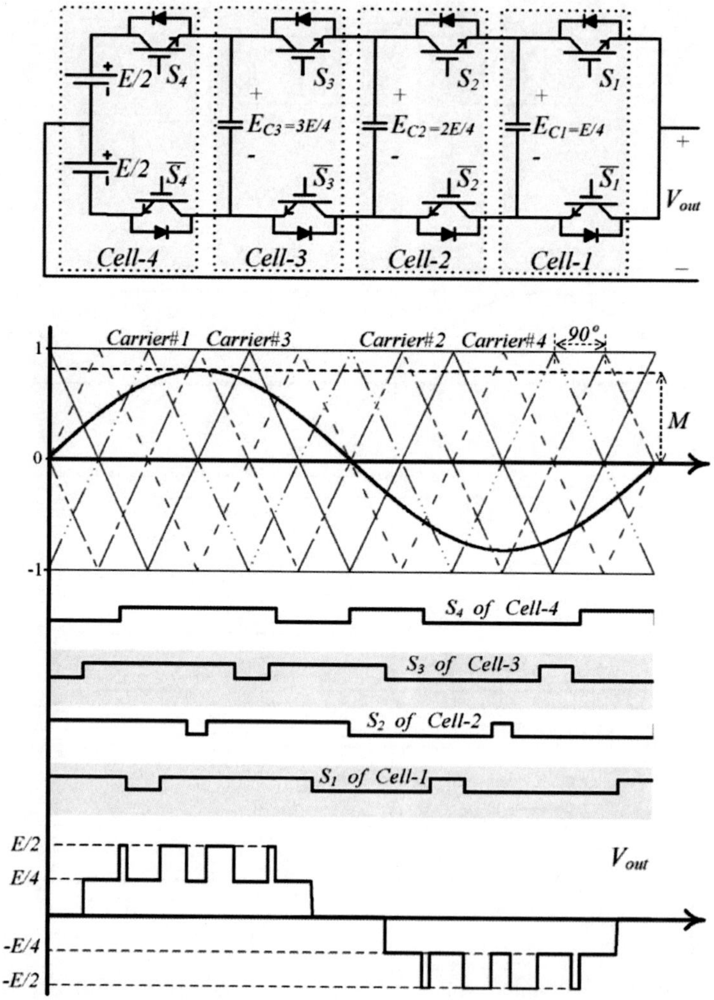

Fig. 9 Switches states and output voltage 4-cell-5-level flying capacitor multicell converter controlled by bipolar PS-SPWM.

As shown in Figs. 8 and 9, the gating for the each switch generated by comparing the related carrier waves, V_{cr}, and the sinusoidal reference signal. In this algorithm, the modulation index is expressed as:

$$m_a = \hat{V}_m / \hat{V}_{cr} \tag{8}$$

Moreover, the device (switch) switching frequency, $f_{sw,dev}$, can be calculated as:

$$f_{sw,dev} = f_{cr} \tag{9}$$

In general, the switching frequency of the converter, $f_{sw,conv}$, using the phase-shifted-SPWM is related to the device switching frequency as:

$$f_{sw,conv} = (m-1) \cdot f_{sw,dev} \tag{10}$$

The harmonics in the output voltage of CM converter controlled by unipolar PS-SPWM is distributed around $(2k \cdot m \cdot f_{cr})^{Hz}$ frequency, where k is an integer number. While the output voltage harmonics of both CM converter and FCM converter controlled by bipolar PS-SPWM are distributed around $(k \cdot m \cdot f_{cr})^{Hz}$.

2.1.2 Level-Shifted SPWM

Level-shifted SPWM (LS-SPWM) is a natural extension of unipolar SPWM technique for multilevel converters [35]. Similar to the phase-shifted modulation, an m-level CM converter using level-shifted multicarrier modulation algorithm requires $(m–1)$ triangular carriers, all having the same frequency and amplitude. The $(m–1)$ triangular carriers are arranged in vertical shifts, instead of the phase-shift used in PS-PWM, so that the bands they occupy are contiguous and each carrier is set between two voltage levels. Since each carrier is set to two levels, the same principle of unipolar SPWM can be applied.

To generate the corresponding levels, the control signal should be directed to the appropriate switches. The carriers span the whole amplitude range that can be generated by the converter. They can be arranged in vertical shifts, with all the signals in phase with each other, called in-phase disposition (IPD-SPWM); with all the positive carriers in phase with each other and in opposite phase of the negative carriers, known as phase opposition disposition (POD-SPWM); and alternate phase opposition disposition (APOD-SPWM), which is obtained by alternating the phase between adjacent carriers [35, 36]. An example of these methods for a five-level converter (four carriers) is illustrated in Fig. 10(a)–(c), respectively.

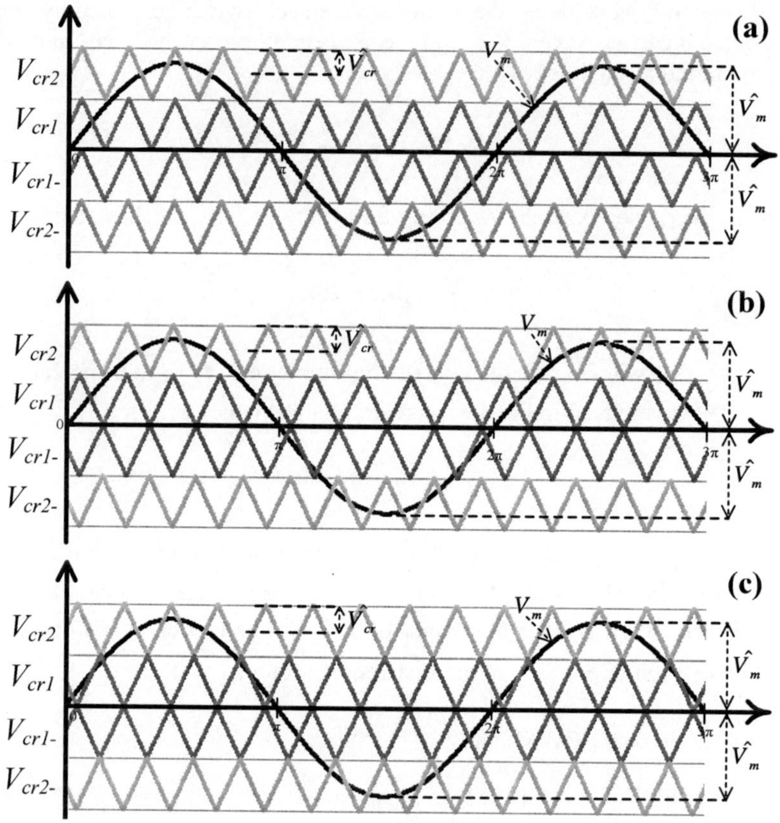

Fig. 10 Carrier arrangement of level-shifted SPWM for five-level converter: (a) in-phase disposition (IPD), (b) phase opposition disposition (POD), (c) alternate phase opposition disposition (APOD).

LS-SPWM leads to less distorted line voltages since all the carriers are in phase compared to PS-PWM [36]. In addition, since it is based on the output voltage levels, the method is so useful for NPC multilevel converter topology. However, the method is not preferred for CM and FCM converters, since it causes an uneven power distribution among the different cells. This generates input current distortion in the CM converter and flying capacitors unbalance voltage in the FCM converter compared to PS-SPWM.

In the following, the IPD-SPWM algorithm is only discussed since it provides the best harmonic profile of all three modulation algorithms. The detailed principle of IPD-SPWM for seven-level CM converter is shown in Fig. 11. The device switching frequency in the phase-shifted SPWM is equal to the carrier frequency. This relationship, however, is not applicable for the level- shifted SPWM. For example, with the carrier frequency of 450 Hz in Fig. 11, the switching frequency of the devices in Cell#1 is only 50 Hz, which is obtained by the number of gating pulses per cycle multiplied by the frequency of the modulating wave

(50 Hz). On the other hand, the switches in Cell#3 are turned on and off only twice per cycle, which means a switching frequency of 100 Hz. Therefore, it can be pointed out that the switching frequency is not same for the devices in different H-bridge cells. In general, the switching frequency of the converter, $f_{sw,\ conv}$, using the level-shifted-SPWM is equal to the carrier frequency, f_{cr}, that is,

$$f_{sw,conv} = f_{cr} \tag{11}$$

The average device switching frequency, $f_{sw,dev}$, is

$$f_{sw,dev} = f_{cr} / (m-1) \tag{12}$$

Besides the unequal device switching frequencies, the conduction time of the devices is not evenly distributed. For example, the device S_5 in Cell#3 conducts much less time than S_1 in Cell#1 per cycle of the fundamental frequency. To evenly distribute the switching and conduction losses, the switching pattern should rotate among the H-bridge cells.

The harmonics in the output voltage of CM converter controlled by unipolar LS-SPWM are distributed around $(k \cdot f_{cr})^{Hz}$, where k is an integer number. The comparison of level-shifted and phase shifted SPWM are illustrated in Table 5. The maximum voltage in multilevel converters as well as two-level converters can be boosted by 15.5% using the third harmonic injection method. This technique can also be applied to the phase- and level-shifted modulation algorithms.

2.2 Space Vector Pulse Width Modulation Technique

The space vector-based pulse width modulation (SVPWM) technique is a digital modulation technique and a well-known method in control the power electronics converters [37]. The SVPWM offers a number of useful features in comparison PWM methods, such as:

- Easy to implement with digital controller (PWM method is an inherently analogue technique),
- Better output waveforms,
- Representing voltage or current in two-dimension reference frames (instated of three-dimensional *abc* frames),
- Reducing the number of switching in each cycle.

In addition, SVPWM has capability to control input and output power factors independently. In the following, the SVPWM will be used to control a voltage source converter.

Fig. 12 illustrates the voltage source inverter (VSI), where input is connected to a dc voltage, and output is connected to a three-phase load with constant current.

Eight possible switching states that can be considered for the six switches of VSI, are shown in Fig. 13. The last two combinations make zero output voltage.

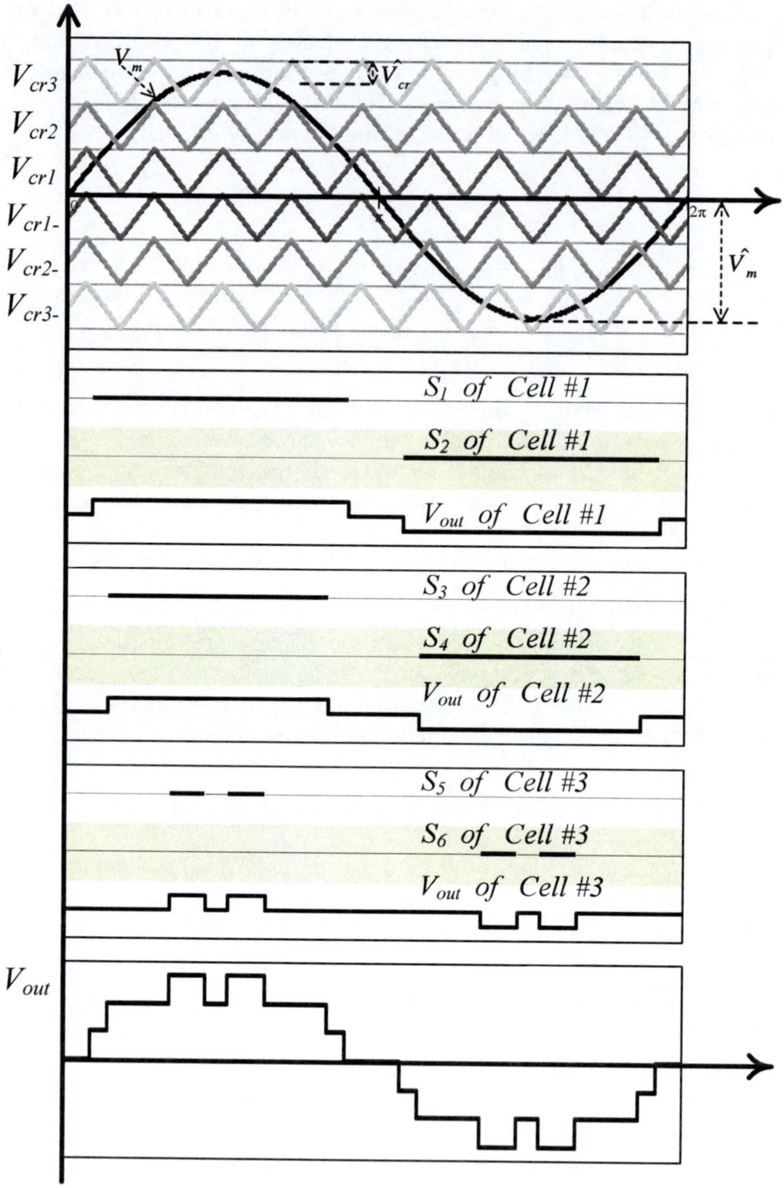

Fig. 11 Level-shifted SPWM for seven-level cascade multicell converter.

Table 5 Comparison of level- and phase-shifted SPWM

Item	Phase-Shifted SPWM	Level-Shifted (IPD) SPWM
Device switching frequency	Same for all devices	Different
Device conduction period	Same for all devices	Different
Rotating of switching patterns	Not required	Required
Line-to-line voltage THD	Good	Better

The desired three-phase output ac voltages V_{AB}, V_{BC} and V_{CA} are a set of quantized values in *abc* coordinate at each state. These voltages can be transferred to a fixed two-dimensional *dq*-frame as follows [38]:

$$\begin{bmatrix} V_{oq} \\ V_{od} \end{bmatrix} = \frac{2}{3} \begin{bmatrix} 1 & \cos(\frac{2\pi}{3}) & \cos(-\frac{2\pi}{3}) \\ 0 & \sin(\frac{2\pi}{3}) & \sin(-\frac{2\pi}{3}) \end{bmatrix} \begin{bmatrix} v_{AB} \\ v_{BC} \\ v_{CA} \end{bmatrix} \tag{13}$$

In complex form, the space vector of the desired quantized output line voltages can be defined by:

$$V_{oL} = \frac{2}{3}(v_{AB} + v_{BC} e^{j\frac{2\pi}{3}} + v_{CA} e^{-j\frac{2\pi}{3}}) \tag{14}$$

This transformation can be used for any of the three-phase quantities in *abc*-frame, *i.e.*, voltages or currents related to the voltage or current source converters.

By calculating the switching state vector for allowed output voltages of the voltage-source inverter in each switching state, six non-zero space switching voltage vectors, $\vec{V}_1, \vec{V}_2, \ldots, \vec{V}_6$, will result, as illustrated in Fig. 14. These vectors can form a hexagon centered at the origin of the *dq* frame. The remaining two zero voltage space voltage vectors, $\vec{V}_0 \& \vec{V}_7$, are located at the origin of the frame [38, 39].

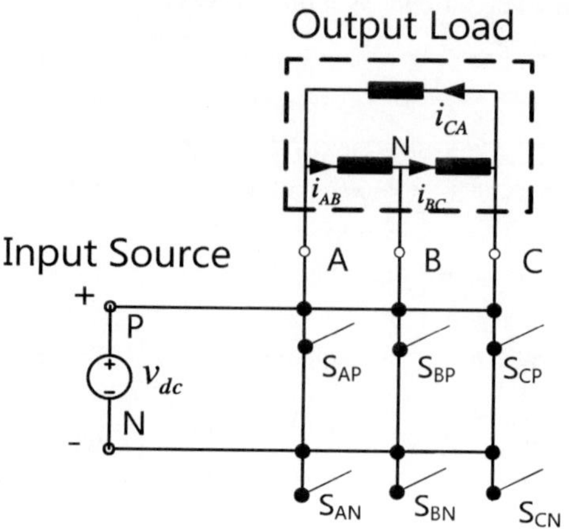

Fig. 12 Voltage-source inverter with fictitious dc-link voltage, V_{dc}.

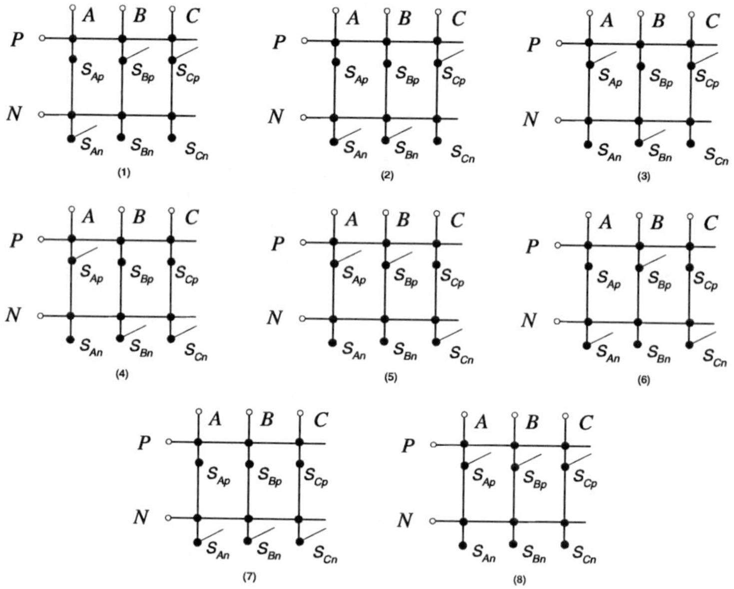

Fig. 13 Eight possible switching states for the six switches of VSI.

The continuous desired output voltages, $v_{AB}(t)$, $v_{BC}(t)$ and $v_{CA}(t)$ can be represented by a space vector given by (14.15).

$$V_{oL\_ref} = \sqrt{3}V_{om}e^{j(\omega_o t+\varphi_o)} \tag{15}$$

This reference vector should be synthesized using the two active vectors adjacent to the reference vector and a zero vector. Fig. 15 illustrates an example of how $\vec{V}_{ol\_ref}$ can be synthesized when it is located in sector 1, between non-zero vectors, $\vec{V}_1$ and $\vec{V}_6$.

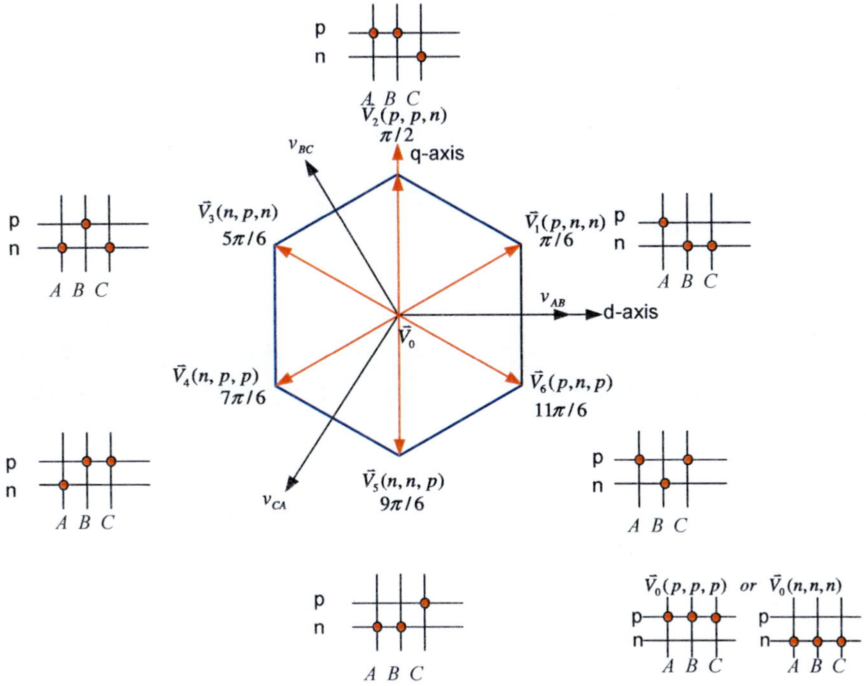

Fig. 14 Output voltage space vector in complex plane.

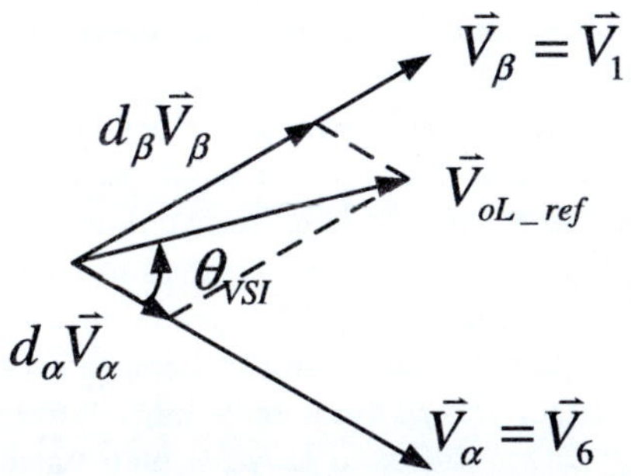

Fig. 15 $\vec{V}_{oL\_ref}$ synthesis using vectors $\vec{V}_1$ and $\vec{V}_6$.

During each switching period, T_s, the V_{oL_ref} is calculated by choosing the time T_α spent in vector $\vec{V}_\alpha$, and time T_β spent in vector $\vec{V}_\beta$. The rest of the time T_0 is dedication to a zero vector. These three time periods can be stated by duty cycles using trigonometric identities as:

$$\begin{cases} T_\alpha = m_v \sin(\frac{\pi}{3} - \theta_{VSI}).T_s = d_\alpha T_s \\ T_\beta = m_v \, sin(\theta_{VSI}).T_s = d_\beta T_s \\ T_{0v} = d_{0v} T_s = T_s - (T_\alpha + T_\beta) \end{cases} \Rightarrow \begin{cases} d_\alpha = m_v \, sin(\frac{\pi}{3} - \theta_{VSI}) \\ d_\beta = m_v \, sin(\theta_{VSI}) \\ d_{0v} = 1 - (d_\alpha + d_\beta) \end{cases} \quad (16)$$

where $0 \le m_v \le 1$ is the modulation index of voltage source inverter, and θ_{SVI} the angle between the reference vector and the closest clockwise state vector.

There are several methods for distributing the time periods (T_α, T_β, T_0) during a switching period in space vector modulation. One possible way is shown in Fig. 16.

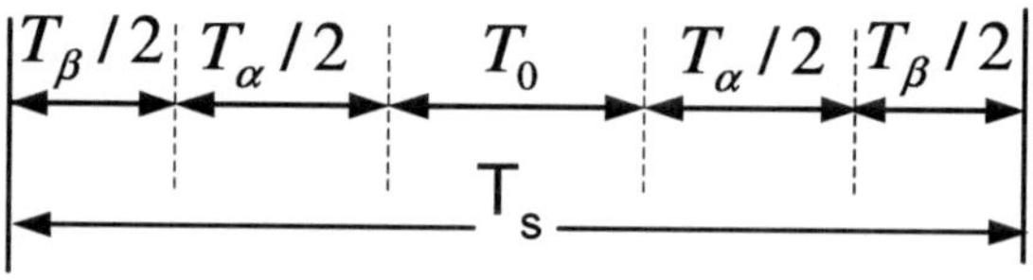

Fig. 16 A possible method of distributing the time periods during one switching period.

2.3 Selective Harmonic Elimination

To control the output voltage and reduce the undesired harmonics, different sinusoidal pulse width modulation (SPWM) and space-vector PWM schemes are explained in previous sections for multilevel inverters; however, PWM techniques are not able to eliminate lower order harmonics completely. The undesirable lower order harmonics of a square wave can be eliminated and the fundamental voltage can be controlled by Selective Harmonic Elimination (SHE) technique in which switching angles should be chosen [40]. SHE technique offer several advantages over other modulation methods including acceptable performance with low switching frequency to fundamental frequency ratios, direct control over output waveform harmonics, and the ability to leave triplen harmonics uncontrolled to take advantage of circuit topology in three phase systems, and therefore have drawn great attention in recent years [40-42]. There are two categories of SHE technique which are SHE staircase modulation and SHE-PWM. In the SHE staircase modulation, only one switching is done at predetermined angle in each level of multilevel output voltage while several switchings are done at predetermined angle in each level of multilevel output voltage controlled by SHE-PWM. A fundamental issue associated with such method is to obtain the arithmetical solution of nonlinear transcendental equations that contain trigonometric terms and naturally present multiple solutions. The principle of SHE staircase modulation and SHE-PWM are illustrated in Figs. 17 and 18, where V_1, V_2, V_3 and V_4 are the output of the H-bridge cells in a nine-level CM converter. In these figures, positive half cycle output voltage is shown with quarter-wave symmetry.

As shown in Fig. 17, the positive half cycle of output voltage controlled by SHE staircase modulation is formed by four-level staircase; thus, there are 4 variables, *i.e.*, θ_1, θ_2, θ_3 and θ_4, which can be determined to eliminate three significant harmonic components and control the fundamental voltage.

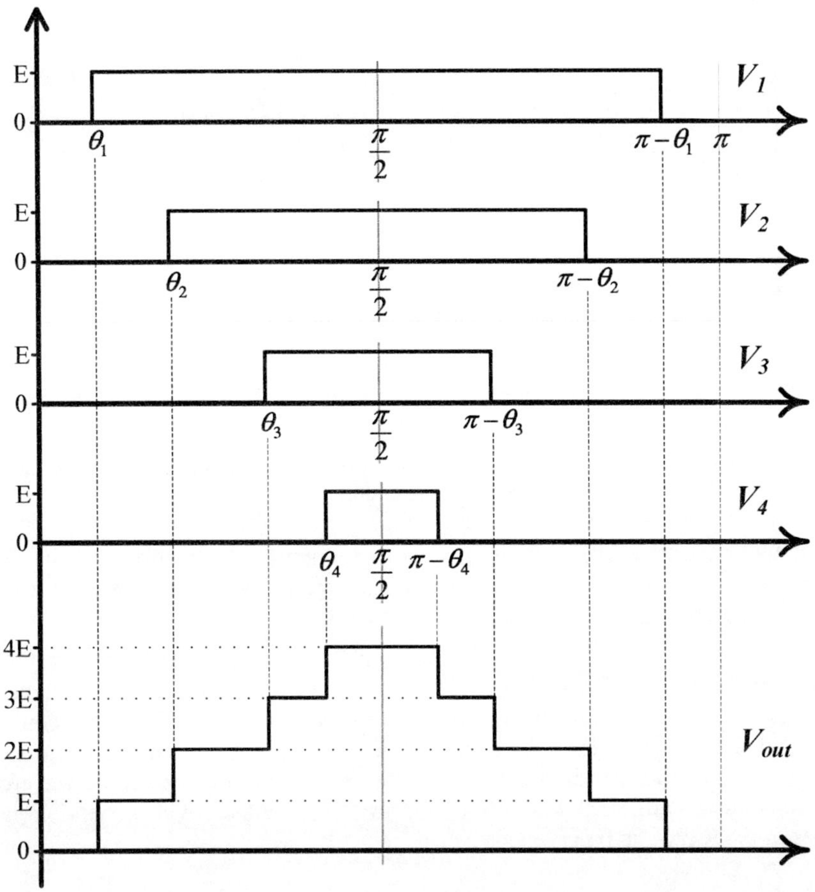

Fig. 17 Principle of selective harmonic elimination staircase modulation for nine-level converter.

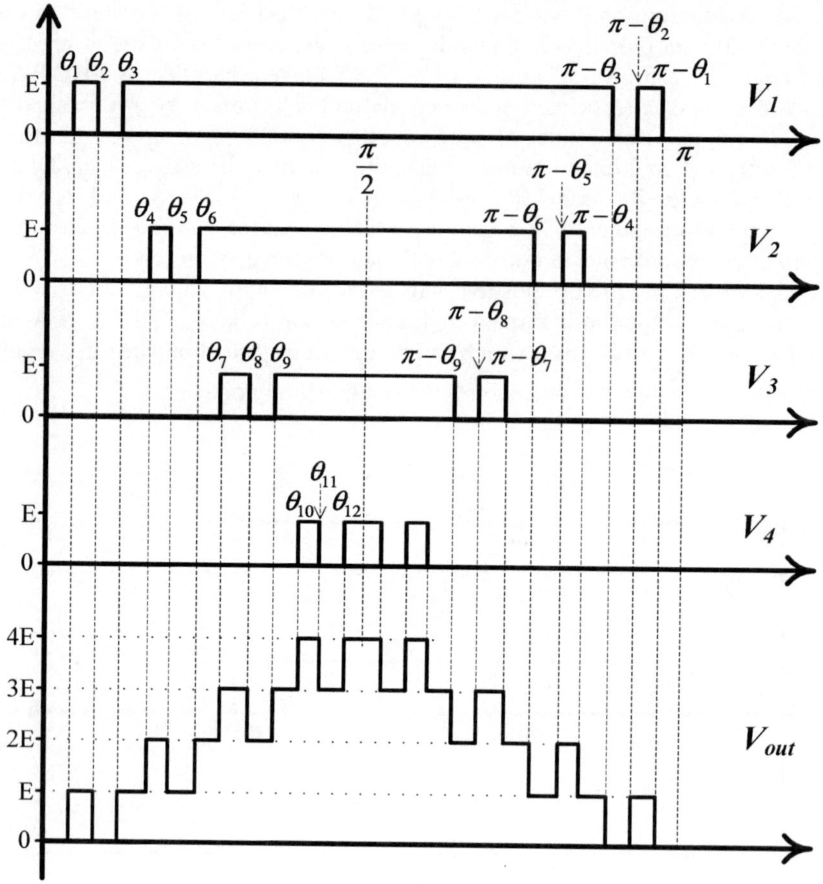

Fig. 18 Principle of selective harmonic elimination pulse width modulation (SHE-PWM) for nine-level converter with twelve switching angles

On the other hand, the positive half cycle of output voltage controlled by SHE-PWM has four levels, while three switchings, can be extended to five, seven, …, (any odd number) switchings, are done in each level, as shown in Fig. 18. Therefore, there are 12 variables, i.e., θ_1, θ_2, … and θ_{12}, which can be determined to eliminate eleven significant harmonic components and control the fundamental voltage. Although the number of switchings in each level in SHE-PWM should be odd, they are not essential to be same. In the SHE staircase modulation and SHE-PWM, a large number of harmonic components can be eliminated if the waveform accommodates additional variables, *i.e.*, switching angles. The general Fourier series of the wave can be given as:

$$V(t)=\sum_{n=1}^{\infty}\left(a_n\cdot\cos(n\omega t)+b_n\cdot\sin(n\omega t)\right) \tag{17}$$

where

$$a_n=\frac{1}{\pi}\int_0^{2\pi}V(t)\cdot\cos(n\omega t)d(\omega t) \tag{18}$$

$$b_n=\frac{1}{\pi}\int_0^{2\pi}V(t)\cdot\sin(n\omega t)d(\omega t) \tag{19}$$

For a waveform with quarter-cycle symmetry, only the odd harmonics with sine components will be present. Therefore,

$$a_n=0 \tag{20}$$

$$V(t)=\sum_{n=1,3,5,\ldots}^{\infty}b_n\cdot\sin(n\omega t) \tag{21}$$

where,

$$b_n=\frac{4}{\pi}\int_0^{\pi/2}V(t)\cdot\sin(n\omega t)d(\omega t) \tag{22}$$

Assuming that the each level has E amplitude, b_n for SHE staircase modulation can be expanded as:

$$b_n=\frac{4E}{\pi}\begin{bmatrix}\int_{\theta_1}^{\theta_2}(+1)\cdot\sin(n\omega t)d(\omega t)+\int_{\theta_2}^{\theta_3}(+2)\cdot\sin(n\omega t)d(\omega t)+\ldots\\ +\int_{\theta_{k-1}}^{\theta_k}(+k-1)\cdot\sin(n\omega t)d(\omega t)+\int_{\theta_k}^{\pi/2}(+k)\cdot\sin(n\omega t)d(\omega t)\end{bmatrix} \tag{23}$$

b_n for SHE-PWM can be expressed as:

$$b_n = \frac{4E}{\pi}\left[\begin{array}{l} \int_{\theta_1}^{\theta_2} (+1)\cdot\sin(n\omega t)d(\omega t) + \int_{\theta_3}^{\theta_4} (+1)\cdot\sin(n\omega t)d(\omega t) + \\ \int_{\theta_4}^{\theta_5} (+2)\cdot\sin(n\omega t)d(\omega t) + \int_{\theta_5}^{\theta_6} (+1)\cdot\sin(n\omega t)d(\omega t) \\ +\int_{\theta_6}^{\theta_7} (+2)\cdot\sin(n\omega t)d(\omega t) \\ +\ldots \\ +\int_{\theta_{3k-2}}^{\theta_{3k-1}} (+k)\cdot\sin(n\omega t)d(\omega t) \\ +\int_{\theta_{3k-1}}^{\theta_{3k}} (+k-1)\cdot\sin(n\omega t)d(\omega t) \\ +\int_{\theta_{3k}}^{\pi/2} (+k)\cdot\sin(n\omega t)d(\omega t) \end{array}\right] \tag{24}$$

Using the general relation

$$\int_{\theta_1}^{\theta_2} (+1)\cdot\sin(n\omega t)d(\omega t) = \frac{1}{n}\left(\cos(n\theta_1) - \cos(n\theta_2)\right) \tag{25}$$

and integrating other components of equation (23) and (24), we can write b_n for SHE staircase modulation as follows:

$$\begin{aligned} b_n &= \frac{4E}{n\pi}\cdot\left[\cos(n\theta_1) + \cos(n\theta_2) + \ldots + \cos(n\theta_k)\right] \\ &= \frac{4E}{n\pi}\cdot\sum_{m=1}^{k}\cos(n\theta_m) \end{aligned} \tag{26}$$

Coefficients b_n for SHE-PWM is as follows:

$$b_n = \frac{4E}{n\pi} \cdot \begin{bmatrix} \cos(n\theta_1) - \cos(n\theta_2) + \cos(n\theta_3) + \ldots \\ \\ \cos(n\theta_{3k-2}) - \cos(n\theta_{3k-1}) + \cos(n\theta_{3k}) \end{bmatrix}$$

$$= \frac{4E}{n\pi} \cdot \sum_{m=1}^{k} \cos(n\theta_{3m-2}) - \cos(n\theta_{3m-1}) + \cos(n\theta_{3m}) \tag{27}$$

where k is number of levels. In the SHE staircase modulation with k number of levels, there are k number of switching angles and k number of simultaneous equations. In this case, the fundamental voltage can be controlled and k-1 harmonics can be eliminated.

For example, consider nine-level CM converter controlled by SHE staircase modulation that the 5th, 7th and 11th harmonics (lowest significant harmonics) should be eliminated and the fundamental voltage is to be controlled. The 3rd and other triplen harmonics can be ignored if the converter has three-phase application. In this case, k=4 and simultaneous equations can be written as follows:

$$b_1 = \frac{4E}{\pi} \cdot \left[\cos(\theta_1) + \cos(\theta_2) + \cos(\theta_3) + \cos(\theta_4) \right] \tag{28}$$

$$b_5 = \frac{4E}{5\pi} \cdot \left[\cos(5\theta_1) + \cos(5\theta_2) + \cos(5\theta_3) + \cos(5\theta_4) \right] \tag{29}$$

$$b_7 = \frac{4E}{7\pi} \cdot \left[\cos(7\theta_1) + \cos(7\theta_2) + \cos(7\theta_3) + \cos(7\theta_4) \right] \tag{30}$$

$$b_{11} = \frac{4E}{11\pi} \cdot \left[\cos(11\theta_1) + \cos(11\theta_2) + \cos(11\theta_3) + \cos(11\theta_4) \right] \tag{31}$$

These nonlinear equations can be solved by iterative methods, such as the Newton–Raphson method, or by the theory of resultant [40-42]. Iterative methods mainly depend on the initial guess which makes a divergence problem especially for high numbers of inverter levels. Also, both techniques are complicated and time consuming. Another approach to deal with the SHE problem is based on modern stochastic search techniques such as genetic algorithm (GA) and particle swarm optimization (PSO) [40-42]. However, by increasing the number of switching angles, the complexity of search space increases dramatically and both the methods trap the local optima of search space. As an example, for a fundamental voltage of $4E$ ($b_1 = 4E$), the θ values are:

$$\theta_1 = 10^\circ \;,\; \theta_2 = 22.1^\circ \;,\; \theta_3 = 40.7^\circ \;,\; \theta_4 = 61.8^\circ \tag{32}$$

The selective harmonic elimination method can be implemented with a microcomputer using a lookup table of switching angles. As the number of switching angles per cycle increases, a higher number of significant harmonics can be eliminated. However, the number of switching angles per cycle, switching frequency, is determined by the converter switching losses. An obvious disadvantage of this scheme is that the lookup table for higher number of switching angles is unusually large.

References

[1] Lezana, P., Rodríguez, J., Oyarzún, D.A.: Cascaded multilevel inverter with regeneration capability and reduced number of switches. IEEE Trans. Industrial Electronics 55(3), 1059–1066 (2008)

[2] Meynard, T.A., Foch, H., Forest, F., Turpin, C., Richardeau, F., Delmas, L., Gateau, G., Lefeuvre, E.: Multicell converters: derived topologies. IEEE Trans. Industrial Electronics 49(5), 978–987 (2002)

[3] Haederli, C., Ladoux, P., Meynard, T., Gateau, G., Lienhardt, A.M.: Neutral point control in multi level converters applying novel modulation schemes. In: IEEE 37th Power Electronics Specialists Conference, PESC, June 18-22, pp. 1–8 (2006)

[4] Lezana, P., Ortiz, G., Rodríguez, J.: Operation of regenerative Cascade Multicell Converter under fault condition. In: 11th Workshop on Control and Modeling for Power Electronics, COMPEL, August 17-20, pp. 1–6 (2008)

[5] Du, Z., Tolbert, L.M., Ozpineci, B., Chiasson, J.N.: Fundamental frequency switching strategies of a seven-level hybrid cascaded H-bridge multilevel inverter. IEEE Trans. Power Electronics 24(1), 25–33 (2009)

[6] Sabahi, M., Hosseini, S.H., Sharifian, M.B.B., Yazdanpanah Goharrizi, A., Ghareh-petian, G.B.: A three-phase dimmable lighting system using a bidirectional power electronic transformer. IEEE Trans. Power Electronics 24(3), 830–837 (2009)

[7] Gateau, G., Fadel, M., Maussion, P., Bensaid, R., Meynard, T.A.: Multicell converters: active control and observation of flying capacitor voltages. IEEE Trans. Industrial Electronics 49(5), 998–1008 (2002)

[8] Lienhardt, A.M., Gateau, G., Meynard, T.A.: Zero-Steady-State-Error input-current controller for regenerative multilevel converters based on single-phase cells. IEEE Trans. Industrial Electronics 54(2), 733–740 (2007)

[9] Turpin, C., Deprez, L., Forest, F., Richardeau, F., Meynard, T.A.: A ZVS imbricated cell multilevel inverter with auxiliary resonant commutated poles. IEEE Trans. Power Electronics 17(6), 874–882 (2002)

[10] Rodríguez, J., Sheng Lai, J., Peng, F.Z.: Multilevel inverters: a survey of topologies, controls and applications. IEEE Trans. Industrial Electronics 49(4), 724–738 (2002)

[11] McGrath, B.P., Holmes, D.G.: Natural current balancing of multicell current source converters. IEEE Trans. Power Electronics 23(3), 1239–1246 (2008)

[12] Nabae, A., Takahashi, I., Akagi, H.: A new neutral point clamped PWM inverter. IEEE Trans. Industry Application IA-17, 518–523 (1981)

[13] Babaei, E.: A Cascade multilevel converter topology with reduced number of switches. IEEE Trans. Power Electronics 23(6), 2657–2664 (2008)

[14] Lezana, P., Rodríguez, J., Aguilera, R., Silva, C.: Fault detection on multicell converter based on output voltage frequency analysis. In: Proc. IEEE Industrial Electronics Conf., pp. 1691–1696 (2006)
[15] Meynard, T.A., Fadel, M., Aouda, N.: Modelling of multilevel converters. IEEE Trans. Industrial Electronics 44(3), 356–364 (1997)
[16] Khoshkbar Sadigh, A., Hosseini, S.H., Barakati, S.M., Gharehpetian, G.: Flying capacitor multicell converter based dynamic voltage restorer. In: Proc. 41st North American Power Symposium (NAPS), USA, October 2009, pp. 1–6 (2009)
[17] Gateau, G., Meynard, T.A., Foch, H.: Stacked Multicell Converter (SMC): properties and design. In: Proc. IEEE PESC Meeting, pp. 1583–1588 (2001)
[18] Khoshkbar Sadigh, A., Babaei, E., Hosseini, S.H., Farasat, M.: Dynamic voltage restorer based on stacked multicell converter. In: IEEE Symposium on Industrial Electronics and Application (ISIEA), Malaysia, October 2009, pp. 1–6 (2009)
[19] Khoshkbar Sadigh, A., Hosseini, S.H., Barakati, S.M., Gharehpetian, G.: Stacked multicell converter based DVR with energy minimized compensation strategy. In: Proc. 41st North American Power Symposium (NAPS), USA, October 2009, pp. 1–6 (2009)
[20] McGrath, B.P., Holmes, D.G.: Analytical modeling of voltage balance dynamics for a flying capacitor multilevel converter. IEEE Trans. Power Electronics 23(2), 543–550 (2008)
[21] McGrath, B.P., Meynard, T., Gateau, G., Holmes, D.G.: Optimal modulation of flying capacitor and stacked multicell Converters Using a State Machine Decoder. IEEE Trans. Power Electronics 22(2), 508–516 (2007)
[22] Feng, C., Liang, J., Agelidis, V.G.: Modified phase-shifted PWM control for flying capacitor multilevel converters. IEEE Trans. Power Electronics 22(1), 178–185 (2007)
[23] Marchesoni, M., Mazzucchelli, M., Tenconi, S.: A non conventional power converter for plasma stabilization. IEEE Trans. Power Electronics 5(2), 212–219 (1990)
[24] Ghiara, T., Marchesoni, M., Puglisi, L., Sciutto, G.: A modular approach to converter design for high power AC drives. In: Proc. of the 4th European Conf. on Power Electronics and Applications (EPE 1991), Firenze, Italy, September 1991, pp. 4-477–4-482 (1991)
[25] Babaei, E., Hosseini, S.H., Gharehpetian, G.B., Tarafdar Haquea, M., Sabahi, M.: Reduction of dc voltage sources and switches in asymmetrical multilevel converters using a novel topology. Elsevier Journal of Electric Power Systems Research 77(8), 1073–1085 (2007)
[26] Hosseini, S.H., Khoshkbar Sadigh, A., Sharifi, A.: Estimation of flying capacitors voltages in multicell converters. In: Proc. 6th International ECTI Conf., Thailand, May 2009, vol. 1, pp. 110–113 (2009)
[27] Meynard, T., Lienhardt, A.M., Gateau, G., Haederli, C., Barbosa, P.: Flying Capacitor MultiCell Converters with Reduced Stored Energy. In: IEEE ISIE, Canada (July 2006)
[28] Shukla, A., Ghosh, A., Joshi, A.: Improved multilevel hysteresis current regulation and capacitor voltage balancing schemes for flying capacitor multilevel inverter. IEEE Trans. Power Electronics 23(2), 518–529 (2008)
[29] Lienhardt, A.M., Gateau, G., Meynard, T.A.: Digital sliding-mode observer implementation using FPGA. IEEE Trans. Industrial Electronics 54(4), 1865–1875 (2007)

[30] Khoshkbar Sadigh, A., Hosseini, S.H., Sabahi, M., Gharehpetian, G.B.: Double flying capacitor multicell converter based on modified phase shifted pulse width modulation. IEEE Trans. Power Electronics 25(6), 1517–1526 (2010)
[31] Khoshkbar Sadigh, A., Gharehpetian, G.B., Hosseini, S.H.: New method for estimation of flying capacitors voltages in stacked multicell and flying capacitor multicell converters. Journal of Zhejiang University SCIENCE-C (Compute & Electron) 11(8), 654–666 (2010)
[32] Hosseini, S.H., Khoshkbar Sadigh, A., Sabahi, M.: New configuration of stacked multicell converter with reduced number of dc voltage sources. Accepted for publication in 5th IET International Conference on Power Electronics, Machines and Drives (PEMD), Brighton, UK (April 2010)
[33] Mohan, N., Undeland, T.M., et al.: Power Electronics—Converters, Applications and Design, 3rd edn. John Wiley & Sons, New York (2003)
[34] Carrara, G., Gardella, S., et al.: A New Multilevel PWM Method: A Theoretical Analysis. IEEE Transactions on Power Electronics 7(3), 497–505 (1992)
[35] Holmes, G., Lipo, T.: Pulse Width Modulation for Power Converters. IEEE Press/Wiley, New York (2003)
[36] van der Broeck, H.W., Skudelny, H.-C., Stanke, G.V.: Analysis and realization of a pulse width modulator based on voltage space vectors. IEEE Tran. on Industry Applications 24, 142–150 (1988)
[37] Masoud Barakati, S.: Applications of Matrix Converters for Wind Turbine Systems. VDM Verlag Dr. Muller, Germany (2008), ISBN-NR: 978-3-639-09860-0
[38] Barakati, S.M.: Modeling and Controller Design of a Wind Energy Conversion System Including a Matrix Converter. PHD Thesis, University of Waterloo, Ontario, Canada (2009)
[39] Hosseini, S.H., Khoshkbar Sadigh, A., Barakati, S.M.: Comparison of SPWM technique and selective harmonic elimination by using genetic algorithm. In: Proc. of 6th International Conference on Electrical and Electronics Engineering (ELECO), Bursa, Turrkey, November 2009, pp. 1–5 (2009)
[40] Tarafdar Hagh, M., Taghizadeh, H., Razi, K.: Harmonic minimization in multilevel inverters using modified species-based particle swarm optimization. IEEE Trans. Power Electronics 24(10), 2259–2267 (2009)
[41] Taghizadeh, H., Tarafdar Hagh, M.: Harmonic elimination of cascade multilevel inverters with non-equal dc Sources using particle swarm optimization. IEEE Trans. Industrial Electronics (in Press)

Modeling and Control of DC-AC Power Converters of Distributed Energy Resources in Microgrids

Mohammadhassan Abdollahi Sofla, Lingfeng Wang, and Roger King

Abstract. The high penetration of power electronic interfaced distributed energy resources in microgrids raises several problems on the stability and power quality of microgrids. In this chapter, the modeling of DC-AC converters is discussed based on the characteristics of the MGs. The control techniques developed for DC-AC converters are briefly discussed. The capabilities and limits of these control methods are presented in order to enable the researchers to find suitable control strategy for a specified microgrid application. Finally, a performance test of microgrids is analyzed based on power quality and stability of microgrids. In this analysis, various scenarios are developed in order to examine the performance of the system. Some of the main issues which must be considered are discussed such as: 1) Power quality and regulation in the steady state operation, 2) Stability under low and medium frequency disturbances, 3) Stability under high frequency disturbances. This chapter is focused on the inner voltage and current controllers rather than power management of microgrids. The performance tests are developed in MATLAB/SIMULINK in which a more realistic performance of the system is tested using the SimPowerSystem Toolbox.

Nomenclature

v	Inverter output voltage (capacitor voltage).
i	Inverter current (inductance current).
L	Filter inductance.
C	Filter Capacitance.
r_L	Inductance resistance.
v_{in}	Inverter input voltage.
L_C	Coupling inductance (transformer inductance).
i_{MG}	Microgrid side current.

Mohammadhassan Abdollahi Sofla · Lingfeng Wang · Roger King
Department of Electrical Engineering and Computer Science, University of Toledo, MS 308, 2801 W. Bancroft St., Toledo, OH, 43606, USA

L. Wang (Ed.): *Modeling and Control*, Green Energy and Technology, pp. 341–366.
springerlink.com

V_{DC}	DC link voltage.
$\boldsymbol{A}, \boldsymbol{B}, \boldsymbol{E}$	System description matrices.
$\boldsymbol{C}, \boldsymbol{D}$	System outputs matrices.
$\boldsymbol{u}$	Input control signal.
d	Disturbance of the system.
$\boldsymbol{x}$	State variables.
k_p	Proportional gain.
k_i	Integral gain.
$\boldsymbol{A}^*, \boldsymbol{B}^*, \boldsymbol{E}^*$	Discredited system matrices.
P	Real power.
Q	Reactive power.
k_i	Integral gain.
$\boldsymbol{f}(x,t), \boldsymbol{B}(x,t)$	Time-variant system matrices.

1 Introduction

One of important applications of smart grids is associated with distribution networks [1]. Future distribution networks integrate various technologies such as distributed energy resources (DERs), smart buildings, Plug-in Hybrid Electric Vehicles (PHEVs). Hence, this kind of distribution networks needs to be controlled and managed smartly. Fig. 1 shows a schematic of a cyber-physical smart distribution network.

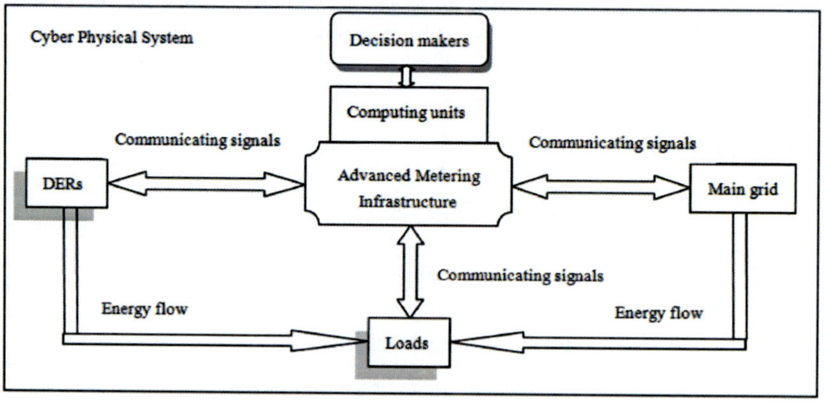

Fig. 1 A cyber physical smart distribution network.

Microgrid (MG) can be defined broadly for any energy system which is a cluster of loads and DERs, able to operate independently if necessary [1]. In MGs, Power Electronic Interfaces (PEIs) has been used as an efficient way to integrate DERs [2]. Hence, proper operation of these interfaces is important in proper operation of MGs or any other energy system. Therefore, this chapter focuses on the modeling and control of DC-AC converters of DERs in microgrid applications. In this study, modeling of Voltage Source Inverters (VSIs) is considered due to the importance of considering characteristics of MGs in designing control methods. In fact, modeling and control strategies of VSI must be adaptable to the characteristics of the MGs.

The control strategy of MGs can be considered in two parts: 1) Control of power flow (power management), and 2) Control of power converters (VSIs). Controllers of VSIs are discussed in this chapter which high and medium frequency dynamics should be controlled [3]. Also, output power quality and reference signals tracking are controlled by these controllers [3]. There are several issues make the modeling and control of power converters for MG applications critical:

- Stability: characteristics of MGs make them vulnerable systems which can easily be led to the instability.
- Power quality: high penetration of DERs in MGs and very low impedance between loads and DERs make the power quality analysis a major issue.

Several techniques have been applied to the control of DC-AC conversion in DERs applications. Recently, Digital Signal Processing (DSP) techniques are widely used instead of the analog approaches. In the MG applications, the current and voltage of DERs must be controlled to stabilize the system. Thus, the inner controllers construct a Current Controlled Voltage Source Converter (CC-VSC). The cascaded control structure has been used mostly to separate the control methods of voltage and current loops based on their respective dynamics [4]. The conventional linear controllers such as Proportional Integral (PI) regulators and resonant controller have been used in several works as voltage or current loop controller [5]. Usually, it has been used as voltage loop controller because it was assumed that voltage dynamics are not as fast as current dynamics [3]. However, more robust control methods have been developed for the voltage loop to enhance its dynamic and transient performance. Sliding mode [6]- [7], H^{∞} [8], robust servomechanism problem [9], repetitive [10], fuzzy [11] control methods are developed to enhance one or more properties of the voltage control loop. A PI controller has been added to a Sliding Mode Control (SMC) for the voltage loop in [12].

The speed and robustness of the controller in the current control loop have been considered much more than in voltage control loop. The conventional linear control methods have been developed in different configurations and reference frames for the inner current loop [10], [13], [14]. Deadbeat (predictive) control method which is a very fast controller has been applied for the current loop and an enhanced one has been presented in [15]. A nonlinear method based on the sliding mode control has been developed in [7], [16] to have a robust current loop against uncertainties and disturbances. Hysteresis current control has been used widely for the current loop in where a fast and robust response was needed [17]. Neural networks method has been also applied to the current controllers [18]. Fuzzy self-tuned PI method which enhances some properties of conventional PI controllers has been applied to the current controllers [19]. These methods are categorized and analyzed in this paper. This analysis is based on the characteristics of different MGs, meaning that each control method is described in a way that readers can find the suitable method for a proposed application of MGs. The modeling will be briefly considered in Section 3. Linear controllers are discussed in Section 3. The nonlinear controllers are described in Section 4. The properties of each control method are described to help in choosing an appropriate method based on the MG characteristics. A control strategy is simulated as an example to illustrate the performance tests in Section 5.

2 Modeling of DC-AC Converters in Microgrids

2.1 Microgrids Description

In MGs, it is supposed there is a considerable penetration of DERs connected to the system (distribution network). As distribution networks are not designed for being active power systems, there might be various interactions between DERs and the distribution system. In the LV distribution networks, low *X/R* ratio is a clear difference from transmission networks. Thus, in the low and medium frequency modes (and almost in the high frequency dynamics) LV distribution networks are resistive dominant. This demands that all dynamics particularly high frequency dynamics be transmitted very quickly. Some fast dynamics that might occur in MGs are listed as follows:

- Any switching action specially capacitors switching.
- Start up or disconnection of DERs.
- Islanding from the main grid.
- Connection or disconnection of a large load.
- Faults in the network.
- Start up or variations of fast dynamic loads such as induction motors.

There are some other medium or low frequency dynamics which are listed bellow:

- Unbalanced network.
- Load variations.
- Energy generation variation (especially those from the renewable energy resources).

All of dynamics influence the voltages and currents in the MG. These dynamics are supposed to be controlled by both voltage and current loops. Fig. 2 shows the configuration of a typical MG.

In a systematic view for the modeling purpose, the following characteristics of MG are obvious:

- MGs might have different operating points.
- Dynamics with higher frequency than system frequency might cause the nonlinearity and saturation of different elements.
- Variation of the modeled signals might be very fast.
- Regarding the network, loads and DERs characteristics, there might be parametric uncertainties.
- The system might have severe disturbances which might cause very high frequency oscillations.

Power quality of power converters needs more improvement since there might be several converters connected to a distribution feeder. These DERs can be sources of harmonics which are located very close to the end customer. Thus, it is necessary to reduce the THD of the output voltage of each DER in order to have a standard power quality in the whole system. A single line microgrid schematic is shown in Fig. 2.

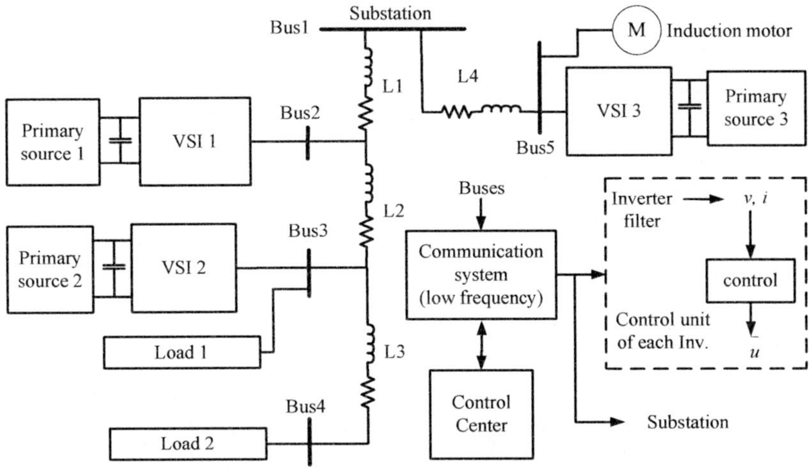

Fig. 2 Single line schematic of a microgrid: Central and local controllers are shown.

Primary source of the DERs may influence the mathematical modeling in both low and high frequency modes of operation. Some DERs have nondispatchable generation (renewable energy units). These generators can be analyzed in the long-term stability as sources of small disturbances however, they may be considered as sources of large disturbances because of existing severe weather disturbance. Moreover, the speed of the DERs in responding to transients should be taken into account in the stability analysis. Dynamics of the primary source was mostly neglected (because of ideal dc link assumption). However, this assumption can be violated in the high frequency modes analysis. Fig. 3 shows various PEI configurations for different DERs. However, it is not limited to only these configurations.

Other characteristics of the MGs come from the possible disturbances in the network. The voltage disturbances with different sources may occur in a MG. They can be considered, as follows: 1) transient over-voltages, which have characteristics of high frequency phenomena, 2) unbalanced voltages, because of different parameters such as nonlinearities of loads, malfunction of measurement devices, signal conversion errors, different phase parameters and single phase loads or generations 3) voltage dips and short interruptions, 4) voltage fluctuations, especially those causing flicker, 5) disturbances in generation units with variations in spectrum, duration and magnitude such as variation in generation of non-dispatchable units, dynamics of primary energy sources and 6) islanding and reconnection of the MG to the utility grid. In a MG system, all types of disturbances are severe and can lead to the voltage instability. Note that in situation when LCL filter is used for Voltage Source Inverters (VSIs) coupling point, high-frequency poles are placed near the instability border, and hence, any disturbance can take the system out of stability.

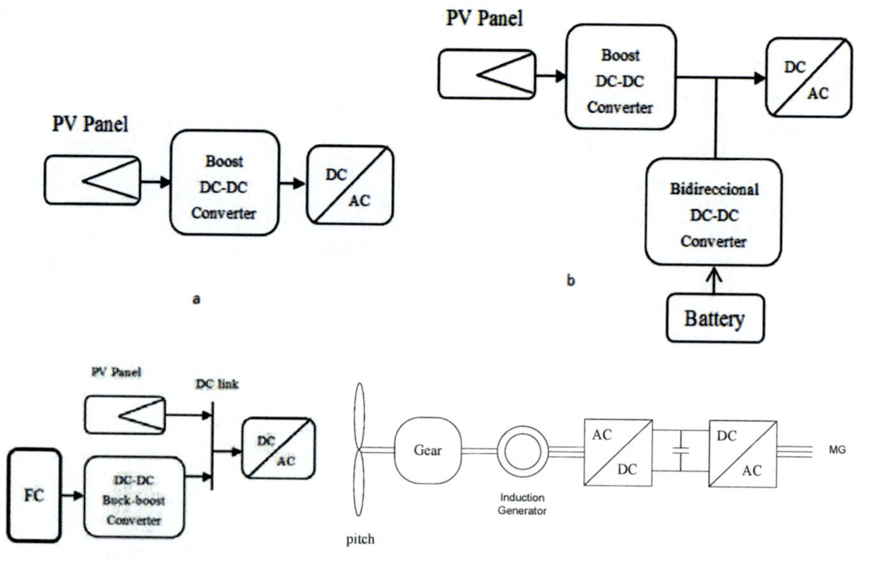

Fig. 3 Various DERs and their PEIs. Top: left) A Photovoltaic (PV) panel, right) Hybrid PV-battery unit. Bottom: left) Hybrid PV-Fuel Cell (FC) unit, right) Wind turbine system.

In designing inner controllers, all transients, the sag or swell of voltage, and the recovery time following a disturbance should be small enough and the transient oscillations must be maximally damped and free of circulating currents [4].

Moreover, VSIs employ passive low pass filters and therefore, they have near-zero inertia and a cut off frequency which is much higher than the fundamental frequency of MG. Inverter-based DERs tend to have faster dynamics but smaller output impedance.

2.2 Microgrids Control Process

High and low frequency modes of MG are controlled by different parts of control process. For high frequency disturbances, large signal analysis has been considered; and for load sharing analysis, small signal analysis has been developed.

In the MG stability analysis, voltage signal including angle is controlled by inner controllers of VSIs. The reference voltage signals come from the load sharing control.

It should be noted that if a DER could not follow proposed voltage angle fast enough, then DER would not remain synchronized with the system. This is important in both grid-connected and islanded modes of operation.

Islanding is a phenomenon in which main grid disconnects for any reason. But, the MGs do not go to the outage and keep working in the islanded mode.

In the islanded mode, severity of disturbances and vulnerability of MG is very important in designing controller for DC-AC converters. Fig. 4 represents the control process in the MGs. There are some control methods providing discrete ON/OFF signals and do not need any type of Pulse-Width-Modulation (PWM).

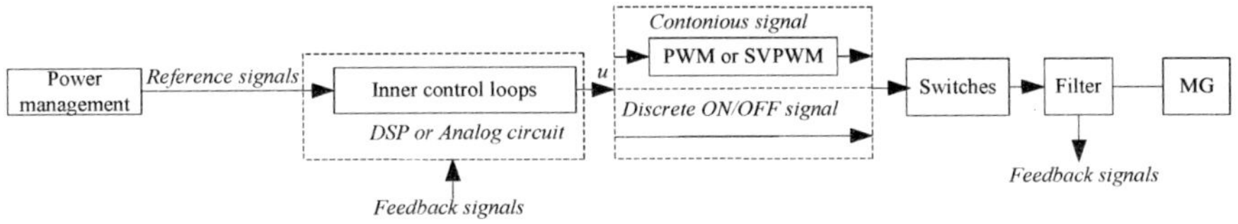

Fig. 4 Control process of inverters includes two main parts: power control and inner control VSIs, a) For continuous input signals, b) For discrete input signals.

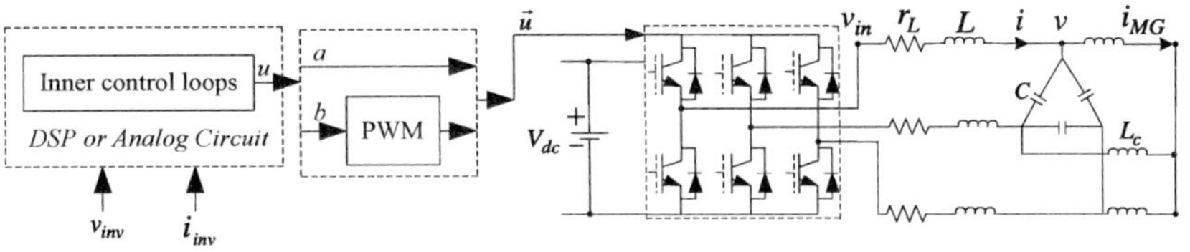

Fig. 5 Three-phase diagram of an inverter interface of a DER connected to the MG, signal a is for discrete u and signal b is for continuous u.

2.3 Modeling of Power Electronic Interfaced DERs in Microgrids

Fig. 5 shows a three-phase diagram of a DER connected to the MG via VSI. LC filter is used to pass the desired fundamental frequency voltages. The coupling inductance models the transformer inductance used for the connection of DERs to MGs.

Modeling of VSIs has been considered in the literature for different purposes. The average model of VSIs is an inaccurate model which is used for small signal analysis. This model sometimes is used to design linear controllers which are model dependent. The nonlinearity of the VSIs comes from the saturation and nonlinear behavior of active elements in high frequency modes. Also, linearized models at a specific operating point are not suitable for the MGs application because the operating point is changing.

There might be some uncertainties in the plant. These uncertainties must be considered in designing the control methods. For example, DC link perturbation in an uncertainty which depends on the type of DERs.

Disturbances in the utility grids are very common. Some of these disturbances can be considered in the model. The switching process is one of possible disturbances in VSIs because it can be a source of undesirable harmonics. But, it has been neglected mostly because the switching frequency is much higher than the system frequency.

Each MG has its own characteristics and aforementioned parameters must be considered in modeling and designing the controllers based on these modeling methods.

For example, in a MG in which the operating points of DERs are almost fixed and the plant parameters are precise enough, a linearized model can be used for designing linear controllers.

The reference frame of the modeling is also important in the performance of the controllers. Stationary $\alpha\beta$ reference frame by Clark's transformation, synchronous dq reference frame by Park's transformation, and positive-negative synchronous sequences reference frame have been developed for inner controllers of VSIs. The use of these methods is dependent on the characteristics of the proposed MGs.

The dynamics of the one phase of a VSI are expressed as follows:

$$\begin{aligned} \frac{dv}{dt} &= \frac{1}{C}i - \frac{1}{C}i_{MG} \\ \frac{di}{dt} &= \frac{1}{L}v_{in} - \frac{1}{L}v - \frac{r_L}{L}i. \end{aligned} \tag{6.1}$$

The state space presentation of this model is represented as follows:

$$\begin{gathered} \begin{cases} \dot{\boldsymbol{x}} = \boldsymbol{A}\boldsymbol{x} + \boldsymbol{B}\boldsymbol{u} + \boldsymbol{E}d \\ \boldsymbol{y} = \boldsymbol{C}\boldsymbol{x} + \boldsymbol{D}\boldsymbol{u} \end{cases} \\ \boldsymbol{x} = [v, i]^T, u = v_{in}, d = i_{MG} \\ \boldsymbol{A} = \begin{bmatrix} 0 & \frac{1}{C} \\ -\frac{1}{L} & -\frac{r_L}{L} \end{bmatrix}, \boldsymbol{B} = \begin{bmatrix} 0 \\ \frac{1}{L} \end{bmatrix}, \boldsymbol{E} = \begin{bmatrix} -\frac{1}{C} \\ 0 \end{bmatrix}, D = 0. \end{gathered} \tag{6.2}$$

The matrix $\boldsymbol{C}$ depends on the controller type. In the cascaded structure, $\boldsymbol{C} = [1, 0]^T$ for the current control loop and $\boldsymbol{C} = [0, 1]^T$ for the voltage control loop.

Some robust control methods use the following representation to develop a robust controller for VSIs:

$$\dot{x} = f(x,t) + B(x,t)\cdot u(x,t) + h(x,t)$$
$$x,\ f(x,t) \in \mathbf{R}^{n},\ \ B(x,t) \in \mathbf{R}^{n\times m},\ \ u(x,t) \in \mathbf{R}^{m}. \tag{6.3}$$

where $f(x,t)$ and $B(x,t)$ are time-variant system matrices. $h(x,t)$ is the disturbance matrix.

The dynamics of the current in (6.2) can be written as follows:

$$\frac{di}{dt} = \frac{u'}{L}V_{DC} - \frac{1}{L}v - \frac{r_L}{L}i.\ \ u' \in \{1,0\} \tag{6.4}$$

The input signal is u' which is a vector. This model is more accurate and also facilitates the modeling of switching disturbances [20].

In (6.2), the coupling inductance current is modeled as a disturbance to the system. Note that for some control methods modeling includes the second coupling inductance dynamics. However, this needs to add more sensors which results in higher cost.

3 Control Techniques for Power Converters of DERs in MGs

For inner Voltage Controller (VC), the following properties are desirable: 1) High loop gain at dc to control the magnitude of the fundamental frequency voltage with low error, 2) High loop gains at fundamental and twice at the fundamental frequency in order to suppress imbalance, 3) High bandwidth to reject harmonic distortions, 4) Robust for uncertainties of the plant model 5) Maximum resonance damping, 6) fast response with smooth and minimum overshoot, 7) Negligible interaction (coupling) between real and reactive powers and 8) Highly stable to any nonlinear behavior toward instability [22]-[23].

For inner Current Controller (CC), the following properties are desirable: 1) Fast transient response with minimum overshoot, 2) Precise current control, 3) Zero steady-state error, 4) Low distortion and current ripple, 6) High bandwidth (this will be limited by the switching frequency and the controller-sampling period), and 7) Less sensitivity to the system uncertain parameters. The response speed of inner current controllers must be much more than that of inner voltage controllers because current dynamics are much faster. The current high frequency modes made by any transient is more common and wider than voltage high frequency modes; thus, the bandwidth of current controller must be much higher to be shown the desired performance.

These properties might be achieved by different techniques for different MGs. For example, those MGs with high uncertainties need a more robust control technique to show the desired performance.

3.1 Linear Control Methods

Linear Controllers have been developed with different structures such as: 1) Conventional/Synchronous PI, 2) Resonant, 3) Predictive (Deadbeat) controllers.

1-1-3. Conventional/Synchronous PI Control

The transfer function of a PI controller is represented as follows:

$$G_c(s) = k_p + \frac{k_i}{s} \tag{6.5}$$

where k_p is a proportional gain, k_i is an integral gain and s is the Laplace operator. Fig. 6 shows a conventional PI control method as VC. The desired performance can be achieved by choosing appropriate gains. These gains are designed by adjusting a Bode plot, crossover frequency and phase margin. Root locus approaches are also used for tuning to improve the transient response. Phase margin is an important parameter to test the robustness of the inner controller. In (6.5), it is shown that PI regulators have a pole (infinite gain) at zero-frequency. Thus, they are not the best solutions to regulate the fundamental frequency and compensate higher harmonic disturbances of inner voltage loops. Also, typical high voltage-recovery times and high voltage variations have been reported [24]. The voltage regulation performance might not be quick enough to mitigate fast voltage disturbances in the subcycle range, such as capacitor switching transients and loads with fast dynamics. Note that conventional PI regulators do not have proper performance for unbalanced systems.

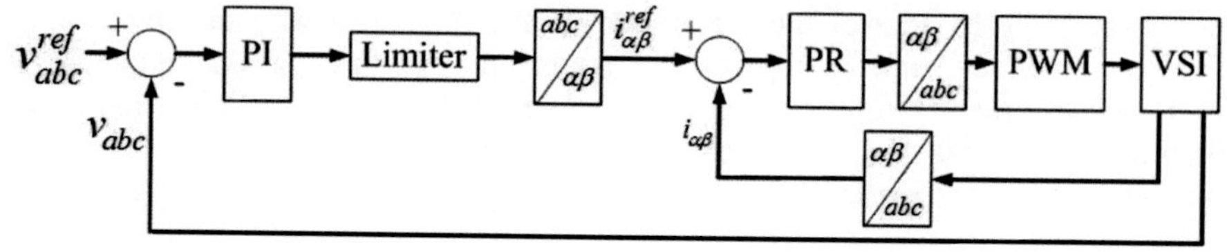

Fig. 6 Conventional PI voltage controller and PR controller of a VSI.

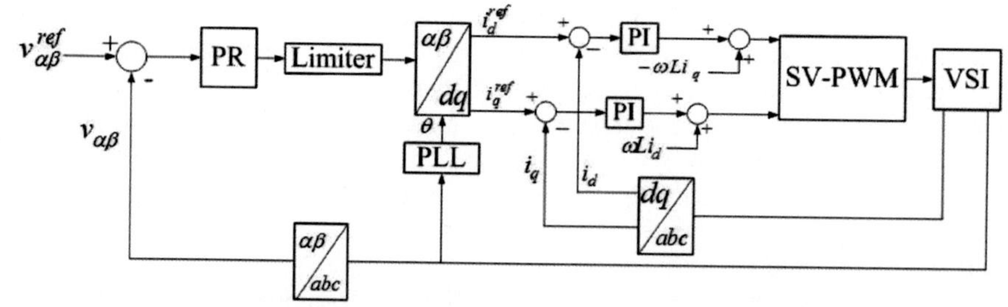

Fig. 7 Synchronous dq PI current and PR voltage controller of a VSI. Space Vector PWM (SV-PWM) is represented to generate control signals for switches.

PI controllers show better performance with some improvements such as using the voltage flux vector [25], or adding a predictive loop [26]. A hybrid PI-SMC controller is also developed for the inner voltage loop which shows an improved performance [12].

Synchronous PI controllers are a kind of PI controllers designed in the *dq* reference frame. Fig. 7 shows this control method as CC. This control method has a better performance than stationary type.

State feedback improves the dynamic behavior of PI controllers. This method improves the bandwidth of the PI controllers and hence improves the robustness.

3.1.1 Resonant Control

Resonant controller is a stationary frame equivalent of the synchronous PI controller. Resonant controller has been developed for inner voltage and current loops of VSIs [27]-[28]. This controller acts on a very narrow band around its resonant frequency ω. Usually, this method has been applied to the voltage control loop. As voltage controller, the implementation of harmonic compensator for low-order harmonics is possible without influencing all the behaviors of the current controller. Transfer function of Proportional Resonant (PR) controller can be as follows:

$$PR(s)=k_p+k_i\frac{s}{s^2+\omega^2}+\sum_{h=5,7}\frac{2k_{ih}s}{s^2+(h\omega)^2} \tag{6.7}$$

where k_{ih} is the resonant gain for the resonant peak adjustment.

This PR controller is also called P+ multi frequency resonant controller. The properties of this method are as follows: 1) Almost zero steady-state error regulation by having significant gains in the vicinity of the controller's resonant frequencies, 2) Actively damping the voltage oscillations, 3) Capability to control harmonics.

Note that the last term of (6.7) is a Harmonic Controller (HC) which can facilitate current harmonics control.

3.1.2 Deadbeat Control or Predictive Control

Predictive control has been mostly used as inner current controller due to its fast response [29]. The basic concept of predictive control is presented here.

The discrete-time dynamics of current can be written as follows:

$$i(k+1)=\boldsymbol{A}^*i(k)+\boldsymbol{B}^*u(k)+\boldsymbol{E}^*d(k). \tag{6.8}$$

$$\boldsymbol{A}^*=\exp(\boldsymbol{A}.T),\quad \boldsymbol{B}^*=\int_0^T e^{\boldsymbol{A}^*.T}\boldsymbol{B}d\tau,\quad \boldsymbol{E}^*=\int_0^T e^{\boldsymbol{A}^*.T}\boldsymbol{E}d\tau.$$

The required control input, $\boldsymbol{u}(k)$ can be obtained as follows:

$$u(k)=\boldsymbol{B}^{*-1}\left[i(k+1)-\boldsymbol{A}^{*}i(k)-\boldsymbol{E}^{*}d(k)\right] \tag{6.9}$$

As it is shown in (6.9), this control method needs a prediction for the parameter with $k+1$ dependent variable. Due to the presence of system delays, a delay compensation method has been applied in [15] to improve its bandwidth characteristics.

This method is a fast discrete-time controller; but, it has some disadvantages such as: 1) High THD for nonlinear loads [16], 2) Precise filter parameters determination [30] and 3) Sensitive to the parametric variations of the controlled system [6]. Though these disadvantages of deadbeat control can be compensated for, a longer control period results in lower control accuracy and lower bandwidth characteristics; hence, some favorable features of the predictive deadbeat control are lost [15].

3.2 Robust Control Methods

MGs may have different levels of uncertainties. Linear Controllers are model-dependent methods; hence, these methods cannot have proper performance in MGs with high uncertainty. Nonlinear control methods have been developed in order to improve the performance of VSIs in such applications. However, linear control methods can be used in MGs whose characteristics fit these methods. These methods are as follows: 1) H^{∞}, 2) Repetitive, 3) Robust servo-mechanism problem and 4) Two nonlinear robust controls: sliding mode and hysteresis control.

3.2.1 H^{∞} Control

H^{∞} control technique is served as a robust controller for VSIs [8]. Fig. 8 shows a standard H^{∞} diagram, in which $K(s)$ is the controller and $G(s)$ is transfer function in frequency domain by model (6). P(s) is the augmented plant by weighting functions, $W_1(s)$ and $W_2(s)$ based on the desired performance indices (third weighting function can be added). The weighting function, $W_1(s)$ is a typical low-pass filter, shaping the sensitivity function S at low frequency to reject disturbance and to reduce tracking errors, and Z_1 is a control variable used to adjust the tracking errors.

The weighting function, $W_2(s)$ is chosen to be a high-pass filter, shaping the complementary sensitivity function T at high frequency to minimize instability effects. Therefore, the steps in designing H^{∞} controller are as follows: 1) Augment the plant $G(s)$ by weighting functions based on the desired performance indices, 2) Suppose that input r can be any sinusoidal signal with amplitude less or equal to 1, performance specification $\|S\|_{\infty} < \varepsilon$ guarantees the tracking error e with amplitude less than ε. Next, we have to find a H^∞ robust controller $K(s)$ to satisfy the following inequality:

$$\left\| \begin{matrix} W_1 S \\ W_2 T \end{matrix} \right\| \le 1, \qquad S(s) = \left(1 + G(s)K(s)\right)^{-1}, \tag{6.10}$$

$$T(s) = G(s)K(s)\left(1 + G(s)K(s)\right)^{-1}$$

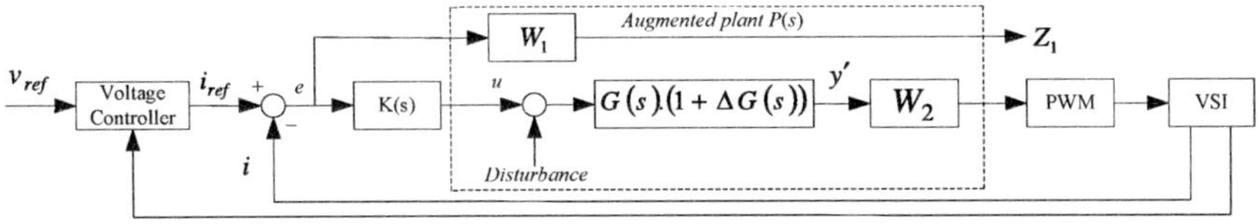

Fig. 8 H^∞ current controller diagram.

and 3) Examine if the design is close to the desired performance indexes based on the evaluation of the singular-value Bode plot; otherwise, step 1 should be repeated.

H^∞ controller can work well with known disturbances though it may not guarantee the voltage tracking capability under parametric uncertainty [24]. However, voltage disturbances are not periodic by nature, and there are some unknown disturbances in the MG [24].

3.2.2 Repetitive Control Method

In the repetitive control method, a repetitive controller is added to the control loop in addition to the conventional regulator. In Fig. 9, $P(z^{-1})$ represents the closed-loop transfer function of the plant which is closed loop regulated by a conventional regulator [31]. Blocks $Q(z^{-1})$ and $S(z^{-1})$are the auxiliary compensators of the repetitive controller.

This system is stable if $P(z^{-1})$ is stable and

$$\left. Q(z^{-1}) - P(z^{-1})S(z^{-1}) \right|_{z=e^{j\omega T}} < 1, \text{ for all } \omega.$$

The design of $Q(z^{-1})$ and $S(z^{-1})$ is a compromise between the relative stability and the convergence rate of the periodic error.

The repetitive control guarantees zero steady-state error at all the harmonic frequencies less than half of the sampling period. This is accomplished because the frequency modes of the periodic error are included in the stable control loops. This

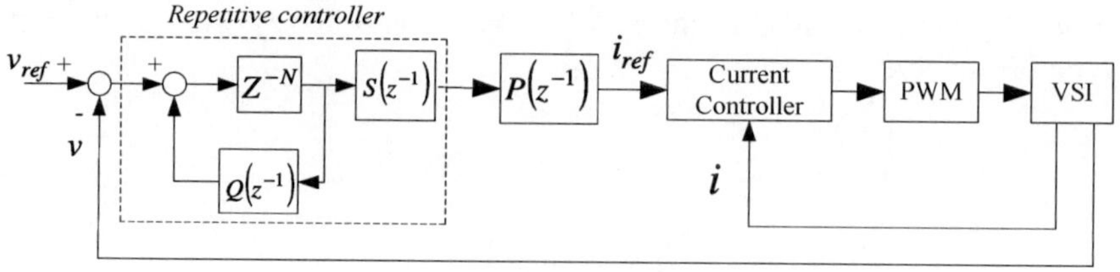

Fig. 9 Repetitive voltage controller schematic.

control method has a suitable performance against periodic disturbances. However, the repetitive control is not easy to stabilize for all unknown load disturbances and cannot obtain very fast response for fluctuating load [16].

3.2.3 Sliding Mode Control and Hysteresis

In the category of nonlinear control methods, hysteresis control is used mostly for the inner current controller. The difference between hysteresis control and SMC is the number of variable states used in the sliding manifold. In the hysteresis control only one variable has been used; however, in SMC, sliding manifold of all system state variables is used in the design.

The system (6.2) is discretized as follows:

$$\boldsymbol{x}(k+1)=\boldsymbol{A}^{*}\boldsymbol{x}(k)+\boldsymbol{B}^{*}\boldsymbol{u}(k)+\boldsymbol{E}^{*}\boldsymbol{d}(k) \tag{6.14}$$

The control signal is obtained from the equivalent control method [32] as follows:

$$\boldsymbol{u}_{eq}(k)=\left(\boldsymbol{C}'\boldsymbol{B}^{*}\right)^{-1}\left(\boldsymbol{x}_{ref}-\boldsymbol{C}'\boldsymbol{A}^{*}\boldsymbol{x}(k)-\boldsymbol{C}'\boldsymbol{E}^{*}\boldsymbol{d}(k)\right) \tag{6.15}$$

This control signal meets the Lyapunov stability criterion in which $S(k+1)$ must be less than or equal to $S(k)$.

Finally, the control law in this discrete-time Sliding Mode Control (SMC) can be stated as follows:

$$\boldsymbol{u}(k)=\begin{cases}\boldsymbol{u}_{eq} & \text{if} \quad \left\|\boldsymbol{u}_{eq}\right\|\leq u_{\max}\\ u_{\max}\dfrac{\boldsymbol{u}_{eq}}{\left\|\boldsymbol{u}_{eq}\right\|} & \text{if} \quad \left\|\boldsymbol{u}_{eq}\right\|>u_{\max}\end{cases}. \tag{6.16}$$

Fig. 10 shows the basic idea of SMC developed by Utkin [32].

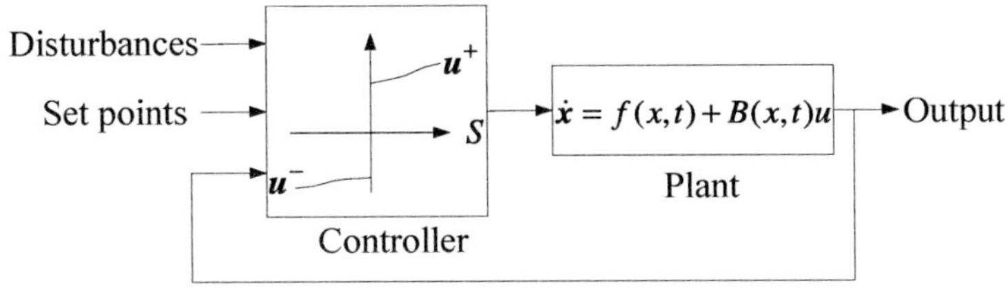

Fig. 10 Sliding mode control diagram.

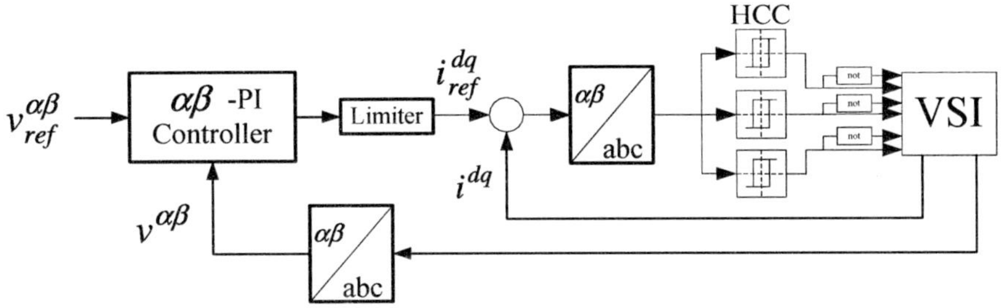

Fig. 11 Stationary $\alpha\beta$ PI voltage controller and hysteresis current controller (HCC).

SMC is an energy-based control mechanism that has been tested for inner control loops of VSIs in [7], [9], [16] and as voltage controller in [6], [33]. SMC is a variable structure controller that exhibits reliable performance during transients which can show acceptable THD if it is designed well. Chattering is a problem in discrete SMC implementation and some methods have been developed to achieve chatter-free SMC.

Fig. 11 represents a control scheme of a synchronous PI voltage controller and a hysteresis current controller (HCC).

3.2.4 Perfect Robust Servomechanism Problem (RSP) Control

Robust Servomechanism Problem (RSP) control is implemented by the internal model principle. This controller which has been used for the inner voltage loop [16] shows robust performance. RSP control ensures perfect tracking of the output voltages under unknown loading. This is accomplished by means of eliminating errors at specified harmonic level while ensuring good transient response and negligible interaction [16], [20]. Suitable transient performance in the RSP control depends on the design of this controller; however, fast response of this method need a proper design of RSP. Also, RSP as inner voltage controller facilitates the desired current harmonics control.

In this control method, the control input is a linear function of the states of the augmented plant x_p^* and servo compensator states η as follows:

$$u = K_0 x_p^* + K_1 \eta.$$

The coefficients are derived by minimizing the performance index function.

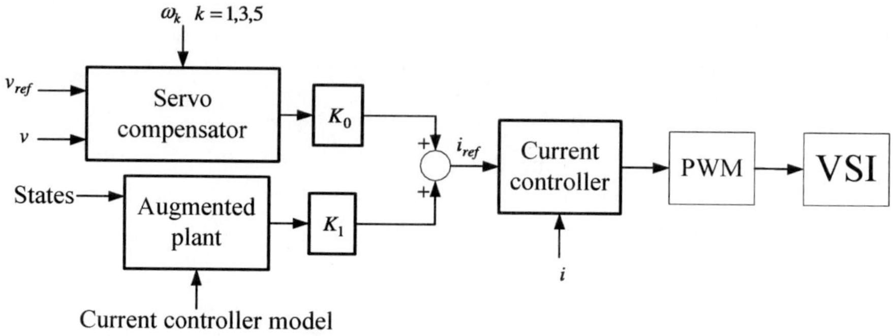

Fig. 12 Servomechanism problem control method schematic.

3.2.5 Neural Networks and Fuzzy Control Methods

Neural networks (NN) control method has been mostly used in current controllers because of its high robustness [34]-[35]. NN control methods are trained in off-line or on-line approaches. In the MG applications, it must be noticed if there is any parametric variation in the plant. In a MG application in which there is any parametric variation, the off-line method lacks a suitable performance. Note that NN control method has been used to modify the performance of other controllers such as deadbeat control in [36].

Fuzzy control has been mostly applied to the current control loops of VSIs [37]-[38]. Fuzzy is used to tune PI controllers to reduce overshoots and improve the tracking error performance. But, fuzzy controller is not a fast control method. In those MG applications in which there are fast dynamics, fuzzy tuned PI control method is not the best solution.

4 Discussion on Control Methods Properties

Depending on the MG application, different control methods can be used for inner controllers of VSIs of DERs. It is discussed which elements should be considered in the MG characteristics for inner controllers design.

Linear controller can have a good performance in an MG with more accurate transfer function and less uncertainty in the equipment and operating points. Proportional resonant control method is a good choice as inner voltage controller. This method can be used with a robust inner current controller to show the desired performance. Deadbeat (predictive) control method needs precise model of the filter to show the desired performance. However, it is a very fast response method and suitable for the current controller.

Among robust methods, H^{∞} control has robust performance for MGs with no parametric uncertainties. Repetitive control is a method which can show robust performance for periodic disturbances. Note that disturbances in the distribution networks are not all periodic. RSP control method has good performance as inner voltage controller in a MG where there might be parametric uncertainty.

SMC is a nonlinear robust controller which has a fast and robust performance. Sometimes, it is difficult to design a SMC to exhibit both suitable transient and zero steady state performance. In this case, a feedforward controller can be employed [6]. Note that adding a well-designed feedforward controller to the SMC improves the zero tracking speed as well as a non-overshoot transient performance.

5 Performance Tests for DC-AC Power Converters in MGs

In this section, power quality and stability analysis of VSIs are considered. This section shows the performance test for a proposed control strategy in an MG application. A MG configuration is proposed which is shown in Fig. 13. In this configuration, there are two feeders connected to the substation. The operation of MG in both of grid-connected and island modes and the transient between these two modes are considered. An islanded MG is a vulnerable system; thus, the dynamic stability analysis is considered when the system is islanded. In Fig. 13, the control center provides reference signals for the local controllers of VSIs. It has been shown that a low frequency communication system is adequate for a suitabl performance of MGs [4]. Table 1 shows the VSIs specifications. MG specifications are shown in Table 2.

The values of the distribution network cable impedences are calculated as follows:

$$L1 = (0.03 + j\omega 50\mu)\Omega\,,\; L2 = (0.012 + j\omega 30\mu)\Omega$$

$$L3 = (0.028 + j\omega 50\mu)\Omega\,,\; L4 = (0.032 + j\omega 55\mu)\Omega$$

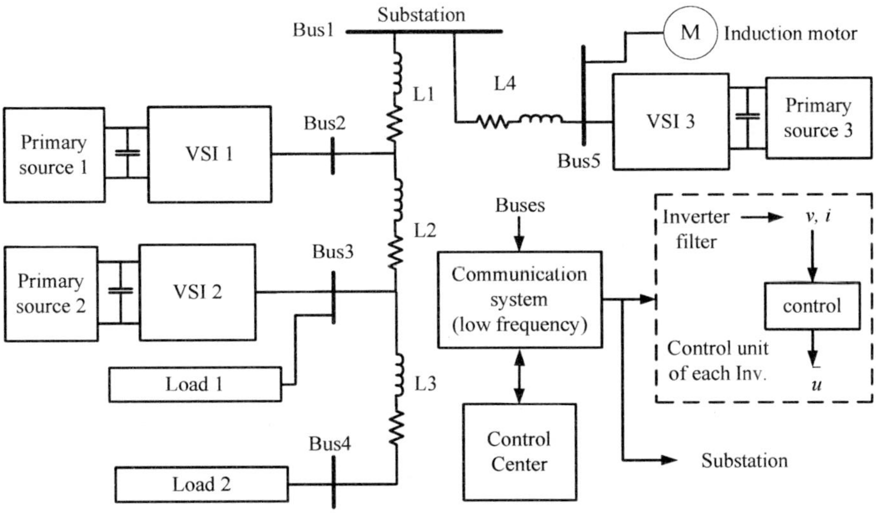

Fig. 13 Single line schematic of the proposed inverter-based microgrid.

Table 1 DC-AC Power Converters Specifications

Parameters	Values
DC Links Voltages	550 V (max)-300 V (min)
VSIs Filters:	L = 300 μ H, r_L = 0.01 Ω , C (Delta Configuration) = 200 μ F
DERs Transformers	50 VA, 200:440 V (rms), 60 Hz, L = 0.02 pu

Table 2 Microgrid Specifications

Substation transformer	440 V: 23 KV (rms), L = 0.03 pu
MG operation	440 V, f = 60 Hz
Static loads	15 kW, pf = 0.85
Induction motor	2 kW, pf = 0.85
DERs nominal power	50 kVA

In the proposed configuration voltage control in the grid-connected mode is necessary to avoid adverse voltage profile. For example, when there is no voltage control, if voltage drops at Bus 4, then the substation will increase the voltage which can result in overvoltage at Bus 2. Voltage control in the grid-connected mode needs to be coordinated to avoid voltage cotrol conflict.

In the islanded mode of operation, it is necessary to have a central controller to share power properly among DERs.

It is assumed that there are some high frequency dynamics in this MG. Induction motor is a fast dynamic load and islanding is a severe disturbance. Regarding these characteristics, a PR voltage controller and a fuzzy sliding mode controller seems to be the best potential candidates. The approximate voltage control loop transfer function diagram and the reference signal track (regulation) are shown in Fig. 14. It is obvious that the voltage controller has achieved good robustness and regulation is excellent.

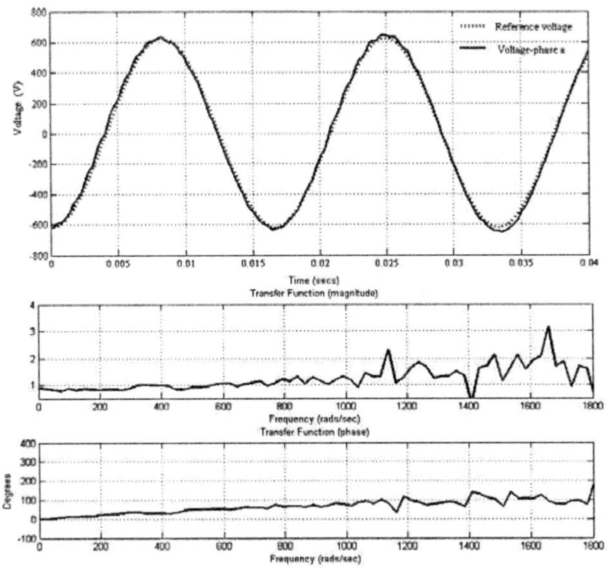

Fig. 14 Voltage regulation and Bode diagram of voltage control loop

The powers of DER 1, DER 2 and DER 3 are set to 10, 30 and 50 kW, respectively. These power are shown in Fig. 15. The output voltage and current of DERs are shown in Fig. 16 and Fig. 17. Substation voltage and injected current to it are shown in Fig. 17. The voltage profiles in the grid-connected mode with and without coordinated voltage control are shown in Fig. 18 and Fig. 19. It is shown that voltage profile can be out of standard voltage level. However, the coordinated voltage control improves the voltage profile in grid-connected mode. The developed algorithm depends on the configuration of the MG. For the propsoed MG, if voltage of Bus 4 drops, DER 2 tries to increase its power as much as possible. If the voltage is still below than desired level, then DER 1 increase. Finally, If voltage is still low substation transformer will increase its voltage. However, voltage of Buses 2 and 3 must not pass the maximum voltage. The THDs of output bus of VSIs are shown in Table 3.

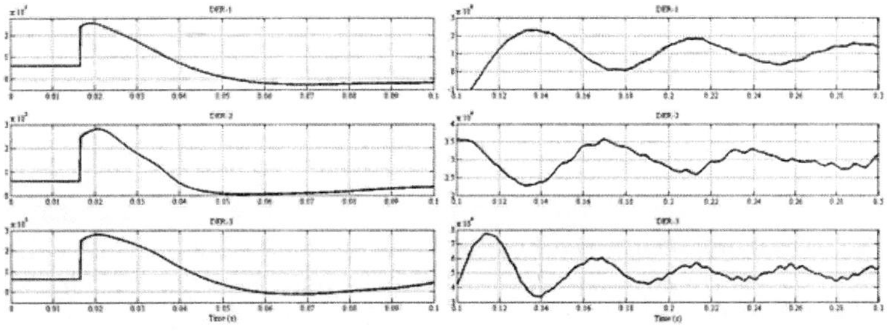

Fig. 15 Setting real power of DERs.

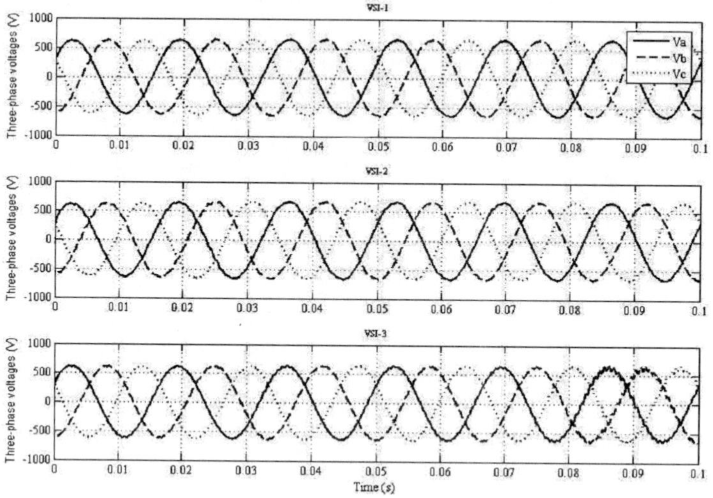

Fig. 16 Voltages of DERs in the steady state.

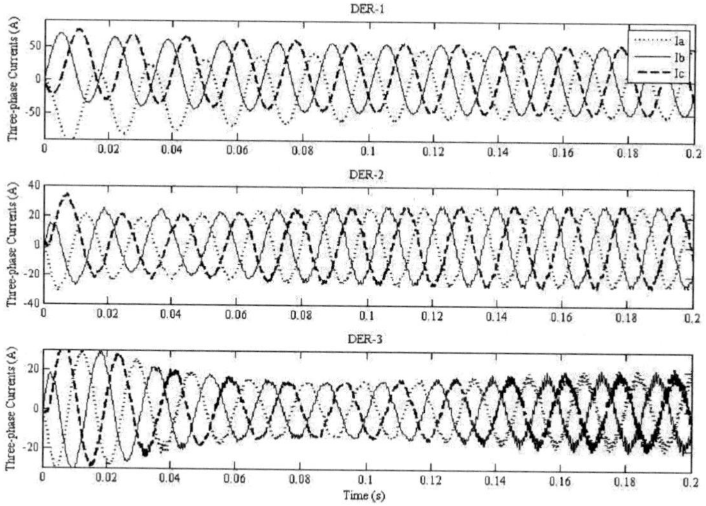

Fig. 17 Currents of DERs in the steady state.

In a dynamic stability analysis of grid-connected mode, a three-phase fault is tested at Bus 3. The current and voltage controller performance is tested in this scenario. The results are shown in Fig. 19-21. These results show that the proposed controller has excellent dynamic performance. This fault has severe impact on DER 2 and then on DER 1. DER 3 is not influenced. Islanding is tested to observe if the proposed control method is able to keep system working uninterruptly. An intentional islanding is tested at t = 2 s under different power flow scenarios. The voltage and frequency of Bus 1 are shown in Fig. 22. To test

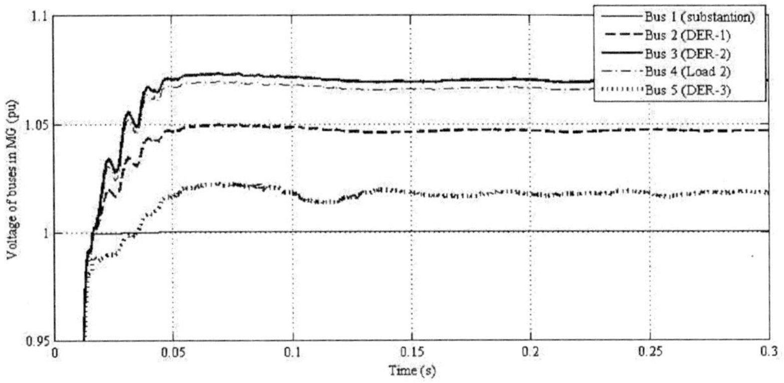

Fig. 18 Voltage of buses without coordinated voltage control.

Table 3 Voltage THD at output Bus of VSIs

	Phase A	Phase B	Phase C
VSI 1	0.96	1.13	2.14
VSI 2	1.10	2.14	2.50
VSI 3	1.40	1.80	2.32

the dynamic stability of the proposed control method in an islanded MG, the induction motor is started at $t = 0.2$ s. The voltage sag resulted from this start-up is well controlled by inner voltage and current controllers. The voltage of Bus 5 is shown in Fig. 23 for this scenario.

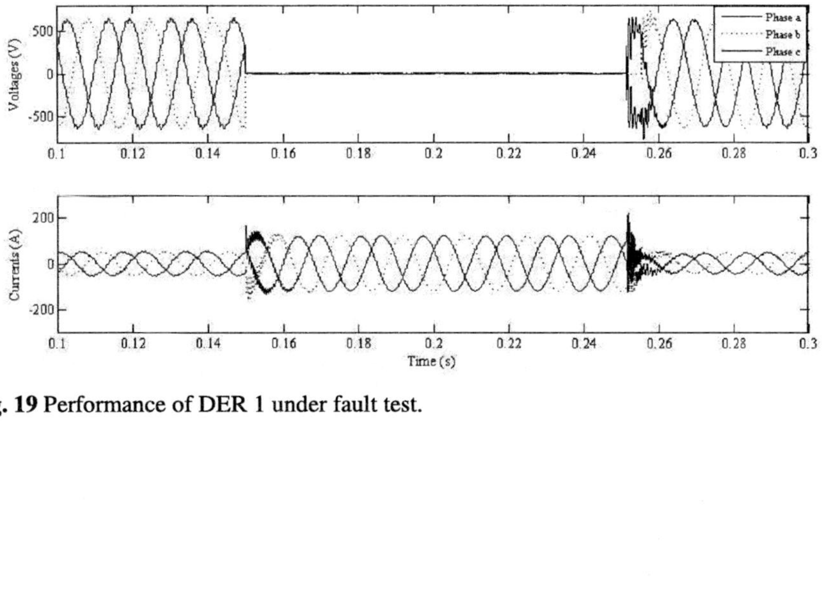

Fig. 19 Performance of DER 1 under fault test.

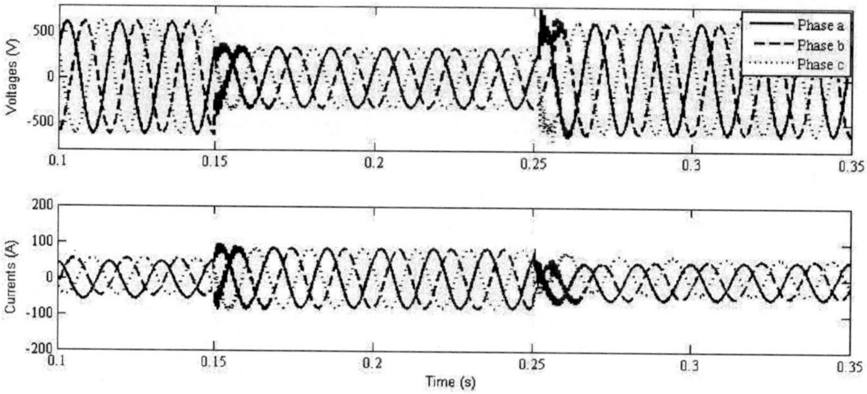

Fig. 20 Performance of DER 2 under fault test.

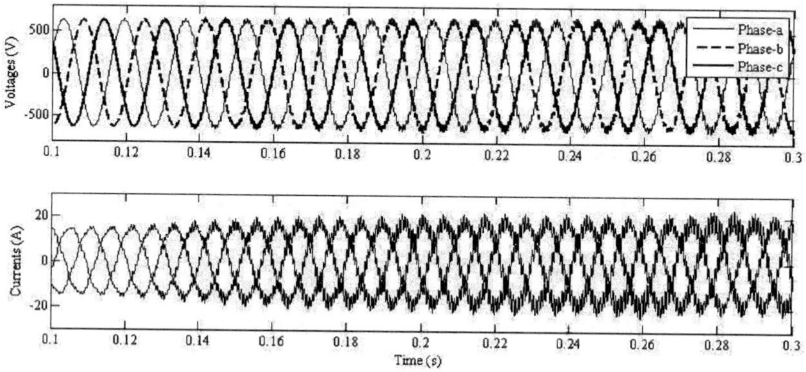

Fig. 21 Performance of DER 3 under fault test.

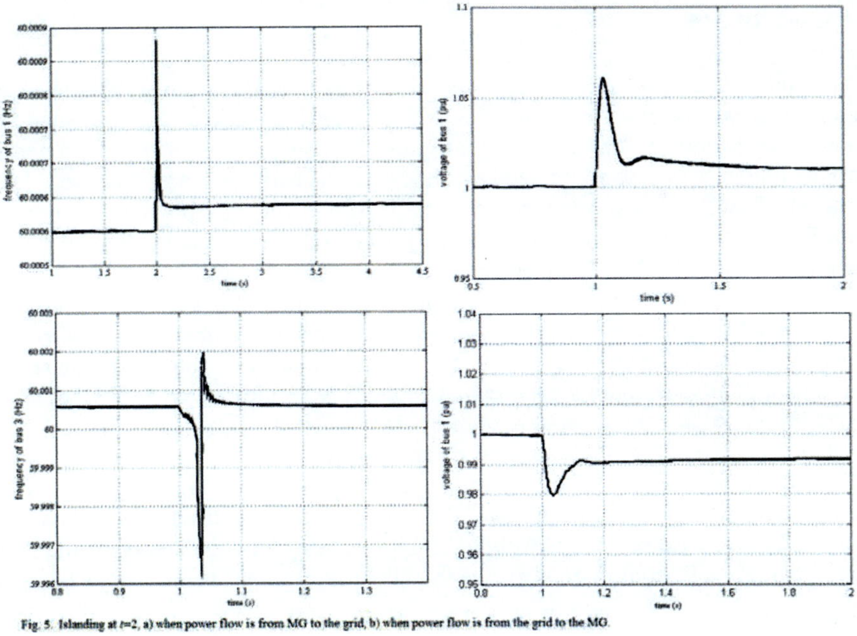

Fig. 22 Islanding at t = 2. Top) Power flow is from MG to the grid. Bottom) Power flow is from the grid to the MG.

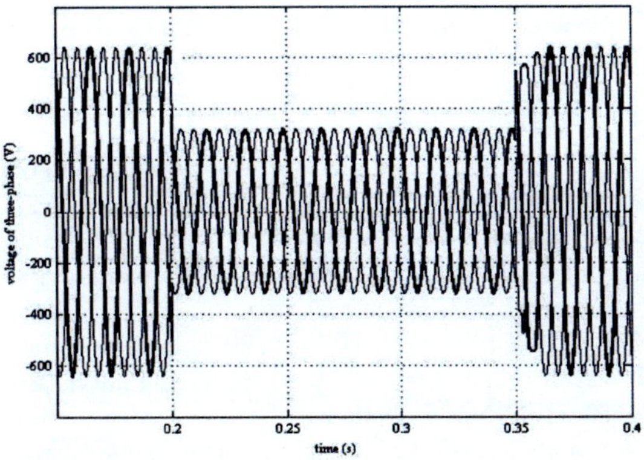

Fig. 23 Dynamic stability test in the island MG with the start-up of the induction motor.

6 Conclusion

The stability and power quality of MGs are discussed in this chapter. The focus of this chapter was on inner controllers of VSIs. Different methods are presented for inner control loops of VSIs. These controllers are categorized as linear and nonlinear methods. The capabilities and weakness of each method are discussed. It is discussed how to choose appropriate control method for a proposed MG application. Each method is useful if it is chosen for an appropriate application. Finally, the performance of a MG is tested for a proposed control method.

It is discussed that voltage stability should be considered in both grid-connected and islanded mode. However, in island MGs, the system is more sensitive to any transient. Different transients may occur in an MG such as load and generation variations, faults, islanding and reconnection. Two poles of each VSI transfer function can easily go to instability for any transient. Thus, more attention is needed to design of the LC filter of VSIs and control methods.

The basic formulation of each control method is presented in order to provide better understanding of its function. The design steps are briefly reviewed and some references are provided. Although, more robust and advance controller can work in where a PI controller works well, but overdesign has its own drawbacks.

References

[1] Lasseter, R.H.: Microgrids. In: Proc. IEEE Power Eng. Soc. General Meeting, August 7, pp. 305–308 (2002)
[2] Blaabjerg, F., Chen, Z., Kjaer, S.B.: Power Electronics as Efficient Interface in Dispersed Power Generation Systems. IEEE Trans. Power Electron 19(5), 1184–1194 (2004)
[3] Green, T.C., Prodanović, M.: Control of inverter-based micro-grids. Electrical Power Energy Syst. 77, 1204–1213 (2007)
[4] Prodanović, M., Green, T.C.: High-quality power generation through distributed control of a power park microgrid. IEEE Trans Ind. Electron 53(5) (2006)
[5] Abdel-Rahim, N., Quaicoe, J.E.: Analysis and design of a multiple feedback loop control strategy for single-phase voltage-source ups inverters. IEEE Transactions on Power Electronics 11(4), 532–541 (1996)
[6] Jung, S.-L., Tzou, Y.-Y.: Discrete sliding-mode control of a pwm inverter for sinusoidal out-put waveform synthesis with optimal sliding curve. IEEE Transactions on Power Electronics 11(4), 567–577 (1996)
[7] Sofla, M.A., Gharehpetian, G.B.: Dynamic performance enhancement of microgrids by advanced sliding mode controller. International Journal of Electrical Power and Energy Systems 33(1), 1–7 (2011)
[8] Lee, T.S., Chiang, S.J., Chang, J.M.: H_∞ loop-shaping controller designs for the single-phase ups inverters. IEEE Transactions on Power Electronics 16(4), 473–481 (2001)
[9] Dai, M., Marwali, M.N., Jung, J.-W., Keyhani, A.: A three-phase four-wire inverter control technique for a single distributed generation unit in island mode. IEEE Trans. Power Electronics 23(1), 322–331 (2008)

[10] Liang, J., Green, T.C., Weiss, G., Zhong, Q.-C.: Power electronics as efficient interface in dispersed power generation systems. In: Proc. IEEE Power Electron. Specialists Conf., pp. 1803–1808 (2002)
[11] Lin, B.-R.: Analysis of neural and fuzzy-power electronic control. IEE Proc.-Sci. Meus. Technol. 144(1), 25–33 (1997)
[12] Mohamed, Y.A.I., El-Saadany, E.F.: Hybrid variable-structure control with evolutionary optimum-tuning algorithm for fast grid-voltage regulation using inverter- based distributed generation. IEEE Trans. Power Electronics 23(3), 1334–1341 (2008)
[13] Timbus, A., Liserre, M., Teodorescu, R., Rodriguez, P., Blaabjerg, F.: Evaluation of current controllers for distributed power generation systems. IEEE Trans. Power Electronics 24(3), 654–664 (2009)
[14] Serban, E., Serban, H.: A control strategy for a distributed power generation microgrid appli-cation with voltage- and current-controlled source converter. IEEE Trans. Power Electronics 25(12), 2981–2992 (2010)
[15] Mohamed, Y.A.I., El-Saadany, E.F.: Robust high bandwidth discrete-time predictive current control with predictive internal model-a unified approach for voltage-source pwm converters. IEEE Trans. Power Electronics 23(1), 126–136 (2008)
[16] Marwali, M., Keyhani, A.: Control of distributed generation systems-part I: Voltages and currents control. IEEE Trans. Power Electron 19(6), 1541–1550 (2004)
[17] Malesani, L., Tenti, P.: A novel hysteresis control method for current controlled VSI pwm inverters with constant modulation frequency. IEEE Trans. Ind. Applicat. 26(1), 88–92 (1990)
[18] Buhl, Lorenz, R.D.: Design and implementation of neural networks for digital current regu-lation of inverter drives. In: Proc. of Conf. Rec. IEEE-IAS Annu. Meeting, pp. 415–421 (1991)
[19] Dzieniakowski, M.A., Kazmierkowski, M.P.: Self-tuned fuzzy pi current controller for pwm-vsi. In: Proc. EPE Conf., Seville, Spain, pp. 1308–1313 (1995)
[20] Karimi, H., Davison, E.J., Iravani, R.: Multivariable Servomechanism Controller for Autonomous Operation of a Distributed Generation Unit: Design and Performance Evaluation. IEEE Trans. Power Syst. 25(2), 853–865 (2010)
[21] Chen, C.-L., Wang, Y., (Jason)Lai, J.-S., Lee, Y.-S., Martin, D.: Design of Parallel Inverters for Smooth Mode Transfer Microgrid Applications. IEEE Trans. Power Electron 25(1), 6–15 (2010)
[22] Li, Y.W., Vilathgamuwa, D.M., Loh, P.C.: Robust Control Scheme for a Microgrid With PFC Capacitor Connected. IEEE Trans. Ind. Applicat. 43(5), 1172–1182 (2007)
[23] Kim, J., Lee, J., Nam, K.: Inverter-Based Local AC Bus Voltage Control Utilizing Two DOF Control. IEEE Trans. Power Electron 23(3), 1288–1298 (2008)
[24] Li, Y., Vilathgamuwa, D.M., Loh, P.C.: Design, Analysis, and Real-Time Testing of a Con-troller for Multibus Microgrid System. IEEE Trans. Power Electron 19(5), 1195–1204 (2004)
[25] Jurado, F., Valverde, M., Almonacid, B.: A Fuzzy Flux Control to Reduce Harmonics in the Utility Interface of Fuel Cell Power Systems. Electric Power Components and Systems 33(7), 781–800 (2005)
[26] Delghavi, M.B., Yazdani, A.: Islanded-Mode Control of Electronically Coupled Distributed-Resource Units Under Unbalanced and Nonlinear Load Conditions. IEEE Trans. Power Deliver 99 (2010)

[27] Liserre, M., Teodorescu, R., Blaabjerg, F.: Multiple Harmonics Control for Three-Phase Grid Converter Systems With the Use of PI-RES Current Controller in a Rotating Frame. IEEE Trans. Power Electron 21(3), 836–841 (2006)
[28] Li, Y.W., Kao, C.-N.: An Accurate Power Control Strategy for Power-Electronics-Interfaced Distributed Generation Units Operating in a Low-Voltage Multibus Microgrid. IEEE Trans. Power Electron 24(12), 2977–2988 (2009)
[29] Huerta, J.M.E., Castelló-Moreno, J., Fischer, J.R., García-Gil, R.: A Synchronous Reference Frame Robust Predictive Current Control for Three-Phase Grid-Connected Inverters. IEEE Trans. Ind. Electron 57(3), 954–962 (2010)
[30] Kim, J., Lee, J., Nam, K.: Inverter-Based Local AC Bus Voltage Control Utilizing Two DOF Control. IEEE Trans. Power Electron 23(3), 1288–1298 (2008)
[31] Tzou, Y.-Y., Jung, S.-L., Yeh, H.-C.: Adaptive Repetitive Control of PWM Inverters for Very Low THD AC-Voltage Regulation with Unknown Loads. IEEE Trans. Power Electron 15(5), 973–981 (1999)
[32] Utkin, V., Guldner, J., Shi, J.: Sliding Mode Control in Electromechanical Systems. Taylor and Francis, Abington (1999)
[33] Sofla, M.A., Radan, A.: Robust nonlinear Method to Control Micro-Grids. International Journal of Integrated Energy Systems 1(1), 65–69 (2009), Serial Publication
[34] Harashima, F., Demizu, Y., Kondo, S., Hashimoto, H.: Application of neural networks to power converter control. In: Conf. Rec. IEEE-IAS Annu. Meeting, San Diego, CA, pp. 1087–1091 (1989)
[35] Trzynadlowski, A.M., Legowski, S.: Application of neural networks to the optimal control of three-phase voltage-controlled inverters. IEEE Trans. Power Electron 9(4), 397–404 (1994)
[36] Mohamed, Y.A.I., El-Saadany, E.F.: Adaptive Discrete-Time Grid-Voltage Sensorless In-terfacing Scheme for Grid-Connected DG-Inverters Based on Neural-Network Identification and Deadbeat Current Regulation. IEEE Trans. Power Electron 23(1), 308–321 (2008)
[37] Hilloowala, R.M., Sharaf, A.M.: A rule-based fuzzy logic controller for a PWM inverter in a stand alone wind energy conversion scheme. IEEE Trans. Ind. Applicat. 32(1), 57–65 (1996)
[38] Naguib, M.F., Lopes, L.: Harmonics Reduction in Current Source Converters Using Fuzzy Logic. IEEE Trans. Power Electron 25(1), 158–167 (2010)

Author Index